Remote Sensing of Soil Salinization

Impact on Land Management

Remote Sensing of Soil Salinization

Impact on Land Management

Edited by
Graciela Metternicht
J. Alfred Zinck

CRC Press
Taylor & Francis Group
Boca Raton London New York

CRC Press is an imprint of the
Taylor & Francis Group, an **informa** business

CRC Press
Taylor & Francis Group
6000 Broken Sound Parkway NW, Suite 300
Boca Raton, FL 33487-2742

First issued in paperback 2019

© 2009 by Taylor & Francis Group, LLC
CRC Press is an imprint of Taylor & Francis Group, an Informa business

No claim to original U.S. Government works

ISBN-13: 978-1-4200-6502-2 (hbk)
ISBN-13: 978-0-367-38622-1 (pbk)

Library of Congress Cataloging-in-Publication Data

Remote sensing of soil salinization : impact on land management / editors, Graciela Metternicht, J. Alfred Zinck.
    p. cm.
    Includes bibliographical references and index.
    ISBN 978-1-4200-6502-2 (alk. paper)
    1. Soils, Salts in--Remote sensing. 2. Soil salinization. 3. Soil management. I. Metternicht, Graciela Isabel. II. Zinck, J. Alfred. III. Title.

S595.R46 2008
631.4'16--dc22
                                            2008044087

Visit the Taylor & Francis Web site at
http://www.taylorandfrancis.com

and the CRC Press Web site at
http://www.crcpress.com

Contents

PART I   Soil Salinity and Remote Sensing: The Object and the Tool

PART II   Trends in Mapping Soil Salinity and Monitoring Salinization with Remote and Proximal Sensing

PART III   *Diversity of Approaches to Modeling*
           Soil Salinity and Salinization

Preface

As most productive land is already under use, agriculture and livestock are increasingly spreading over to marginal lands that have limited suitability for crop production and grazing. As a result of this, naturally salt-affected soils, among other soils of constrained suitability, are being used for farming in sprawling new agricultural frontiers. On the other hand, land already under use is often exposed to increasing salinity, especially irrigated land, because of inadequate soil and water management. This calls for a careful inventory of salt-affected soils before land use starts and frequent monitoring of ongoing salinization in land already under use. Remote sensing has been advocated as a powerful tool to play an important role in identifying, mapping, and monitoring soil salinity and salinization.

The integration of remote sensing technologies in salinity studies is an expanding field of research and publication, taking advantage of steadily improving technology for remote and proximal sensing of land features at the terrain surface as well as in subsurface layers. Over the last two decades, several review papers have described the usefulness of remote sensing for salinity mapping and hazard assessment (Mougenot et al., 1993; Metternicht and Zinck, 2003; Farifteh et al., 2006). Now might be the appropriate time to offer a comprehensive volume on the subject along with a large spectrum of recent and ongoing research using technological developments in sensors and platforms and new methodological approaches. This book emphasizes on the contribution of remote sensing to assessing soil salinization, which is considered an ongoing land degradation process, and to managing salt-affected lands.

This book consists of 17 contributions describing a variety of sensors, ranging from ground-based to airborne and satellite borne, and their use in diverse geographical regions worldwide and environmental settings from coastal to inland saline areas. The contributions are grouped in three parts.

Part I focuses on the relationships between landscape salinity and remote sensing tools. Chapter 1 reviews the various source-areas of salts in nature generating primary salinity and analyzes the causes of human-induced (secondary) salinity. The relevance of remote sensing for monitoring salinization and for hazard prediction is highlighted. In Chapter 2, the spectral behavior of salt types most commonly found in soils is addressed. Reflectance of pure minerals is controlled mainly by salt mineralogy that determines the presence or absence of absorption features in the electromagnetic spectrum. However, in natural conditions, salts often distribute very gradually across the landscape and intermingle with other soil constituents. Solutions to these issues that control the spectral response of salts in nature are analyzed by proposing fuzzy logic–based mapping and the use of contextual indicators to discriminate variations in soil salinity. Chapter 3 provides an overview of the remote sensing methods and techniques most commonly used for mapping and monitoring salt-affected soils. The full variety of sensors and data is covered. A relevant distinction is made between techniques used to detect soil salinity at the surface and subsurface levels.

Part II is concerned with mapping soil salinity (primary and secondary) and monitoring the process of salinization. A variety of case studies from different dry regions of the world are presented. The performances of the most commonly used sensors are assessed, from multispectral to hyperspectral, from passive to active, and from remote to proximal. Approaches vary from the use of one kind of remotely sensed data, including multispectral, hyperspectral, microwave, ground-penetrating radar, and electromagnetic induction, to the combinations of several types of data. Some contributions emphasize the relevance of hyperspectral data for salt identification and mapping because salt types have characteristic spectral features in narrow bands. Salinity in dryland farming areas and increasing salinization of irrigated land are addressed. Landscape settings covered by the studies range from coastal areas to dry inland depressions (saladas, sabkhas, and playas).

Part III shows the diversity of approaches used in modeling soil salinity and salinization in space and time. For the purpose of modeling, all contributions use a combination of remotely sensed data and statistical techniques including descriptive, adaptive, stochastic, and spatial statistics. Modeling is geared toward inferring salinity parameters at unvisited sites in grid surveys as well as assessing salinity hazards in areas potentially exposed to salinization. Expert knowledge and data mining are shown to be essential for calibrating predicted salinity data.

As editors we are very grateful to all the contributors who worked hard to make this project a success, providing and sharing with readers their experiences in the application of remote sensing to salinity mapping, monitoring, and modeling. This book would not be a reality without their contributions. We are indebted to the team at Taylor & Francis for believing in the need to publish this book. Lastly, Graciela Metternicht would like to acknowledge the School of Natural and Built Environments of the University of South Australia for its support in the preparation of this book.

Graciela Metternicht
Panama City, Panama

J. Alfred Zinck
Enschede, Netherlands

Contributors

Diganta Adhikari
Center for Irrigation Technology
California State University
Fresno, California

Uri Basson
Geosense Ltd.
Tel Aviv, Israel

Eyal Ben-Dor
Department of Geography and Human
  Environment
Tel Aviv University
Tel Aviv, Israel

Paula D. Blanco
Centro Nacional Patagónico
Consejo Nacional de Investigaciones
  Científicas y Técnicas
Puerto Madryn (Chubut), Argentina

Norma Fernández Buces
Instituto de Geología
Universidad Nacional Autónoma de México
Ciudad Universitaria
México City, México

Florence Cassel S.
Center for Irrigation Technology
California State University
Fresno, California

Carmen Castañeda
Unidad de Suelos y Riegos
Centro de Investigación Technología
Agroalimentaria del Gobierno de Aragón
Zaragoza, Spain

Héctor F. del Valle
Centro Nacional Patagónico
Consejo Nacional de Investigaciones
  Científicas y Técnicas
Puerto Madryn (Chubut), Argentina

Ahmed Douaik
Research Unit on Environment and
  Conservation of Natural Resources
National Institute of Agricultural Research
Rabat, Morocco

Anna Dutkiewicz
Department of Water, Land and Biodiversity
  Conservation
Adelaide, South Australia

and

Future Farm Industries Cooperative
  Research Centre
University of Western Australia
Crawley, West Australia

Ravi Shankar Dwivedi
National Remote Sensing Agency
Hyderabad, India

Peter Eklund
Department of Information Systems and
  Technology
The University of Wollongong
Wollongong, New South Wales, Australia

Néstor O. Elissalde
Estación Experimental Agropecuaria
Instituto Nacional de Tecnología
  Agropecuaria
Trelew (Chubut), Argentina

Jamshid Farifteh
Institute Geologico y Minero de
  España (IGME)
Madrid, Spain

and

Geo-Research and Development
Enschede, Netherlands

Abbas Farshad
International Institute for Geo-Information
  Science and Earth Observation (ITC)
Enschede, Netherlands

David Fraser
School of Mathematical and Geospatial
  Sciences
RMIT University
Melbourne, Victoria, Australia

Naftaly Goldshleger
Soil Erosion and Conservation
  Research Station
Ministry of Agriculture
Tel Aviv, Israel

Philip C. Goodell
Department of Geological Science
University of Texas at El Paso
El Paso, Texas

Dave Goorahoo
Center for Irrigation Technology
California State University
Fresno, California

and

Department of Plant Science
California State University
Fresno, California

José Gumuzzio
Departmento de Geología y Geoquíca
Universidad Autónoma de Madrid
Madrid, Spain

Juan Herrero
Departamento de Suelo y Agua
Estación Experinental de Aula Dei, Conseijo
Superior de Investigaliones Científicas
Zaragoza, Spain

Fares M. Howari
Bureau of Economic Geology
Jackson School of Geosciences
J.J. Pickle Research Campus
University of Texas at Austin
Austin, Texas

Stephen D. Kirkby
Maxamine Inc.
Lafayette, California

Magaly Koch
Center for Remote Sensing
Boston University
Boston, Massachusetts

Ramana Venkata Kothapalli
National Remote Sensing Agency
Hyderabad, India

Megan Lewis
School of Earth and Environmental
  Sciences
University of Adelaide
Glen Osmond, South Australia

and

Future Farm Industries Cooperative
  Research Centre
University of Western Australia
Crawley, West Australia

Graciela Metternicht
Division of Early Warning and Assessment
  United Nations Environment Programme
Regional Office for Latin America and the
  Caribbean
Panama City, Panama

and

School of Natural and Built Environments
University of South Australia
Adelaide, South Australia

Vladmir Mirlas
Soil Erosion and Conservation
  Research Station
Ministry of Agriculture
Tel Aviv, Israel

Eshel Mor
Department of Geography and Human
  Environment
Tel Aviv University
Tel Aviv, Israel

Contributors

Contributors

Bertram Ostendorf
School of Earth and Environmental
  Sciences
University of Adelaide
Glen Osmond, South Australia

and

Future Farm Industries Cooperative
  Research Centre
University of Western Australia
Crawley, West Australia

José Luis Palacio Prieto
Instituto de Geografía
Universidad Nacional Autónoma de México
Ciudad Universitaria
México City, México

César M. Rostagno
Centro Nacional Patagónico
Consejo Nacional de Investigaciones
  Científicas y Técnicas
Puerto Madryn (Chubut), Argentina

Thomas Schmid
Departmento de Medio Ambiente
CIEMAT (Centro de Investigaciones
  Energéticas, Medioambientales y
  Tecnológicas)
Madrid, Spain

Dhruba Pikha Shrestha
International Institute for Geo-Information
  Science and Earth Observation (ITC)
Enschede, Netherlands

Christina Siebe
Instituto de Geología
Universidad Nacional Autónoma de México
Ciudad Universitaria
México City, México

Amarendra Narayana Singh
Uttar Pradesh State Remote Sensing
  Application Centre
Lucknow, India

Walter Sione
Centro Regional de Geomática
Facultad de Ciencia y Tecnología
Universidad Autónoma de Entre Ríos
Oro Verde (Entre Ríos), Argentina

and

Programa de Desarrollo e Investigación en
  Teledetección
Universidad Nacional de Luján
Luján (Buenos Aires), Argentina

Tibor Tóth
Research Institute for Soil Science and
  Agricultural Chemistry
Hungarian Academy of Sciences
Budapest, Hungary

Marc Van Meirvenne
Department of Soil Management
  and Soil Care
Ghent University
Ghent, Belgium

Richard Webster
Rothamsted Research
Harpenden, Hertfordshire, United Kingdom

J. Alfred Zinck
International Institute for Geo-Information
  Science and Earth Observation (ITC)
Enschede, Netherlands

David Zoldoske
Center for Irrigation Technology
California State University
Fresno, California

Part I

Soil Salinity and Remote Sensing:
The Object and the Tool

1 Soil Salinity and Salinization Hazard

J. Alfred Zinck and Graciela Metternicht

CONTENTS

1.1 INTRODUCTION

Soil salinity is in the first instance a natural feature, while soil salinization is mainly a human-induced process. The origin of the salts determines the geographic distribution of the natural source areas, the pathways of further salt dissemination, and the location of the areas potentially prone to human-induced salinization. This knowledge indicates where the contribution of remote sensing can be useful. Knowing the origin of the salts also informs about their nature, mineralogy, variety of crystallization patterns, physical structure, modes of occurrence in the soil mantle, and characteristic figures they usually form at the soil surface. This indicates how remote sensing can contribute to identifying and monitoring salt-affected areas. After describing the source areas of salt in nature and the particular landscape positions where salts mainly concentrate, the severity and causes of human-induced salinization are addressed. Finally, the relevance of remote sensing for monitoring the salinization process and hazard prediction is highlighted.

1.2 ORIGIN OF SALT IN NATURE

In natural conditions, the presence and concentration of salts at the terrain surface and in the soil mantle are controlled by geologic, geomorphic, climatic, and hydrologic factors. All salts originate from the weathering of the primary minerals contained in crystalline rocks. From this starting point, salts are redistributed at the earth surface and become afterward constituents of sedimentary rocks and unconsolidated sedimentary deposits. They enter the soil system after weathering of the hard rocks, crystalline and sedimentary, or release from the salt-rich sediments (Figure 1.1).

1.2.1 RELEASE OF SALTS FROM WEATHERING OF PRIMARY MINERALS

The primary minerals of igneous and metamorphic rocks, mainly silicates and aluminosilicates, are the original sources of all salt found in nature. Upon weathering, primary minerals release cations and anions that combine to form a variety of salts such as chlorides, sulfates, carbonates, and bicarbonates of sodium, calcium, magnesium, and potassium, among others. Neoformed salts seldom stay in the sites of origin, except in dry regions where saline soil horizons can form atop the rock-weathering zone, although weathering is usually slow because of insufficient rainfall. The most soluble salts are geographically disseminated by water moving on the terrain surface or through the soil mantle from their original areas to lower landscape positions, where they naturally contaminate soils that might have been free of salts. Upon arrival, salts are redistributed in the soil cover according to their level of solubility under prevailing climatic, topographic and hydrologic conditions, and increasingly under the influence of the land management practices. Ultimately, large amounts of salt accumulate in the oceans, lakes, lagoons, and other inland depressions, as end reservoirs of the salt cycle on earth.

1.2.2 RELEASE OF SALTS FROM SALINE SEDIMENTARY ROCKS

Saliferous sedimentary rocks are lagunary formations, most often gypso-saline, thus rich in calcium sulfate and sodium chloride, associated with variegated (pink, blue, red, green) strata of shales and marls also containing evaporites. Frequently, the lagunary formations dating back to the Permian, Trias, Oligocene, and Miocene are now found in hilly relief or even more rugged landscapes following terrain uplift and folding. Because of their relative fluidity, gypso-saline rocks usually undergo a special tectonic style (diapirism) that causes the formation of salt domes, nearly vertical pipe-like masses of salt that appear to have pierced covering rock strata and punched their way upward to the surface from an underlying salt bed. As a consequence, these saline formations occur now on higher landscape positions, having impregnated the contact formations and being centers of redistribution of salts over the surrounding landscapes through surface runoff and groundwater flow. Anhydrite, gypsum, and native sulfur are commonly associated with salt domes (Hurlbut and Klein, 1977). The terrains affected by this kind of geologic salinization, related to paleo-lagunary sedimentary rocks, are the most frequent and most extensive in nature (Gaucher and Burdin, 1974).

Saliferous sedimentary rocks are the main sources of salt found in dryland seeps. Seeps are resurgence sites of underground water flows fed by recharge areas. In a recharge area, the water

FIGURE 1.1   Source areas of salt. (Photos courtesy of Zinck.) (a) Salt release from weathering of low-grade metamorphic rocks (slates) in the region of Potosi (Bolivia); (b) mud volcano as a source of salt; salt is deflated by wind from the dry lakeshores and spread over the surrounding dry-farming areas (Aq-Qala, Golestan, NE Iran); (c) salt lake of Salar de Uyuni on the Bolivian altiplano at 3650 m elevation, the world's largest salt flat measuring 10,600 km^2; (d) large chott depression with gypsum-rich salt concentration at the terrain surface and briny water table at shallow depth; the grayish-brown surface color is due to desert dust (Chott-el-Djerid, Southern Tunisia); (e) playa landscape with no conspicuous evidence of salt accumulation at the soil surface but high salt concentration in the subsoil related to the water table; light-colored terrain surface around the dug hole corresponds to silty topsoil sealing (Aq-Qala, Golestan, NE Iran); (f) saline soil (Aquisalid) of the playa landscape represented in (e), showing capillary rise of salt from the water table at approximately 80 cm depth to 20 cm from the soil surface (Aq-Qala, Golestan, NE Iran).

that is not used by plants and not retained by the soil moves to the water table and flows downslope through subsoil and bedrock. The salts dissolved along the pathway accumulate in and around the discharge area after evaporation of the seep flow. In Alberta (Canada), several types of dryland saline seeps have been recognized according to the origin of the salts and the way seeps form, including seeps related to contact salinity, slope-change salinity, outcrop salinity, artesian salinity, depression bottom salinity, coulee bottom salinity, and slough ring salinity (Wentz, 2000).

1.2.3 Release of Salts from Volcanic Activity

During volcanic eruptions, wind-blown ash spreads salt-containing minerals over large areas. The weathering of these minerals often releases sodium carbonate that may lead to soil alkalinization (sodication). After eruption, volcanic activity often provides sources of salts, especially chlorine and sulfur components in the proximity of fumaroles and thermal-water springs. Also mass movements can take place on the slopes of ash cones, redistributing sulfur-rich materials on the surrounding piedmonts. Such events are frequent in the high Andes where the melting of snowcaps on semiactive volcanoes originates sulfurous lahar and mudflows that reach down the valleys and bury towns and fertile alluvial soils. Sulfur-rich sediments evolve into acid sulfate soils after drying up (e.g., Armero area, Colombia, after the eruption of the Nevado del Ruiz volcano in 1985).

1.2.4 Deposition of Salts in Coastal Marshes and Estuaries

In low-lying coastal areas, marshes that maintain an inlet connection with the open sea are submerged regularly at high tide and enriched in salts that concentrate at the marsh bottom at low tide (Figure 1.2). During exceptional events such as heavy storms, very high tides, and tsunamis that often cause the breaking of protection dams, coastal areas much larger than marshes can be flooded with seawater and become saline. In estuary environments, the electrolytes of the seawater cause the flocculation of the fine-grained sediments (silt and clay) transported by the rivers, and these saline electrolytes remain fixed on the flocculated particles. Salts also accumulate in water-bearing interdune areas and tropical mangroves.

1.2.5 Deposition of Salts in Inland Depressions

In low-lying inland areas, runoff water and dissolved salts converge toward depressions where high temperature causes water to evaporate and salts to precipitate. Such positions are frequent in enclosed basins (playa, chott, sabkha) and have a lagunary water regime, with evaporation exceeding water influx. In general, lagunary deposits are gypso-saline evaporites, with associated calcium sulfate and sodium chloride, provided by saline water from surrounding saliferous rock outcrops. In inland sabkhas, the terrain is covered by salt crust, and gypsiferous mud layers with well-developed gypsum crystals are common sediments. During long dry periods, the salt crust cracks into large polygons and salt is deflated by the wind. Some salt is dissolved by rainwater, inflowing wadi water or groundwater, and part thereof recrystallizes as salt lenses within the underlying muddy sediments (Reineck and Singh, 1980). Also surface salt beds formed by gradual evaporation and ultimate drying up of enclosed bodies of salt water may have subsequently been buried by other rock strata. The extreme condition of pure salt cover is represented by the dried up salt lakes, with hundreds of meters of salt beds, that occur in arid intermountain valleys and high plateaus (e.g., Salar de Uyuni on the Bolivian altiplano, the world's largest salt flat with 10,600 km^2 at 3,650 m elevation).

1.2.6 Aeolian Influx of Salts

In coastal areas, salt is transported by the wind from the sea to the land through the spray of breaker foam that results from the wave uprush on the strand and more characteristically against rock outcrops along rocky coasts exposed to storms. This type of primary salinization through marine aerosols can affect areas located relatively far away from the coast line, as is the case in Western Australia where salt is believed to have been carried inland by the wind over millions of years and deposited by rainfall and dust fallout in the present West Australian wheat-belt (National Land and Water Resources Audit, 2000). Similarly, the main source of soluble salts found in desert soils of dry inland areas such as in Israel is often the slow concentration from the small amount of airborne salts in the rainwater and dust fallout deposition (Dan and Yaalon, 1982). The deposition of

FIGURE 1.2 Coastal salt marsh. (Photos courtesy of Zinck.) (a) Coastal marsh with white salt cover at low tide; at high tide, the marsh is flooded by seawater crossing the sand bar through a connecting channel (to the left of the picture); as the tide regime is diurnal in this region, reflectance changes twice a day (west of Barcelona, NE Venezuela); (b) rim of the coastal marsh shown in (c), covered by salt-tolerant *Salicornia* (west of Barcelona, NE Venezuela); (c) bare core area of a coastal marsh no longer submitted to tide flooding, after the channel connection with the open sea was clogged by beach-sand encroachment; polygonal surface pattern controlled by the columnar structure of the underlying saline–sodic soil (Natrargid); dark surface material dispersed because of high ESP; circular pattern of white saline spots (west of Barcelona, NE Venezuela); (d) saline-sodic soil (Aquic Natrargid) of the coastal marsh shown in (c); clayey topsoil with columnar structure (10–25 cm) covered by a sodium-dispersed, platy surface layer (0–10 cm); loamy sand subsoil; briny water table at 70 cm depth (west of Barcelona, NE Venezuela).

soluble salts of aeolian origin can result in considerable accumulations, as thick as 3–8 m in the soils of the Atacama Desert, Chile (Prellwitz et al., 2006).

1.3 HUMAN-INDUCED SALINIZATION

Salinization is first of all a natural process resulting from the migration and dissemination of water-soluble salts from a source area to an area originally free of salt. Hereafter, salinization is considered mainly as a human-induced process, also called secondary salinization, which contributes to increase the salt concentration in soils already salt-affected or leads to contaminate salt-free soils because of inadequate water and land management.

1.3.1 EXTENT AND SEVERITY OF SALINIZATION

Salt-affected soils occur in all continents and under almost all climatic conditions, but they are more extensive in dry regions than in humid ones. Dry areas are naturally prone to salinization because of low rainfall and high evapotranspiration that restrict the leaching of the salts. This effect is magnified when the land is used with inadequate farming practices.

1.3.1.1 Global and Regional Extent

Statistics on the worldwide extent of salt-affected areas vary according to data sources. On the basis of data collected about 30 years ago and reported in Abrol et al. (1988), 932 million ha are salt-affected, with 38.4% in Australasia, 33.9% in Asia, 15.8% in the Americas, 8.6% in Africa, and 3.3% in Europe (Table 1.1). According to Food and Agriculture Organization (FAO, 2000), the global salt-affected surface area is 831 million ha, including 397 million ha of saline soils and 434 million ha of sodic soils. Other estimates are closer to 1 billion ha, which represents about 7% of the earth's continental surface area (Ghassemi et al., 1995) or approximately 10 times the size of a country like Venezuela and 20 times the size of France. Saline soils are common in semiarid and arid regions, where low rainfall and high evapotranspiration cause Na, Mg, and Ca salts to concentrate, mainly in the form of chloride and sulfate. In less arid climates, sodium carbonate and bicarbonate dominate and sodium ions get adsorbed on the soil exchange complex, causing the formation of sodic soils.

Besides the naturally salt-affected areas, 76.3 million ha have been salinized as a consequence of human activities, according to the global assessment of human-induced soil degradation (GLASOD) carried out by the GLASOD project in the late 1980s (Oldeman et al., 1991). This represents 3.9% of the 1964 million ha affected by human-induced soil degradation worldwide (Table 1.2). It is much less than the proportions of land damaged by the loss of topsoil through water and wind erosion (46.8% and 23.1%, respectively), but more than the soil degradation caused by other processes such as overblowing, pollution, acidification, compaction, waterlogging, and subsidence of organic soils. The largest proportion of soils suffering secondary salinization is found in Asia (69% of the global area), followed by Africa, America, Europe, and Australasia (Table 1.3). About 58% of all salinized areas concentrates in irrigation schemes. On average, 20% of the world's irrigated lands is affected by salts (45 million ha out of the 230 million ha irrigated worldwide), but this figure increases to more than 30% in countries such as Egypt, Iran, and Argentina (Ghassemi et al., 1995). Of the almost 1500 million ha of dryland agriculture, 32 million ha (2.1%) are salt-affected, with varying degrees of human-induced degradation (FAO, 2000).

TABLE 1.1
Worldwide Distribution of Salt-Affected Areas (Million ha)

Area	Saline Soils	Sodic Soils	Total	Percent
Australasia	17.6	340.0	357.6	38.4
Asia	194.7	121.9	316.5	33.9
America	77.6	69.3	146.9	15.8
Africa	53.5	26.9	80.4	8.6
Europe	7.8	22.9	30.8	3.3
World	351.2	581.0	932.2	100

Source: Data summarized from Szabolcs (1974) for Europe and Massoud (1977) for the other continents, as reported by Abrol et al. (1988) in FAO Soils Bulletin 39.

TABLE 1.2
Worldwide Human-Induced Soil Degradation

Type	Million ha	Percent
Loss of topsoil	920.3	46.8
Terrain deformation	173.3	8.8
Water	1093.6	55.6
Loss of topsoil	454.2	23.1
Terrain deformation	82.5	4.2
Overblowing	11.6	0.6
Wind	548.3	27.9
Loss of nutrients	135.3	6.9
Salinization	76.3	3.9
Pollution	21.8	1.1
Acidification	5.7	0.3
Chemical	239.1	12.2
Compaction	68.2	3.5
Waterlogging	10.5	0.5
Subsidence of organic soils	4.6	0.2
Physical	83.3	4.2
Total	1964.4	100

Source: Data summarized from Oldeman, L.R., Hakkeling, R.T.A., and Sombroek, W.G., World map of the status of human-induced soil degradation: An explanatory note (GLASOD project). ISRIC, Wageninen, the Netherlands, and UNEP, Nairobi, Kenya, 1991.

There are no recent data on global soil salinization, but it can be assumed that, since the earlier data gathering in the 1970s and 1980s, salinization has expanded as newly affected areas most probably exceed reclamation and rehabilitation ones. At country level, data are generally more accurate and sometimes more up-to-date. For instance, Australia, a country with large naturally

TABLE 1.3
Human-Induced Salinization (Million ha)

Area	Light	Moderate	Strong	Extreme	Total	Percent
Asia	26.8	8.5	17.0	0.4	52.7	68.8
Africa	4.7	7.7	2.4	—	14.8	19.3
America	2.1	1.8	0.5	—	4.4	5.7
Europe	1.0	2.3	0.5	—	3.8	5.0
Australasia	—	0.5	—	0.4	0.9	1.2
World	34.6	20.8	20.4	0.8	76.6	100

Source: Data summarized from Oldeman, L.R., Hakkeling, R.T.A., and Sombroek, W.G., World map of the status of human-induced soil degradation: An explanatory note (GLASOD project). ISRIC, Wageninen, the Netherlands and UNEP, Nairobi, Kenya, 1991.

saline areas, had in 2001 an estimated 2.5 million ha of land salinized since the introduction of European farming, compared to the 0.9 million ha identified by the GLASOD project in the late 1980s (Oldeman et al., 1991). Salinization may affect a total of 17 million ha of valuable farmland within the next 50 years (National Land and Water Resources Audit, 2000). In Alberta (Canada), secondary salinization by 1980 affected about 82,000 ha in irrigation districts and about 648,000 ha in nonirrigated areas. The rate of increase in soil salinity is estimated to range from 0.8% to 16% annually (Hecker, 2007).

1.3.1.2 Severity and Effects

Human-induced salinization is by all means a severe environmental hazard, although it affects less surface area than other land degradation processes. Salinization is a main resource concern because salts hinder the growth of crops by limiting their ability to take up water. Excess salt has the same effect on plants as drought. Salt is the savor of foods but the scourge of agriculture (Donahue et al., 1977). Salt concentration in the soil environment has strong impact on crop yields and agricultural production in both dry and irrigated areas due to poor land and water management and the expansion of the agricultural frontier into marginal drylands (Figures 1.3 and 1.4). In the future, more drylands will be put into agricultural production because of increasing population pressure. This will mainly be achieved with irrigation and will thus expand the salinization hazard. Furthermore, salinity also affects other major soil degradation processes such as soil dispersion and compaction, increased soil erosion, and corrosion of engineering structures.

Crop productivity and production losses caused by salinization have considerable impact on farm and irrigation project economics. For instance, the economic damage caused by secondary salinization was estimated at US$750 million per year for the Colorado River Basin in the United States, US$300 million per year for the Punjab and Northwest Frontier Provinces in Pakistan, and US$208 million per year for the Murray–Darling Basin in Australia (Ghassemi et al., 1995). Frequently, these costs do not consider the losses on property values of farms with degraded land, and other indirect costs such as eutrophication of rivers and estuaries, damage to infrastructure (including roads and buildings), and the social cost of farm businesses. A study focusing on dryland salinity risk assessment in Australia (National Land and Water Resources Audit, 2000) estimates that some 20,000 km of major roads and 1,600 km of railways are already at risk, with a potential increase to 52,000 and 3,600 km, respectively, by 2050, unless proper measures to halt degradation are taken. Degradation assessment and hazard prediction require that the causes of salinization be clearly identified.

1.3.2 Causes of Salinization

Salinization is controlled by a set of factors related to environmental conditions (climate, hydrology), water supply and control systems (irrigation, drainage), and cropping practices (type and density of plant cover and rooting characteristics). These factors affect the soil–water balance and therefore the movement and accumulation of salts in the soil. In general, salinization develops in places where the following conditions occur together: presence of soluble salts, high water table, high rate of evaporation, and low annual rainfall. The ways these factors combine vary according to whether the land is irrigated or not, but in any case the presence of salts in the soil makes water less available for uptake by plant roots.

1.3.2.1 Irrigated Areas

1.3.2.1.1 Rising Groundwater Table under Irrigation
Groundwater tends to rise in overirrigated areas that lack natural drainage or have no artificial drainage to evacuate excess water. Similarly, cropping systems based on shallow-rooted crops that do not extract excess subsoil moisture may lead to the rise of the water table. In both situations, salts tend to remain or concentrate on the terrain surface or in the upper soil horizons. Groundwater levels

FIGURE 1.3 Effects of salinization in irrigated land. (Photos courtesy of Zinck.) (a) Salt accumulation in a drainage ditch at parcel level, reflecting the advanced stage of salinization of fields irrigated with low-quality water (Aq-Qala, Golestan, NE Iran); (b) salt-laden material extracted from the drainage channel shown in Figure 1.3d (Golestan, NE Iran); (c) saline seeps at the escarpment of a plateau cultivated with irrigated orchards (Neuquen region, Southern Argentina); (d) deep drainage channel at the periphery of an irrigation scheme, recently excavated to improve the evacuation of salt-laden leaching water from salinized fields; salt exudation spots and lines on the channel walls (Golestan, NE Iran); (e) abandoned farmer village located in the Sharra valley shown in Figure 1.3f, after long-time flood irrigation had fully salinized the cropland (Hamadan region, center-west of Iran); (f) valley bottom and lower alluvial terraces (white) fully salinized by century-long irrigation; irrigated farming has moved to the piedmont landscape on both sides of the valley (red); false-color composite from a Landsat TM image (Sharra valley, Hamadan region, center-west of Iran). (Courtesy of Farshad.)

that fluctuate at less than 1.5–2 m depth cause the concentration of salts in the root zone and can contribute to evaporation from the soil surface.

1.3.2.1.2 Use of Saline Groundwater for Irrigation

Saline groundwater used for irrigation can cause the concentration of salts in the root zone, especially if the internal drainage is restricted and leaching insufficient. This process of salinization occurs in dry environments and covers small areas. Even when irrigation water contains only small amounts of dissolved salt, long-term use of such water in areas of high evapotranspiration can result in building-up salt concentration in the soil.

FIGURE 1.4 Effects of salinization on crops and soils. (Photos courtesy of Zinck.) (a) Cropland abandoned because of generalized salinization after long-term flood irrigation (Marvdasht plain, Shiraz, south-central Iran); (b) Sunflower growing on saline soil derived from lake sediments; salt affects crop density and health; salt lake in the background (Fuente de Piedra, Antequera region, SE Spain); (c) salt patch impeding alfalfa to grow in the distal part of a flood-irrigated field (Sacaba, Cochabamba region, Bolivia); (d) dryland salinity in an area used for extensive sheep grazing (Marvdasht plain, Shiraz, south-central Iran); (e) flood-irrigated field showing the formation of spotty saline crust in furrows and microdepressions (Golestan, NE Iran); (f) salinity maps established by kriging of ECe data determined at three soil depths on a 500 m interval grid. The same 40 sites sampled in 1972 were sampled again in 1995, allowing retrospective monitoring by comparison of two historical maps. Irrigation starting in 1972 has contributed to salt leaching from the topsoil but caused salt accumulation in the subsoil because of insufficient drainage (Bandar-e-Turkman, Golestan, NE Iran). (From Naseri, M.Y., Characterization of salt-affected soils for modelling sustainable land management in semi-arid environment: A case study in the Gorgan region, northeast Iran. ITC dissertation 52, International Institute for Aerospace Survey and Earth Sciences, Enschede, the Netherlands, 2001. With permission)

1.3.2.1.3   Saline Seeps
Seepage is frequent along irrigation canals and in the proximity of water reservoirs and farm ponds, causing salt crusting in and around the seep discharge areas. Together with the excess of irrigation water applied above crop requirements, and the poor maintenance of drainage ways, seeps contribute to raising the subsoil water level and may form a perched water table (Abrol et al., 1988).

1.3.2.1.4   Intrusion of Seawater or Fossil Saline Groundwater
The overexploitation of fresh groundwater for irrigation or urban uses in the proximity of salty water bodies (sea or lake) favors the intrusion of saline water in the freshwater-depleted aquifers. The subsequent rise of the salty groundwater level causes salinization of the subsoil, while the use of the same water for irrigation leads to the salinization of the surface soil. For instance, the rise of the Caspian Sea water level starting in the late 1980s is causing damages to houses and coastal structures, but also favors the encroachment of sea water into the regional aquifers that are used for irrigation in the Golestan province, Iran (Naseri, 2001).

1.3.2.2   Dryland Areas

1.3.2.2.1   Saline Seeps
In low-lying semiarid areas such as rims of playas and shallow water bodies, soils often receive water from below the surface, the evaporation of which causes salts to accumulate close to or on the soil surface. Farming such areas to benefit from scarce water resources in an otherwise dry context increases often surface salt concentration, and this leads ultimately to the abandon of fully salinized spots.

1.3.2.2.2   Effect of Increased Evapotranspiration
In dryland areas, water recharge and discharge are more or less in balance under forest or shrub cover, keeping salts dispersed in the soil mantle. Upon clearing the native vegetation, excessive loss of soil moisture under rainfed agriculture in areas where evapotranspiration exceeds rainfall, leads to salt concentration in soils developed on salt-containing parent material or with saline groundwater.

1.3.2.2.3   Effect of Reduced Evapotranspiration
In contrast to the former situation, evapotranspiration is reduced following the change from a natural forest to a cereal grain crop, or, the introduction of summer fallow management practices in a grain farming system. Such land cover and land use changes, from deep-rooting trees to shallow-rooting annual crops, cause soil moisture content to increase to the point that excess water leaches the soil and the salt-laden water moves to seeps located at the base of hillslopes or in landscape depressions, where salts accumulate as water evaporates. When excess water cannot flow freely through the landscape, water not consumed by the shallow-rooted crops and pasture causes the water table to rise. As the water table rises, the salt lying dormant in the soil mantle dissolves and concentrates, and eventually moves upward to the upper soil horizons and the soil surface.

1.4   MONITORING AND HAZARD PREDICTION: THE CONTRIBUTION OF REMOTE SENSING

In dry and irrigated areas, salts tend to concentrate on the soil surface. As salinity increases, more salts will appear at the soil surface, favoring the use of conventional remote sensing tools. To keep track of changes in salinity and anticipate further degradation, monitoring is needed so that proper and timely decisions can be made to modify the management practices or undertake reclamation and rehabilitation. In general, monitoring salinity changes corresponds actually to the monitoring of the salinization process itself, as facts show that soil salinization resulting from poor soil and water management has considerably increased over time, while reclamation and rehabilitation of salt-affected soils through leaching are much less frequent. Monitoring salinity means first identifying the places where salts concentrate and, second, detecting the temporal and spatial changes in this occurrence. Both largely depend on the peculiar way salts distribute at the soil surface and within the soil mantle, and on the capability of the remote sensing tools to identify salts (Chapters 2 and 3).

1.4.1 ISSUES IN MONITORING SOIL SALINITY

1.4.1.1 Salinity Indicators

Conventional ways of keeping track of changes in soil salinity are based on field observation and laboratory analysis of crops and soils. Usual indicators include: (a) the invasion of salt-tolerant weeds; (b) irregular patterns of crop growth, lack of plant vigor, and other indicators of crop stress that result from direct effect of high osmotic pressure on plant growth and yield; (c) the presence of white crust and efflorescence on saline soils and dark spots on alkaline soils; (d) the presence of white spots and streaks in the soil, even where no surface crusting is visible; and (e) the determination of soil parameters such as pH, electrical conductivity (EC), exchangeable sodium percentage (ESP), and sodium adsorption ratio (SAR). These means are widely used in the conduction of commercial dry-farming, the management of irrigation schemes, and the reclamation and rehabilitation of salt-affected areas for agricultural production. They are best applicable at farm level, although laboratory determinations might be too expensive for frequent replications that would allow accurate monitoring of salinity changes. Remote sensing offers complementary data and constitutes often a less costly, more versatile and timely alternative to the conventional monitoring ways, especially at regional and smaller scales.

1.4.1.2 Data Comparability

Monitoring is based on the comparison of salinity data between at least two dates in a given place to assess changes in soil salinity over time, using remote sensing, field and laboratory data or combinations thereof. This has a number of spatial and temporal implications to secure data comparability (Zinck, 2001; Metternicht and Zinck, 2003).

- As salt-related surface features drastically change with seasons, time series of remote sensing data must be captured in similar periods of the year to be comparable, preferably at the end of the dry season if passive remote sensors are used.
- Monitoring salinity changes from past to present faces the difficulty that, in general, there is no ground-truth information available for past situations. Thus, validating historical remote sensing data involves uncertainties. To overcome this difficulty, fusion of multisource remote sensing data and their integration with field and laboratory data have been advocated.
- Ground-truth data must be collected contemporaneously to validate the spectral signatures from the satellite images. These include radiometric field measurements, surface features' description and mapping, as well as soil sampling for laboratory analyses.
- Georeferencing and coregistration of multitemporal data are essential to enable ground sites to be traced over time and satellite data to be compared with ancillary data. Approaches, such as those developed by the Land Monitor project in Western Australia, require calibration between images to "like-values" so that digital numbers from different dates can be compared (Campbell et al., 1994).
- Remotely sensed data are particularly efficient for the identification and mapping of salt-related surface features. However, soil salinity can substantially change with depth, and surface salinity is frequently controlled by subsoil or substratum salinity and by the seasonal fluctuations of the water table. It is thus important for salinity monitoring to go beyond the soil as a 2D surface area and consider it a 3D body.

1.4.2 APPROACHES TO MONITORING SOIL SALINITY

Mapping, monitoring, and early warning of salinization are best served by a solid synergy between remote sensing, field and laboratory data, and the use of geographic information system (GIS) facilities for processing, transforming, and displaying the data (Metternicht, 1996; Tóth et al., 1998). The reclamation, rehabilitation, and management of salt-affected areas require that not only classes but also severity levels of salinity–alkalinity be determined. This also needs full synergy of multi-source data (Mongkolsawat et al., 1991; Younes et al., 1993; Khalil et al., 1995; De Dapper et al., 1996; Metternicht and Zinck, 1997; Peng, 1998). A variety of approaches to monitoring soil salinization using specific tools has been developed, but case studies with full data integration are still the exception. Farifteh et al. (2006) have proposed an integrated approach that combines remote sensing data, geophysical survey, and solute transport modeling to assess and map salt-affected soils.

1.4.2.1 Comparing Historical Soil Salinity Maps

One effective monitoring approach consists in comparing historical and current soil salinity maps, when historical field and laboratory data are available. This was the case for a region of the Golestan province in northeast Iran (Naseri, 2001). Soil salinity–alkalinity maps of several irrigation schemes were available for 1972, providing EC and pH data from composite samples taken at three depths (0–50, 50–100, and >100 cm) on a regular grid of 500 m intervals. The same points were visited again in 1995 and sampled for EC and pH determinations. The values of the two dates were compared and the differences expressed as discrete map units, in terms of percent salinity or alkalinity increase/decrease and in terms of class changes (Figure 1.4f). This method requires accurate interpolation techniques to generate continuous salinity maps from field observation points and the design of adequate sampling strategies to cater for the spatial variations proper of salt-affected soils (Utset et al., 1998). It also requires consistency of the analytical methods used during the sampling period to secure data comparability (Herrero and Pérez-Coveta, 2005). This kind of information, resulting from monitoring along the third dimension of the soil cover as a volume, still lies largely beyond the current technical capabilities of satellite-based remote sensing but is increasingly provided by proximal subsurface sensing using electromagnetic induction (EM) and ground penetrating radar (GPR).

1.4.2.2 Coping with Salinity Change Uncertainties

Monitoring errors can arise from inaccurate image registration, poor pixel identification and matching, and feature misclassification. There are also error sources when the two images do not represent similar field conditions at the selected dates. This creates uncertainties concerning the likelihood, nature, and magnitude of the changes. To cope with these uncertainties, expert knowledge can be mobilized and formalized as "if–then" rules in a monitoring model, as discussed by Metternicht (2001). In particular, expert judgment allows assessing whether given changes, as detected from the difference image, might really occur or are rather artifacts resulting from inaccurate map overlay. For example, increase of two degrees in salinity or change from full saline to full alkaline are unlikely to take place in a short time interval. This information is incorporated in the model as degrees of likelihood (Figure 1.5). The expert grades the likelihood of occurrence in terms of certainty factors within a range of 0–1, from absolutely unlikely to very likely. The values so obtained are further used to generate a map of likelihood of changes between two selected reference dates. Changes unlikely to occur are corrected to validate two difference maps. One shows the nature of the changes: for example, from nonaffected to saline or from saline to saline–alkaline. The other shows the magnitude of the changes: for example, one class in alkalinity or two degrees in salinity.

1.4.2.3 Incorporating Contextual Landscape Information

Accuracy in mapping and monitoring soil salinity is improved when incorporating contextual landscape information. Geomorphic mapping provides the best informational substrate layer for salinity mapping and monitoring. For example, in the Cochabamba region, Bolivia, highly saline–alkaline areas are strongly correlated with specific geomorphic positions, in particular with playas and lagunary flats, while nonaffected areas are mainly on piedmont glacis. Inclusion of this kind of

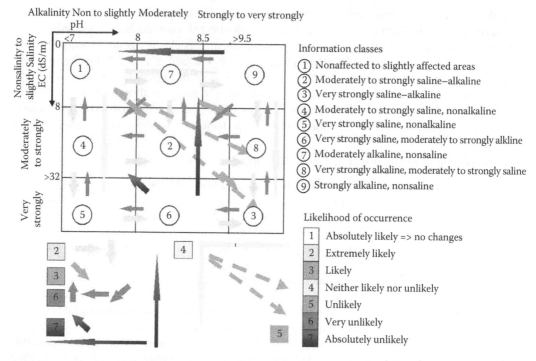

FIGURE 1.5 Graphical representation of the likelihood of changes in salinity–alkalinity. Numbers in the matrix boxes refer to information classes; arrows indicate the likelihood and direction of the changes. (From Metternicht, G.I., Detecting and Monitoring land degradation features and processess in the Cochabamba Valleys, Bolivia: a synergistic approach. ITC Dissertation 36, International Institute for Aerospace Survey and Earth Sciences, Enschede, the Netherlands, 1996. With permission.)

relationships in spatial analysis and modeling led to better spatial discrimination of salinity (Metternicht and Zinck, 1996, 1997). Similarly, in the Quibor area of northern Venezuela, a large semiarid intermountain depression covered by alluvial deposits lying atop lacustrine sediments, the distribution of salts and exchangeable sodium was strongly related to the nature and distribution of the geomorphic units. Low-lying decantation basins had higher pH, EC, and ESP values than the other depositional units, including overflow basins, overflow mantles, and splay mantles (Zinck and Suárez, 1972).

1.4.2.4 Recycling Ancillary Data in GIS Context

Another effective approach for mapping and monitoring salt-affected soils combines digital Landsat TM imagery and ancillary data such as topographic maps, geologic maps, soil maps, land use maps, groundwater quality, and historical data on salinity (Sah et al., 1995). GIS-assisted data integration was found effective in classifying low-salinity and potential saline areas, as well as correcting areas misclassified as extremely or moderately saline. Likewise, Eklund et al. (1998) used GIS modeling for monitoring salinization, by isolating appropriate indicators and determining their weight. They found certain attributes to be more significant than others for discriminating salinity classes. Groundwater depth, in particular, was a prime discriminator of saline discharge.

1.4.2.5 Incorporating Geophysical Data

Airborne EM geophysics has revealed to be a promising technique to measure soil salinity in the third dimension (i.e., variations with depth), especially in situations with extensive saline areas, such as in Australia, where ground determination is less practical (Street and Duncan, 1992; George et al., 1998). According to McGowen and Mallyon (1996), highest accuracy in detecting dryland salinity in New South Wales, Australia, was obtained when using multitemporal Landsat TM imagery and ground salinity data measured with an EM31 instrument providing information on subsoil salinity. Data were georeferenced with a GPS and integrated within a GIS. A similar approach was implemented by Mackenzie et al. (1989) to map dryland salinity recharge areas from the combination of EM34 measurements, vegetation information, infiltration rates, and airborne scanner imagery. Irrigated areas are usually exposed to frequent and rapid variations in salinity resulting from changes in cropping systems and/or in soil and water management practices. In such conditions, a handheld electromagnetic sensor is a cost-effective tool for repetitive measurements of soil salinity. Amezketa (2006) used an EM38 instrument coupled with a software program to assess and map soil salinity in an irrigated plot in the Ebro valley, Spain. Working in another saline irrigation district of Spain, Lesch et al. (1998) developed a statistical monitoring strategy that was able to quantify the temporal changes in the soil salinity patterns from EM induction data and repetitive soil sampling.

1.4.2.6   Integrating Soil Salinity and Hydrology

Separate modeling of soil chemistry and hydrology has led to a good understanding of the processes that control soil and groundwater salinity. However, the intimate relationship between soil and groundwater calls for a better integration of the hydrology and salt chemistry of the vadose zone with the groundwater system. Schoups et al. (2005) have developed a hydrosalinity model able to reconstruct historical changes in soil and groundwater salinity in the western San Joaquin Valley, California.

1.4.3   Prediction of Salinity Hazard

The identification of temporal changes in soil salinity via monitoring allows for extrapolation of spatial trends and constitutes thus an appropriate basis for hazard prediction. Retrospective monitoring, based on time series of satellite images, shows trends that can be extended into the future, within given probability ranges. Thus, change detection can be the basis for change prediction. Metternicht (2001) used a monitoring procedure based on fuzzy logic to highlight areas where major changes in salinity and/or alkalinity have taken place between the situation at the investigation date and the situation at some date in the past for the Cochabamba region in Bolivia. Areas that had undergone significant changes in the recent past were considered as particularly prone to potential increase in salinity and presented, therefore, a severe hazard of salinization (Metternicht, 1996, 2001). The degree of probability for such an evolution to occur can be attached to the mapped areas, as was done for the lower delta of the Amudarya in Russia (Vinogradov, B.V. 1996).

Other methods for predicting areas at risk of dryland salinity use a combination of multitemporal satellite imagery, information derived from large-scale digital elevation models (DEMs), and subjective probability (Evans and Caccetta, 2000; Florinsky et al., 2000). Evans and Caccetta (2000) used decision trees based on the relationships between salinity risk and variables describing landscape features to reproduce expert opinions about the future extent of salinity in a certain area, whereas Florinsky et al. (2000) assumed that the build-up of salts at macrotopographic scale occurs preferably in depressions. De Nys et al. (2005) used a model that calculates soil water and salt balances in the topsoil to predict soil salinity for different on-farm management strategies in an irrigation scheme in northeast Brazil.

1.5   CONCLUSION

All salt in nature originates from the weathering of the primary minerals contained in igneous and metamorphic rocks. Departing from these original source areas, salts geographically disseminate through the landscapes, settle in coastal and inland depressions, and ultimately accumulate in oceans and lakes. Inadequate or poorly adapted land and water management practices in irrigation schemes and dryland farming contribute to increase salt concentration in areas already naturally affected and to contaminate areas originally free of salts. About 76 million ha have been salinized as a consequence of human activities, and this secondary salinization bears high economic costs at farm and development project levels.

Human-induced salinization is the process of increasing the original status of salt content in the soil to levels that are harmful to soils, crops, and infrastructures. Assessing the severity of this process and the related changes in soil salinity in time and space requires monitoring. Data comparability for monitoring purpose is affected by a number of issues including seasonal variability of soil salinity, lack or insufficient ground truth, errors in georeferencing multitemporal data, and poor inventory of subsoil and substratum salinity. Remote sensing, when supported by field and laboratory data and other ancillary information, has proven to be effective in detecting spatial and temporal changes in soil salinity. A variety of approaches combining such data has been implemented to monitor soil salinity and predict salinization hazard.

REFERENCES

Abrol, I.P., Yadav, J.S.P., and Massoud, F.I. 1988. Salt-affected soils and their management. FAO Soils Bulletin 39, Food and Agriculture Organization of the United Nations, Rome, Italy.

Amezketa, E. 2006. An integrated methodology for assessing soil salinization, a pre-condition for land desertification. *Journal of Arid Environments* 67:594–606.

Campbell, N., Furby, S., and Fergusson, B. 1994. Calibrating different images from different dates. Report to LWRRDC, Project CDM1, CSIRO Division of Mathematics and Statistics, Perth, WA, Australia.

Dan, J. and Yaalon, D.H. 1982. Automorphic saline soils in Israel. In: *Aridic Soils and Geomorphic Processes*, Ed. D.H. Yaalon, Catena Supplement 1. Cremlingen, FRG, pp. 103–115.

De Dapper, M., Goossens, R., Gad, A., and El Badawi, M. 1996. Modelling and monitoring of soil salinity and waterlogging hazards in the desert-delta fringes of Egypt, based on geomorphology, remote sensing and GIS. In: *Geomorphic Hazards*, Ed. O. Slaymaker. Wiley, Chichester, UK, pp. 169–182.

De Nys, E., Raes, D., Le Gal, P.Y., Cordeiro, G., Speelman, S., and Vandersypen, K. 2005. Predicting soil salinity under various strategies in irrigation systems. *Journal of Irrigation and Drainage Engineering* 131(4):351–357.

Donahue, R.L., Miller, R.W., and Shickluna, J.C. 1977. *Soils: An Introduction to Soils and Plant Growth*. Prentice-Hall, Englewood Cliffs, NJ.

Eklund, P., Kirkby, S., and Salim, A. 1998. Data mining and soil salinity analysis. *International Journal of Geographical Information Science* 12:247–268.

Evans, F. and Caccetta, P. 2000. Broad-scale spatial prediction of areas at risk from dryland salinity. *Cartography* 29:33–40.

FAO. 2000. Extent and causes of salt-affected soils in participating countries. *Global Network on Integrated Soil Management for Sustainable Use of Salt-Affected Soils*. FAO-AGL website.

Farifteh, J., Farshad, A., and George, R.J. 2006. Assessing salt-affected soils using remote sensing, solute modelling, and geophysics. *Geoderma* 130:191–206.

Florinsky, I., Eiler, R., and Lelyk, G. 2000. Prediction of soil salinity risk by digital terrain modelling in the Canadian prairies. *Canadian Journal of Soil Science* 80:455–463.

Gaucher, G. and S. Burdin. 1974. *Géologie, géomorphologie et hydrologie des terrains salés*. Presses Universitaires de France, Paris, France.

George, R., Beasley, R., Gordon, I., Heislers, D., Speed, R., Brodie, R., McConnell, C., and Woodgate, P. 1998. National airborne geophysics project, Final Report. AFFA, NDSP.

Ghassemi, F., Jakeman, A.J., and Nix, H.A. 1995. *Salinisation of Land and Water Resources: Human Causes, Extent, Management and Case Studies*. The Australian National University, Canberra, Australia, and CAB International, Wallingford, Oxon, UK.

Hecker, F. 2007. Using remote sensing to map soil salinity. Alberta Government, Canada. Available at http://www1.agric.gov.ab.ca./$department/deptdocs.nsf/all/irr7184.

Herrero, J. and Pérez-Coveta, O. 2005. Soil salinity changes over 24 years in a Mediterranean irrigated district. *Geoderma* 125:287–308.

Hurlbut, C.S. and Klein, C. 1977. *Manual of Mineralogy*. John Wiley & Sons, New York.

Khalil, K.A., Fahim, M., and Hawela, F. 1995. Soil reflectances as affected by some soil parameters. *Egyptian Journal of Soil Science* 35:477–491.

Lesch, S.M., Herrero, J., and Rhoades, J.D. 1998. Monitoring for temporal changes in soil salinity using electromagnetic induction techniques. *Soil Science Society of America Journal* 62:232–242.

Mackenzie, M., Bellamy, G., Fraser, D., and Ellis, G. 1989. Mapping dryland salinity recharge using EM conductivity measurements and airborne scanner imagery. In *Proceedings of the 5th Australian Soil Conservation Conference*, Vol. 1, Perth, WA, Australia, pp. 41–48.

McGowen, I. and Mallyon, S. 1996. Detection of dryland salinity using single and multi-temporal Landsat imagery. In *Proceedings of the 8th Australasian Remote Sensing Conference*, Canberra, Australia, pp. 26–34.

Metternicht, G.I. 1996. Detecting and monitoring land degradation features and processes in the Cochabamba Valleys, Bolivia: A synergistic approach. ITC dissertation 36, International Institute for Aerospace Survey and Earth Sciences, Enschede, the Netherlands.

Metternicht, G. 2001. Assessing temporal and spatial changes of salinity using fuzzy logic, remote sensing and GIS: Foundations of an expert system. *Ecological Modelling* 144:163–177.

Metternicht, G.I. and Zinck, J.A. 1996. Modelling salinity–alkalinity classes for mapping salt-affected topsoils in the semi-arid valleys of Cochabamba (Bolivia). *ITC Journal* 2:125–135.

Metternicht, G.I. and Zinck, J.A. 1997. Spatial discrimination of salt- and sodium-affected soil surfaces. *International Journal of Remote Sensing* 18:2571–2586.

Metternicht, G.I. and Zinck, J.A. 2003. Remote sensing of soil salinity: Potentials and constraints. *Remote Sensing of Environment* 85:1–20.

Mongkolsawat, C., Thirangoon, P., and Eiumnoh, A. 1991. A practical application of remote sensing and GIS for soil salinity potential mapping in Korat basin, northeast Thailand. In: *Proceedings International Workshop on Conservation and Sustainable Development*. Asian Institute of Technology, Bangkok, Thailand, pp. 290–297.

Naseri, M.Y. 2001. Characterization of salt-affected soils for modelling sustainable land management in semi-arid environment: A case study in the Gorgan region, northeast Iran. ITC dissertation 52, International Institute for Aerospace Survey and Earth Sciences, Enschede, the Netherlands.

National Land and Water Resources Audit. 2000. Australian dryland salinity assessment 2000: Extent, impacts, processes, monitoring and management options. National Land and Water Resources Audit, Canberra, Australia.

Oldeman, L.R., Hakkeling, R.T.A., and Sombroek, W.G. 1991. World map of the status of human-induced soil degradation: An explanatory note (GLASOD project). ISRIC, Wageninen, the Netherlands and UNEP, Nairobi, Kenya.

Peng, W.L. 1998. Synthetic analysis for extracting information on soil salinity using remote sensing and GIS: A case study of Yanggao Basin in China. *Environmental Management* 22:153–159.

Prellwitz, J.S., Rech, J.A., and Buck, B.J. 2006. Geochemistry of soil salts in the hyper-arid core of the Atacama Desert, Chile. In: *Geological Society of America, Philadelphia Annual Meeting*, Philadelphia (PA), Abstracts, 38(7):544.

Reineck, H.E. and Singh, I.B. 1980. *Depositional Sedimentary Environments*. Springer-Verlag, Berlin, Germany.

Sah, A., Apisit, E., Murai, S., and Parkpian, P. 1995. Mapping of salt-affected soils using remote sensing and geographic information systems: A case study of Nakhon Ratchasima, Thailand. In *Proceedings of the 16th Asian Conference of Remote Sensing*, Thailand, G.3.1–G.3.6.

Schoups, G., Hopmans, J.W., Young, C.A., Vrugt, J.A., Wallender, W.W., Tanji, K.K., and Panday, S. 2005. Sustainability of irrigated agriculture in the San Joaquin Valley, CA. PNAS website, 102(43):15352–15356. Available at http://www.pnas.org_cgi_doi_10.1073_pnas.0507723102.

Street, G. and Duncan, A. 1992. The application of airborne geophysical surveys for land management. In: *Proceedings of the 7th International Soil Conservation Organisation*, Vol. 2, Sydney, Australia, pp. 762–770.

Tóth, T., Kertesz, M., and Pásztor, L. 1998. New approaches in salinity/sodicity mapping in Hungary. In: *Proceedings of the 16th International Congress of Soil Science*. International Soil Science Society, Montpellier, France, 6 pp.

Utset, A., Ruiz, M., Herrera, J., and Deleon, D. 1998. A geostatistical method for soil salinity sample site spacing. *Geoderma* 86:143–151.

Vinogradov, B.V. 1996. Remote sensing monitoring of Saline soils. *Eurasion Soil Science* 29(11):1260–1267.

Younes, H.A., Gad, A., and Rahman, M.A. 1993. Utilization of different remote sensing techniques for the assessment of soil salinity and water table levels in the Serry Command Area, Egypt. *Egyptian Journal of Soil Science* 33:343–354.

Wentz, D. 2000. Dryland saline seeps: Types and causes. Alberta Government, Canada. Available at http://www1.agric.gov.ab.ca./$department/deptdocs.nsf/all/agdex167?opendocument.

Zinck, J.A. 2001. Monitoring salinity from remote sensing data. In: *Proceedings of the 1st Workshop of the EARSeL Special Interest Group on Remote Sensing for Developing Countries*, Eds. R. Goossens and B.M. De Vliegher. Ghent University, Belgium, pp. 359–368.

Zinck, J.A. and C. Suárez. 1972. Condiciones de salinidad y alcalinidad en la depresión de Quibor, Estado Lara. *Revista Agronomía Tropical* XXII (4):405–428.

2 Spectral Behavior of Salt Types

Graciela Metternicht and J. Alfred Zinck

CONTENTS

2.1 INTRODUCTION

Remote sensing of salt-affected soils is favored by a set of factors and conditions peculiar to halomorphic soils, especially the concentration of salts at the soil surface in the form of white crusts with high reflectance and the characteristic reflectance features controlled by salt mineralogy. However, the identification of salt-affected soils through remote sensing is also hampered by the peculiar way salts distribute at the soil surface, in particular their patchy occurrence, gradual spatial variations, and mixture with other soil properties that interfere with salt reflectance. This calls for an approach for classifying and mapping salt-affected soils that is able to cope with their discontinuous and fuzzy distribution on the landscape.

Mapping individual salinity–alkalinity parameters, without taking into consideration the concept and typology of salt-affected soils as defined by soil and agronomic sciences, lacks practical interest. Therefore, this chapter starts reviewing the kind of information classes handled by the conventional approaches to the classification of salt-affected soils, as they provide value ranges and thresholds of the salinity–alkalinity parameters to which the reflectance values are to be compared and correlated. To overcome the constraint of the conventional crisp class boundaries, an approach based on fuzzy set is proposed. Similarly, remote sensing of salt-affected soils is constrained by the types of salt commonly occurring in soils, mainly chlorides, sulfates, and carbonates. As the reflectance of salt-affected soils largely depends on salt mineralogy that determines the presence or absence of absorption features in the electromagnetic spectrum, emphasis is given in this chapter to the chemical composition and mineralogy of the salt types as the main factors controlling their spectral behavior. Finally, we address the advantages and limitations that remote sensing of salt-affected soils can derive from terrain surface features and elements such as the spatial patterns created by salt precipitation at the soil surface, the presence of native halophytic vegetation, and the cultivation of salt-tolerant crops.

2.2 CLASSIFICATION OF SALT-AFFECTED SOILS

The presence of high amounts of water-soluble salts and exchangeable sodium in the soil reduces soil quality and restricts plant growth and crop yield. Determining a salt content threshold to separate saline from nonsaline soils is not straightforward, since some cultivated crops and native vegetation species are more tolerant than others to the effects of high salt content (Driessen and Schoorl, 1973). Thus, proper information classes are required to successfully identify and discriminate saline and alkaline areas from remotely sensed data. A variety of approaches has been implemented to determine such information classes, which can be reduced to two main alternatives based either on strict or fuzzy class boundaries.

2.2.1 CRISP CLASSIFICATIONS

The most common approaches to identify and classify salt-affected soils are based on strict information classes with crisp boundaries, drawn either from soil properties such as soil reaction (pH), electrical conductivity (EC), exchangeable sodium percentage (ESP), and sodium adsorption ratio (SAR), or from the ratios of anions present in the soil saturation extract (Metternicht and Zinck, 2003).

2.2.1.1 Classification Based on Total Salt Content

According to the conventional scheme developed by the U.S. Salinity Laboratory (Richards, 1954), three types of salt-affected soils, namely saline, alkaline (or sodic), and saline–alkaline, are differentiated on the basis of selected values of soil pH, EC, and ESP as shown in Table 2.1. Basic thresholds commonly accepted are >4 dS m^{-1} EC to discriminate saline soils, $>15\%$ ESP to discriminate alkaline soils, and a combination of the former values to discriminate saline–alkaline soils. This approach considers only the total salt level in the soil. This can be a shortfall for some land management applications, as not all salts are equally harmful to soils and plants. For instance, chlorides are more harmful than carbonates or bicarbonates (Table 2.2). This limitation is partially corrected in the classification proposed by Szabolcs (1989), which distinguishes five types of salt-affected soils, including saline soils, alkali soils, magnesium-rich soils, gypsiferous soils, and acid-sulfate soils (Table 2.1).

2.2.1.2 Classification Based on Anion Ratios

A classification of salt-affected soils based on the nature of the salts, in terms of chloride, sulfate, and carbonate anion ratios present in the saturation extract, was first proposed by Russian soil scientists (Plyusnin, 1964). Four categories, namely sulfate soils, chloride–sulfate soils, sulfate–chloride soils, and chloride soils, are recognized according to Cl/SO$_4$ ratios (Table 2.3). A dominance of sulfates over chlorides results in lower anion ratios.

Sulfate-type soils contain the least toxic-soluble salts, are not alkaline, and have about five times more sulfate than chloride anions of sodium, magnesium, and calcium. Chloride–sulfate soils are saline (2%–5% salt content), nonalkaline to slightly alkaline, and characterized by the dominance of sulfates over chlorides. Chloride proportions are up to five times higher in the sulfate–chloride soils,

TABLE 2.1
Chemical and Physical Parameters of Salt-Affected Soils

Soil Types	Chemical Indicators	Predominant Anions	Predominant Cations	Other Properties	Effects on Soil Properties
Saline soils	EC >4 dS m^{-1}	Chlorides, sulfates, and sometimes nitrates; small amounts of bicarbonates	Na: not more than half of the soluble cations	Generally flocculated. Permeability equal to or higher than that of similar nonsaline soils	Higher osmotic pressure
	ESP <15%	Carbonates absent	Ca and Mg: considerable amounts	Presence of white crust on the soil surface	
	pH <8.5		K: not common; sometimes gypsum and lime		
Alkaline (sodic) soils	EC <4 dS m^{-1}	Chlorides, sulfates, and bicarbonates	Na: dominant	Organic matter dispersion and dissolution	Changes in structure
	ESP >15%	Small amounts of carbonates	K: sometimes (exchangeable and soluble)	Clay deflocculation	Decrease in permeability and porosity
	pH 8.5–10		Ca and Mg: small amounts. At high pH and in the presence of carbonates, Ca and Mg precipitate	Columnar or prismatic structure	Changes in soil biological activity pH increases beyond 9 or 10
Saline–alkaline (sodic) soils	EC >4 dS m^{-1}	If excess of salts: appearance and properties similar to those of saline soils (i.e., pH <8.5, clay particles remain flocculated)	Soils become unfavorable for the entry and movement of water and for tillage		
	ESP >15%	If soluble salts are leached downward: appearance and properties similar to those of sodic soils (i.e., pH >8.5 and soil particles are dispersed)			
	pH: variable				

Source: From Richards, L., Ed., *Agriculture Handbook 60*, U.S. Department of Agriculture, Washington, DC, 1954.

TABLE 2.2

Harmful (Gray Cells) and Harmless (White Cells) Salts

Chlorides	Sulfates	Carbonates	Bicarbonates
NaCl	Na_2SO_4	Na_2CO_3	$NaHCO_3$
$MgCl_2$	$MgSO_4$	$MgCO_3$	$Mg(HCO_3)_2$
$CaCl_2$	$CaSO_4$	$CaCO_3$	$Ca(HCO_3)_2$

Source: After Plyusnin, I., *Reclamative Soil Science*, Foreign Languages Publishing House, Moscow, 1964.

TABLE 2.3

Classification of Salt-Affected Soils Based on Anion Ratios

Soil Types	Ratio	Plyusnin	Rosanov	Sadovnikov
Sulfate soils	Cl/SO_4	<0.5	<0.2	<0.2
Chloride–sulfate soils	Cl/SO_4	0.5–1.0	0.2–1.0	0.2–1.0
Sulfate–chloride soils	Cl/SO_4	1.0–5.0	1.0–2.0	1.0–5.0
Chloride soils	Cl/SO_4	>5.0	>2.0	>5.0
Soda soils	CO_3/SO_4			<0.05
Sulfate–soda soils	CO_3/SO_4			0.05–0.16
Soda–sulfate soils	CO_3/SO_4			>0.16

Source: After Plyusnin, I., *Reclamative Soil Science*, Foreign Languages Publishing House, Moscow, 1964.

which are referred to as saline–alkaline soils, with large quantities of soluble salts in the top horizons (3%–5%). Chloride soils contain considerable amounts of harmful salts (e.g., sodium and magnesium chlorides), and chloride anions largely dominate over sulfates.

Using the proportion of carbonates to sulfates present in the soil, the classification further considers soda soils, sulfate–soda soils, and soda–sulfate soils (Table 2.3). Soda soils contain less than 1.5% soluble salts and have a dominance of carbonates, bicarbonates, and sulfates of sodium with pH values ranging from 9.5 to 12. Therefore, these soils are classified as alkaline, nonsaline to slightly saline (Szabolcs, 1989). The presence of sodium carbonate (Na_2CO_3) in the sulfate–soda soils also prompts high alkalinity (pH values between 8.7 and 9.5), but with comparatively higher levels of salinity than those of soda soils. Thus, these soils are referred to as saline–alkaline soils.

In addition to the chemical and physical soil properties mentioned in Tables 2.1 and 2.3, diagnostic horizons are useful to identify salt-affected soils, including salic horizon with soluble salts, natric horizon with exchangeable sodium, gypsic and petrogypsic horizons, sulfuric horizon, and calcic and petrocalcic horizons (Mougenot et al., 1993).

2.2.2 FUZZY CLASSIFICATION

In nature salt contents vary gradually within the soil mantle, horizontally as well as vertically, and thus broad zones of gradual transition may be misrepresented because of the arbitrary assignment of sharp class boundaries (Metternicht, 1998). Crisp set theory is driven by a logic that allows a proposition to have one out of two values: 1 or 0 (true/false). Using crisp classification criteria, an area is considered saline if the EC values are equal to or higher than 4 dS m^{-1}. The question then rises as to whether a soil with an EC value of 3.9 dS m^{-1} is more similar to a saline soil than to a nonsaline one. The classification of such a soil, lying close to a crisp salinity class threshold, can be best addressed using a fuzzy logic approach (Zadeh, 1965). Fuzzy logic is useful for describing the vagueness of entities in the real world (e.g., saline soils), where belonging to a set is really a matter of degree (Malczewski, 1999). An object or a property thereof can have a continuum of membership grades to a set ranging from 0 to 1, instead of a crisp 0 or 1 (no/yes). The membership grade can be interpreted as the possibility for a soil to be saline, rather than a definite inclusion or exclusion based on rigid class boundaries.

A fuzzy set (class) A in X is characterized by a membership function $fA(x)$, which associates a real number in the interval [0, 1] with each point in X. The value of $fA(x)$ represents the grade of membership of x in A. Points belong to a fuzzy set to a greater or minor degree, as indicated by a larger or smaller membership grade. This grade is not a probability measure but an admitted possibility related to the degree to which a given point is compatible with the concept represented by the fuzzy set (e.g., saline soil). In essence, fuzzy logic is a multivalued logic enabling inter-mediate values to be determined between conventional evaluations such as yes/no or true/false (Malczewski, 1999). According to this kind of reasoning, a soil with an EC value of 3.9 dS m^{-1} could now have the possibility of being saline and nonsaline at the same time, defined by the degree of membership to the fuzzy sets saline and nonsaline soils, as shown in Figure 2.1. Thus, such a proposition enables coping with the uneven spatial distribution of salts within the soil profile and across the landscape, better than mapping soil salinity based on crisp classification.

The degree of membership to a fuzzy set is usually determined using a fuzzy membership function. Like with probability distribution functions (e.g., normal, lognormal, rectangular, hyperbolic, etc.), there are different kinds of fuzzy membership function (e.g., lineal, bell-shaped, trapezoidal, etc.). These functions, selected by the user, control the edges and boundary values of the set and determine the transition width and type (e.g., abrupt, smooth). Robinson (2003) provides an excellent summary on the use of fuzzy sets and fuzzy logic in a geographic information system (GIS) context, including a discussion on the types of fuzzy membership function. In some instances, membership functions to fuzzy sets may be drawn from a group of experts (e.g., soil scientists), using an appropriate method for scaling the expert perceptions and assessment.

Metternicht (1998) proposed the definition of transitional fuzzy class boundaries, derived from continuous salinity classes that intermingle gradually, to better represent the real-world situation of saline landscapes. Figure 2.2 illustrates the contrast between the concepts of crisp and fuzzy regions, where a fuzzy region can be defined as a geographical entity with imprecise boundaries (Altam, 1994). Crisp regions present sharp boundaries, and the properties are considered to have homogeneous values within the boundaries of the unit, so that point observations can be extrapolated to characterize the entire unit. In this context, a salinity attribute can take the value of either 1 or 0

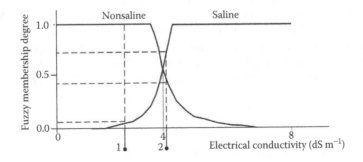

FIGURE 2.1 Membership functions for nonsaline and saline fuzzy sets, defined using EC values. Soil sample 1 has a fuzzy membership degree of 1 to the set "nonsaline soils" and a degree of 0.1 to the set "saline soils." Soil sample 2 belongs to the fuzzy set "saline soils" with a degree of 0.7 and to the fuzzy set "nonsaline soils" with a degree of 0.45.

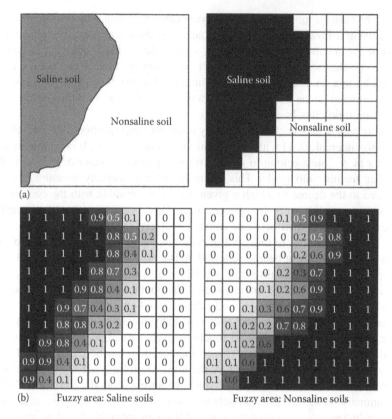

(b)   Fuzzy area: Saline soils   Fuzzy area: Nonsaline soils

FIGURE 2.2   (a) Crisp regions delineating saline and nonsaline soils; (b) fuzzy regions assuming a smooth transition between areas of saline and nonsaline soils.

at each location (i.e., the cells in Figure 2.2a). In a fuzzy region, the concentration of the attribute soil salinity takes a grade of membership that varies according to location (Figure 2.2b). This replicates what occurs in nature, where salinity levels vary vertically and horizontally because of the dynamic nature of the process of salinization and salt accumulation in the landscape.

2.3   SALT TYPES OCCURRING IN SOILS

Salt-affected soils vary according to the chemical composition of the salts. In the processes of salinization, the various salts accumulate in a proportion inverse to their solubility. In nonsaline or slightly saline soils, Ca^{2+}, Mg^{2+}, and HCO_3^- predominate in the soil solution. As salinity increases, the more soluble salts become dominant. Strongly saline soils contain large quantities of NaCl and Na_2SO_4. Sodium salts capable of alkaline hydrolysis, such as Na_2CO_3 and $NaHCO_3$, prevail in alkaline soil-forming processes. The main characteristics of the salt compounds are described hereafter.

2.3.1   CARBONATES

Carbonates exert different effects on soils, depending on the cation the carbonate is bound to, the amount accumulated in the soil and the solubility. For instance, calcium carbonate may occur as a pure chemical but is also found as a calcium–magnesium mixture. The latter has very low solubility in water, which depends on the carbonate concentration and the pH of the solution. The precipitation/dissolution equilibrium of carbonates in the soil is governed by two factors: the CO_2

pressure of the soil atmosphere and the concentration of dissolved ions in the soil solution (Driessen and Dudal, 1991). The equilibrium can be represented as follows:

$$CO_2 + H_2O \rightarrow H_2CO_3$$
$$CaCO_3 + H_2CO_3 \rightarrow Ca(HCO_3)_2$$

An increase in carbon dioxide of the soil air leads to a decrease in the soil pH, thus increasing the solubility of calcium carbonate. Precipitation occurs if the CO_2 pressure lowers and the pH increases.

Magnesium carbonate occurs in soils in the form of magnesite ($MgCO_3$). A saturated solution of magnesium carbonate has a pH of around 10. Its formation results from the chemical decomposition of crystalline rocks. It might be toxic, especially when the ratio Ca:Mg in the soil solution and exchange complex is unbalanced, with more magnesium than calcium cations.

Sodium carbonate is a highly soluble compound, which, as a result of alkaline hydrolysis, reaches pH values between 9 and 11. The formation of bicarbonate reduces soil alkalinity regardless of the anion bound. The transformation of carbonate into bicarbonate is a reversible process, and a decrease in carbon dioxide causes the formation of carbonate from bicarbonate. This happens when the activity of microorganisms is weak and the content of organic matter low. Sodium carbonate accumulations result from carbonates and bicarbonates. The absence of calcium sulfate compounds allows the formation of soda–saline soils.

Sodium carbonate can develop under different processes and conditions, in particular through biological reduction, chemical weathering due to the interaction of silicates and water containing CO_2, clayey texture, and the interaction between NaCl or Na_2SO_4 and $CaCO_3$ (Szabolcs, 1989). When the soil reaches a total salt concentration of 0.3%–0.5%, Na_2CO_3 precipitates in preference to the other salts, as illustrated in Figure 2.3 (Kovda, 1947; Szabolcs, 1989).

2.3.2 SULFATES

Magnesium and sodium sulfates are highly soluble and mobile salts with harmful effects on soils. Magnesium sulfate precipitates in the form of epsomite that does not accumulate in soils as a pure salt; it mixes with other readily soluble salts, such as sodium sulfate, sodium chloride, and magnesium chloride (Szabolcs, 1989). As the concentration of sodium sulfate increases, the saturation of magnesium sulfate decreases. Sodium sulfate crystallizes in different ways according to temperature. Under relatively low temperature, it precipitates in the form of needle-shaped mirabilite,

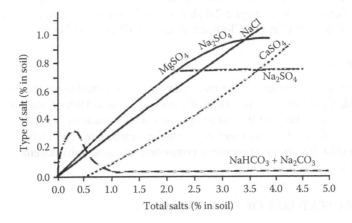

FIGURE 2.3 Relationship between the degree of alkalinization of the soil and the accumulation of alkaline salts. (Modified from Kovda, V., *The Genesis and Regime of Salt-Affected Soils*, Izd. Ak., Moscow-Leningrad, 1947.)

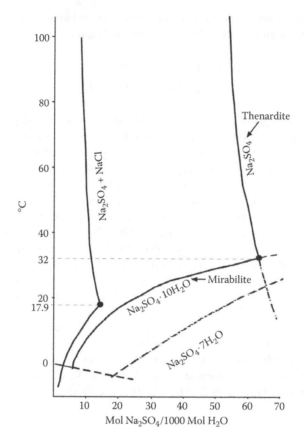

FIGURE 2.4 Na_2SO_4–H_2O system, where thenardite precipitates as needle-shaped mirabilite in the absence of NaCl (right-hand side curve) and at saturation in the presence of NaCl (left-hand side curve). (Modified from Braitsh, O., *Salt Deposits, Their Origin and Composition*, Springer-Verlag, New York, 1971.)

which dehydrates and turns into white powdery thenardite when temperature increases. In the Na_2SO_4–H_2O system, the stable phases are mirabilite, also known as Glauber's salt, thenardite and heptahydrite, the latter not definitely identified in nature. The solubility of mirabilite and thenardite is reduced by the addition of NaCl. Figure 2.4 shows that the mirabilite–thenardite transition point is lowered from 32°C to 17.9°C in a NaCl-saturated solution (Braitsh, 1971).

2.3.3 CHLORIDES

Together with sulfates, chlorides are the main compounds responsible for the formation of saline soils. They are highly soluble and may be highly toxic (Szabolcs, 1989). Calcium chloride ($CaCl_2$) is rarely present in soils, because it commonly reacts with sodium sulfate and sodium carbonate, precipitating in the form of calcium sulfate or calcium carbonate. The highly soluble and toxic sodium chloride (NaCl) is the most common component of saline soils, together with sodium and magnesium sulfates.

2.4 SPECTRAL FEATURES OF SALT MINERALS

The preceding section describes the common salt types in terms of their chemical composition, their effects on soils, and the kinds of mixing that usually occur in salt-affected landscapes. This section presents a summary of the spectral behavior of common salt types as derived from research works conducted in the period 1970–2007, mostly in the spectral regions of the visible, near-infrared and shortwave infrared (V-NIR-SWIR), and in the thermal infrared to a lesser extent. It is noteworthy that many of the reported spectral reflectance curves were derived from measurements taken under controlled laboratory conditions. In natural conditions, salts intermingle and the reflectance properties of salt minerals are affected by factors such as impurities, elemental composition, and crystalline structure (Hunt and Salisbury, 1970), in addition to the effect of soil particle size distribution and the interference with other soil properties. Departures from the ideal reflectance curves can be expected when ground-based, airborne, or satellite-borne sensors are used for spectral measurements of salinity in field conditions.

The mineralogy of the carbonate, sulfate, and chloride salts determines the presence or absence of absorption features in the electromagnetic spectrum. Hunt and Salisbury (1970, 1971) found that the V-NIR-SWIR absorption bands in the spectra of saline minerals were largely associated with internal vibration modes due to the excitation of overtones and combination tones of the fundamental anion groups (e.g., HOH, OH^-, CO_3^{2-}, and SO_4^{2-}). Using laboratory spectrometry, Csillag et al. (1993) identified six spectral ranges useful to characterize the salinity status of soils undergoing different salinization and alkalinization processes. These ranges are located in the visible (550–770 nm), NIR (900–1030 and 1270–1520 nm), and mid-infrared (1940–2150, 2150–2300, and 2330–2400 nm). Hereafter, we briefly discuss the location of the spectral features of common salt compounds found in soils.

2.4.1 CHLORIDES

Pure halite (NaCl) does not induce absorption bands in the visible, NIR, and thermal infrared (Hunt et al., 1971). Several authors report absorption bands near the 1400, 1900, and 2250 nm spectra of halite (Figure 2.5) that are attributed to moisture and fluid inclusions (Crowley, 1991; Mougenot et al., 1993; Howari et al., 2002; Farifteh, 2007). A continuum removal reflectance spectrum of soil samples treated with halite and sylvite (KCl) shows consistent absorption features at 1440 and 1933 nm; while soil samples treated with a $MgCl_2$ solution (bischofite) exhibit absorption features at wavelengths of 1190 and 1824 nm (Farifteh, 2007). Studies by Howari et al. (2002) and Farifteh (2007) show a trend of increased feature absorption depth with increased salt concentration. Calcium chloride rarely occurs in soil because of its tendency to react with sodium sulfates and sodium carbonates (see Section 2.3.3).

2.4.2 SULFATES

Nearly all vibration features in the V-NIR-SWIR spectra of sulfate minerals are due to OH^- vibrations (Hunt et al., 1971). Mulders (1987) reports the location of absorption bands around 1500–1730 nm for soil surface features rich in gypsum ($CaSO_4 \cdot 2H_2O$). Howari et al. (2002) identified an additional absorption feature at 1978 nm, that is similar to an absorption band recorded for thenardite (Na_2SO_4), making it thus difficult to distinguish between gypsum and thenardite in soils. Mulders (1987) also signals an absorption band for sulfate anions near 10,200 nm (thermal IR), caused by overtones or combination tones of internal vibrations of constitutional water molecules. Epsomite ($MgSO_4 \cdot 7H_2O$) presents several absorption features in the V-NIR-SWIR (i.e., 793, 999, 1240, 1490, 1631, 1760, 1946, 3270, and 3400 nm), as reported by Moenke (1962) and Farifteh (2007).

2.4.3 CARBONATES AND BICARBONATES

Carbonates present absorption features in the thermal range (11,000–12,000 nm) due to internal vibrations of the CO_3^{2-} group. Carbonate absorption features were also reported at 2340 nm (Siegal and Gillespie, 1980). Calcareous soils have several absorption bands in the SWIR at 2250, 2350, 2380, and 2465 nm due to the internal vibrations of the carbonate ion in calcite ($CaCO_3$) (Farifteh et al., 2006). Howari et al. (2002) identified additional calcite absorption features at 1052, 1479, and

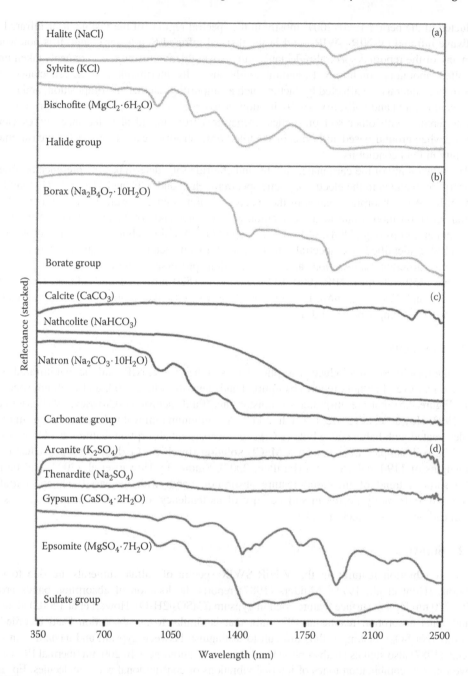

FIGURE 2.5 Laboratory-derived spectra of salt minerals: (a) halide group, (b) borax, (c) carbonate group, and (d) sulfate group. (From Farifteh, J., Imaging spectroscopy of salt-affected soils: Model-based integrated method. ITC dissertation 143, Enschede, the Netherlands, 2007. With permission.)

2100 nm. In soils with mixtures of nathcolite and calcite ($NaHCO_3$–$CaCO_3$), nathcolite-related spectral features (i.e., absorption bands at 1334, 1472, and 1997 nm) dominate because the less-soluble calcite precipitates first, leaving the more soluble salts to dominate in the spectral contribution (Howari et al., 2002).

Compared to these laboratory-based measurements, field-based spectral discrimination between types of salt-affected soils using optical remote sensing data (V-NIR-SWIR) is more complicated, because variations in soil reflectance cannot be attributed to a single soil property or salt type. The salinization process is per se complex in nature, affecting both physical and chemical soil properties (Csillag et al., 1993).

2.5 FACTORS AFFECTING THE REFLECTANCE OF SALINE AND SODIC SOILS

Surface salinity is a highly dynamic process. This constrains the identification of salt-affected soils and the monitoring of the salinization process because it influences the spectral, spatial, and temporal behavior of the salt features. Ground observations and radiometric measurements show that the main factors affecting the reflectance of salt-affected soils are quantity and mineralogy of salts, together with soil moisture, color, and surface roughness. Soil surface salinity can be detected from remotely sensed data, obtained by ground-based, airborne, or space-borne sensors, through direct indicators that refer to salt features visible at the soil surface and indirect indicators that refer to contextual features, such as the presence of native halophytic vegetation or the performance level of salt-tolerant crops (Metternicht and Zinck, 2003; Farifteh, 2007). Figures 2.6 and 2.7 illustrate some examples of direct and indirect field indicators of soil salinity.

2.5.1 SALT FEATURES AT THE SOIL SURFACE

Soil surface indicators are used when mapping bare soil areas, including features such as efflorescence figures, crusts, surface sealing, deflation spots, and moist pans with distinctive spectral characteristics (Mabbut, 1986; Mougenot et al., 1993). Figure 2.6 illustrates soil surface-related indicators of salinity.

* Saline efflorescence and crusting occur in places where salts precipitate directly on the soil surface, after capillary rise of salt solutions from the subsoil or water table. Crusts vary largely in spatial continuity, thickness, roughness, and color. Usually, cracking and curling occur upon desiccation. Salt mineralogy produces distinctive macromorphological features at the soil surface. For instance, puffy crusts form because of the abundance of sodium sulfates, while smooth salt crusts are due to the presence of chlorides (Driessen and Schoorl, 1973).
* Surface sealing takes place in the presence of high sodium content that contributes to weaken soil aggregation and causes soil dispersion in sodic soils. The dispersive effect of exchangeable Na is due to the highly hydrated nature of this ion and its thick water shell, preventing clay particles to flocculate. Sealing is commonly associated with the dissolution of soil organic matter in presence of high pH and the concentration of the dispersed humates as dark incrustation on the soil surface.
* Salt scalds form by wind or water erosion. Aeolian deflation is particularly active in areas affected by surface cracking or soil dispersion. On the margin of the scalds, rills and gullies may form, while the breakdown of puffy surfaces over the loose topsoil increases wind erosion.
* Saline seeps in areas of active water discharge cause the formation of moist horizons above perched saline water tables, which increase capillary rise of salts to the soil surface.

The tendency of salts to concentrate locally, in patchy spots, creates variable mixtures of features within individual pixels. Structure and color of saline crusts can vary considerably, resulting in large variations of reflectance (Escadafal, 1989; Epema, 1992; Metternicht and Zinck, 1997). The alteration of a saline crust by trampling, or the curling of an alkaline crust by desiccation, increases surface roughness, which causes reflectance to decrease in comparison to that of a smooth crust with similar salt content (Metternicht and Zinck, 2003).

FIGURE 2.6 Soil surface indicators of salinity. (Photo courtesy of Zinck and Metternicht.) (a) Glassy, nonsalty soil surface sealed by fine particles that have been detached from soil aggregates by heavy rains; reflectance of a sealed silty surface might be comparable to or higher than that of a salt crust (Sacaba, Cochabamba region, Bolivia); (b) regular polygon and dot pattern of the crust covering the Uyuni salt lake at 3650 m elevation (Southern Bolivia); (c) puffy crust formed from soil material rich in sodium sulfate; microtopography caused by crust curling influences the radar backscattering (Cliza, Cochabamba region, Bolivia); (d) smooth salt crust (Cliza, Cochabamba region, Bolivia); (e) gypsum-rich efflorescence crust at the surface of a chott depression (Chott-el-Djerid, Southern Tunisia).

2.5.2 Presence of Halophytic Vegetation

The presence of salts and/or sodium ions in high proportions creates conditions adverse to vegetation growth, resulting usually in bare terrain surfaces. However, lower salt concentrations allow specialized halophytic plant communities (e.g., *Salicornia* spp.) to form that can be used as indirect indicators for the cartography of saline and/or alkaline areas. Likewise, the presence of dead

FIGURE 2.7 Variation in surface patterns. (Photo courtesy of Zinck and Metternicht.) (a) White salt efflorescence pattern on cracked yellow brown soil surface; color mixture decreases reflectance as compared to that of a pure salt crust (Cliza, Cochabamba region, Bolivia); (b) powdery efflorescence pattern, exposed to selective deflation by wind that lets appear spots of the underlying dark topsoil horizon and causes mixed pixels (Cliza, Cochabamba region, Bolivia); (c) thin salt crust atop a dark topsoil horizon; resists wind deflation better than a powdery efflorescence pattern (Cliza, Cochabamba region, Bolivia); (d) mixed pixel with intermingled patches of *Cynodon dactylon* grass and sealed surface areas (Cliza, Cochabamba region, Bolivia); (e) patched crop development in a paddock of the agricultural belt of Western Australia; (f) presence of dead trees in a highly saline area, located in the southwest of the Western Australia agricultural belt; (g) playa landscape with a mixture of surface patterns, including mounds covered by *C. dactylon* grass in the middle ground, mounds showing wind grooves carving the grass cover in the foreground, and salt crust in the intermediate depressions; short-distance association of contrasting surface features causes mixed pixels (Cliza, Cochabamba region, Bolivia).

trees and shrubs (Figure 2.7f) and the invasion of salt-tolerant pioneer plants colonizing previously vegetated areas are landscape indicators of ongoing salinization.

Halophytic plants are common in saline and alkaline areas, but not all have revealed to be good remote sensing indicators of soil salinity. For instance, the spectral curve of salt-tolerant Chenopodiaceae, obtained from ground-based radiometric measurements in the Cochabamba area, Bolivia, was found to be comparable to that of chlorophyll-rich vegetation, with strong absorption in the visible range of the spectrum and higher reflectance in the NIR (Metternicht, 1996). In contrast,

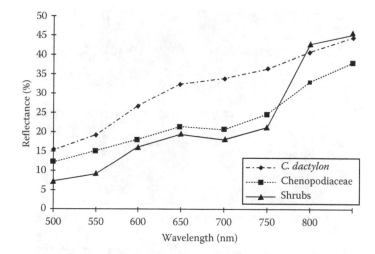

FIGURE 2.8 Reflectance curves of halophytic vegetation obtained with a Cropscan multiband radiometer. Salt-tolerant Chenopodiaceae shrubs follow a trend comparable to chlorophyll-rich vegetation, while *C. dactylon*, a salt-tolerant grass, presents a spectral curve that increases monotonically through the V-NIR bands.

the spectral curve of *C. dactylon*, also a halophyte, increased monotonically throughout the visible and NIR wavebands, indicating lower chlorophyll content (Figure 2.8). In this Bolivian study area, the remote sensing of halophytic vegetation allowed distinguishing saline–alkaline areas from nonaffected ones.

Vegetation indicators can be used in different ways for remote mapping of salinized landscapes. Spatial distribution of soil salinity can be detected by using contrasted associations of vegetation and bare soils (Metternicht and Zinck, 1997; Zhang et al., 1997; Szabo et al., 1998) or by taking into consideration vegetation types at macroscale level and vegetation status, as influenced by salinity stress, at microscale level (Dale et al., 1986; Wiegand et al., 1994a,b).

2.5.3 Crop Performance Indicators

Some crops such as alfalfa, barley, and cotton can be grown on salt-affected soils with variable success, and their level of performance reflects thus the severity of salinity–alkalinity. The degree to which the performance of these crops is diminished by the presence of salt or exchangeable sodium can be inferred from a set of indicators that include features such as

- Sparse germination of crop plants and patched crop development, with significant variations in growth at paddock level and farm level (Figure 2.7e)
- Evidence of plant moisture stress because of high osmotic concentration of the soil solution
- Decrease in plant cover and invasion of salt-tolerant weeds
- Reduction in crop yield (Madrigal et al., 2003; Lobell et al., 2007)
- Leaf angle orientation (leaf roll) and increased chlorosis (Bastiaanssen et al., 2000)

Cotton appears to be the ideal performance-oriented indicator because it is largely cultivated in irrigated drylands. The health status of cotton plants and cotton yield are strongly correlated with EC. The health condition of the crop can be estimated from color infrared aerial photographs, video images, and multispectral images using spectral brightness coefficients, photo densities, and the normalized difference vegetation index (NDVI) (Golovina et al., 1992). High correlations between the NDVI of cotton and sugarcane crops and the soil EC have been used to separate saline–alkaline soils from nonaffected ones (Wiegand et al., 1994a,b).

However, performance-oriented indicators should be used with caution. Lobell et al. (2007) warn on the use of interannual crop yield changes as an indicator of soil salinity at regional level. Their study, based on randomly sampled EC data that were correlated with wheat yield data derived from remote sensing, shows that many factors besides soil salinity contribute to yield losses throughout an agricultural region. Thus, yield mapping of a single year does not generally provide a reliable estimator of soil salinity, especially when highly saline soils are infrequent. In contrast, Madrigal et al. (2003) report successful correlation between remote sensing-derived yield of wheat and salinity, in a case study where soil salinity values were as high as 20 dS m^{-1}. Lobell et al. (2007) obtained consistent correlation between yield losses and soil salinity using 6-year temporal series of yield images derived from remote sensing data.

2.5.4 Spectral Confusions with Nonsaline Surface Features

Surface features common in drylands, such as braided stream beds, eroded terrain surfaces with truncated soils, and nonsaline silt-rich structural crusts, can generate high levels of reflectance, similar to those of areas with high salt concentration. Structural crusts in particular can cause spectral confusion with saline crusts. The former result from the concentration of the finer soil particles at the terrain surface after destruction of the soil aggregates during a rainstorm or an irrigation event. This kind of surface sealing or crusting causes an overall increase in reflectance, which might be up to 12% in the NIR and SWIR spectral regions according to De Jong (1992) and Goldshleger et al. (2001).

Using multispectral field measurements in the V-NIR spectral range in the Cochabamba area of Bolivia, Metternicht (1996) found spectral confusions between salt crusts and bright silt-loam structural crusts in the blue and green regions of the spectrum (450–550 nm) (Figure 2.9). Surface brightness due to high silt content determines higher reflectance than that of a puffy or smooth salty crust. In general, nonsaline crusts have lower reflectance and slightly different curve

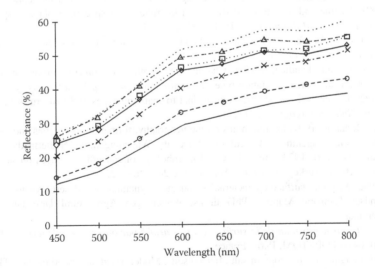

FIGURE 2.9 Soil surface reflectance spectra in similar textural conditions but with varying crust patterns and salt contents. (········) silt loam bright crust; (−·-△-·-) silt loam salty crust; (··-□-··) silt loam crust; (—○—) silt loam puffy surface; (-·×-·-) silt loam dull crust; (—◇—) silt loam smooth crust (0.01% O.M., 60% silt, 22% clay contents); (———) silt loam smooth crust (0.87% O.M., 56%, silt 14% clay contents). (From Metternicht, G. and Zinck, J.A., *Rem. Sens. Environ.*, 85, 1, 2003. Farifteh, J., Imaging spectroscopy of salt-affected soils: Model-based integrated method. ITC dissertation 143, Enschede, the Netherlands, 2007. With permission.)

shape than salt crusts. The puffy, sodium-sulfate-rich crusts seem to respond more to the influence of silt than to salt content, as the curve shape and level of reflectance are similar to those of the bright silt-loam crust. A clear decrease in reflectance is observed as the crusts become nonsaline, together with lower silt and clay contents.

2.6 CONCLUSION

Remote detection and mapping of salt-affected soils are favored by a set of factors and conditions proper to halomorphic soils, especially the concentration of salts at the soil surface in the form of white crusts with high reflectance in the V-NIR regions of the spectrum, and the characteristic reflectance features controlled by salt mineralogy. However, the spatial patterns of salt distribution at the soil surface can also restrict remote identification of salt-affected soils. Remote sensing-based approaches to map and monitor salinized landscapes should take into consideration: (1) the discontinuous way in which salts tend to distribute on the landscape, so that appropriate classification schemes can be developed, and (2) the mineralogy of salt types, as this controls the occurrence of spectral absorption features in specific regions of the electromagnetic spectrum, while also influencing the appearance of soil surface indicators of salinity (e.g., puffy crusts, and land cover patterns). Salinity surface indicators can be direct, based on salt features clearly visible at the soil surface, or indirect, referring to contextual features such as the presence of native halophytic vegetation or the performance level of salt-tolerant crops.

REFERENCES

Altam, E.I. 1994. Fuzzy set theoretic approaches for handling imprecision in spatial analysis. *International Journal of Geographical Information Systems* 8: 271–289.

Bastiaanssen, W.G.M., Molden, D.J., and Makin, I.W. 2000. Remote sensing for irrigated agriculture: Examples from research and possible applications. *Agricultural Water Management* 46: 137–155.

Braitsh, O. 1971. *Salt Deposits, Their Origin and Composition.* Springer-Verlag, New York.

Crowley, J. 1991. Visible and near-infrared (0.4–2.5 μm) reflectance spectra of playa evaporate minerals. *Journal of Geochemical Research* 96 (B10): 16231–16240.

Csillag, F., Paztor, L., and Biehl, L. 1993. Spectral band selection for the characterization of salinity status of soils. *Remote Sensing of Environment* 43: 231–242.

Dale, P., Hulsman, K., and Chandica, A. 1986. Classification of reflectance on color infrared aerial photographs and sub-tropical salt marsh vegetation types. *International Journal of Remote Sensing* 7: 1783–1788.

De Jong, S. 1992. The analysis of spectroscopical data to map soil types and soil crusts of Mediterranean eroded soils. *Soil Technology* 5: 199–211.

Driessen, P. and Dudal, R. 1991. Lecture notes on the geography, formation, properties and use of the major soils of the world. Wageningen Agricultural University, Wageningen, the Netherlands.

Driessen, P. and Schoorl, R. 1973. Mineralogy and morphology of salt efflorescences on saline soils in the Great Konya Basin, Turkey. *Journal of Soil Science* 24: 436–442.

Epema, G.F. 1992. Mapping surface characteristics and their dynamics in a desert area in southern Tunisia with Landsat Thematic Mapper. PhD thesis, Wageningen Agricultural University, Wageningen, the Netherlands.

Escadafal, R. 1989. *Caractérisation de la surface des sols arides par observation de terrain et par télédétection.* Editions de l'ORSTOM, Paris, France.

Farifteh, J. 2007. Imaging spectroscopy of salt-affected soils: Model-based integrated method. ITC dissertation 143, Enschede, the Netherlands.

Farifteh, J., Farshad, A., and George, R. 2006. Assessing salt-affected soils using remote sensing, solute modelling and geophysics. *Geoderma* 130: 191–206.

Goldshleger, N., Ben-Dor, E., Benyamini, Y., Agassi, M., and Blumber, D. 2001. Characterization of soil's structural crust by spectral reflectance in the SWIR region (1.2–2.5 μm). *Terra Nova* 13: 12–17.

Golovina, N.N., Minskiy, D., Pankova, Y., and Solovyev, D.A. 1992. Automated air photo interpretation in the mapping of soil salinization in cotton-growing zones. *Mapping Sciences and Remote Sensing* 29: 262–268.

Howari, F., Goodell, P., and Miyamoto, S. 2002. Spectral properties of salt crusts formed on saline soils. *Journal of Environmental Quality* 31: 1453–1461.

Hunt, G. and Salisbury, S. 1970. Visible and near-infrared spectra of minerals and rocks. I. Silicate minerals. *Modern Geology* 1: 283–300.

Hunt, G. and Salisbury, J. 1971. Visible and near-infrared spectra of minerals and rocks. II. Carbonates. *Modern Geology* 2: 23–30.

Hunt, G., Salisbury, J., and Lenhoff, C. 1971. Visible and near-infrared spectra of minerals and rocks. IV. Sulphides and sulfates. *Modern Geology* 3: 1–4.

Kovda, V. 1947. *The Genesis and Regime of Salt-Affected Soils*, Vol. 2. Izd. Ak. Moscow-Leningrad, USSR.

Lobell, D., Ortiz-Monasterio, J., Cajigas-Gurrola, F., and Valenzuela, L. 2007. Identification of saline soils with multiyear remote sensing of crop yields. *Soil Science Society of America Journal* 71: 777–783.

Mabbutt, J.A. 1986. Desertification indicators. *Climatic Change* 9: 113–122.

Madrigal, L.P., Wiegand, C.L., Merz, J.G., Rubio, B.D., Estrada, X.C., and Ramirez, O.L. 2003. Soil salinity and its effect on crop yield: A study using satellite imagery in three irrigation districts. *Ingeniería Hidráulica en México* 18: 83–97.

Malczewski, J. 1999. *GIS and Multicriteria Decision Analysis*. John Wiley & Sons, New York.

Metternicht, G.I. 1996. *Detecting and Monitoring Land Degradation Features and Processes in the Cochabamba Valleys, Bolivia: A Synergistic Approach*. ITC Publication 36, Enschede, the Netherlands.

Metternicht, G.I. 1998. Fuzzy classification of JERS-1 SAR data: An evaluation of its performance for soil salinity mapping. *Ecological Modelling* 111: 61–74.

Metternicht, G.I. and Zinck, J.A. 1997. Spatial discrimination of salt- and sodium-affected soil surfaces. *International Journal of Remote Sensing* 18: 2571–2586.

Metternicht, G.I. and Zinck, J.A. 2003. Remote sensing of soil salinity: Potentials and constraints. *Remote Sensing of Environment* 85: 1–20.

Moenke, H. 1962. *Mineralspektren I*. Akademie-Verlag, Berlin.

Mougenot, B., Epema, G., and Pouget, M. 1993. Remote sensing of salt affected soils. *Remote Sensing Reviews* 7: 241–259.

Mulders, M. 1987. *Remote Sensing in Soil Science*. Elsevier, Amsterdam, the Netherlands.

Plyusnin, I. 1964. *Reclamative Soil Science*. Foreign Languages Publishing House, Moscow.

Richards, L., Ed. 1954. Diagnosis and improvement of saline and alkali soils. *Agriculture Handbook 60*. U.S. Department of Agriculture, Washington, DC.

Robinson, V. 2003. A perspective on the fundamentals of fuzzy sets and their use in geographic information systems. *Transactions in GIS* 7: 3–30.

Siegal, B. and Gillespie, A.R. 1980. *Remote Sensing in Geology*. John Wiley & Sons, New York.

Szabo, J., Pásztor, L., Suba, Z., and Varallyay, G. 1998. Integration of remote sensing and GIS techniques in land degradation mapping. In *Proceedings of the 16th International Congress of Soil Science*, Montpellier, France.

Szabolcs, I. 1989. *Salt Affected Soils*. CRC Press, Boca Raton, FL.

Wiegand, C.L., Anderson, G., Lingle, S., and Escobar, D. 1994a. Soil salinity effects on crop growth and yield. Illustration of an analysis and mapping methodology for sugarcane. *Journal of Plant Physiology* 148: 418–424.

Wiegand, C.L., Rhoades, J.D., Escobar, D.E., and Everitt, J.H. 1994b. Photographic and videographic observations for determining and mapping the response of cotton to soil salinity. *Remote Sensing of Environment* 49: 212–223.

Zadeh, L. 1965. Fuzzy sets. *Information and Control* 8: 338–353.

Zhang, M., Ustin, S., Rejmankova, E., and Sanderson, E. 1997. Monitoring Pacific-Coast salt marshes using remote sensing. *Ecological Applications* 7: 1039–1053.

3 Review of Remote Sensing-Based Methods to Assess Soil Salinity

*Eyal Ben-Dor, Graciela Metternicht, Naftaly Goldshleger,
Eshel Mor, Vladmir Mirlas, and Uri Basson*

CONTENTS

3.1 INTRODUCTION

Remote sensing is the process of collecting data about the earth's surface and environment from a distance, usually by sensors mounted on ground equipment, aircraft, or satellite platforms. Depending on the spectral position of the bands, sensors collect energy that is reflected (visible [VIS] and infrared), emitted (thermal infrared [TIR]), or backscattered (microwave) by a landscape surface and/or the atmosphere (Metternicht, 2007).

There are different ways to classify data acquired by remote sensors, according to the source of energy (passive or active sensors), the type of platform (ground-, air, or space-borne), the region of the spectrum used to image the earth's surface (optical, infrared, and microwave), the platform trajectory distinguishing between sun-synchronic and geostationary satellites, the number and width of spectral bands (panchromatic, multispectral, hyperspectral), the spatial resolution (high, medium also known as Landsat-like, and low), the spatial coverage (point or image view), the temporal resolution (hourly, daily, and weekly revisiting frequency), and the radiometric resolution (8, 12, and 16 bits) (Metternicht, 2007). Passive remote sensors use the sun's irradiation and/or the earth's emission in the visible, near-infrared, and shortwave infrared (VIS–NIR–SWIR) (400–2500 nm) and in the TIR (2,500–14,000 nm) spectral regions. These sensors provide information on spatial and spectral domains from elevations above the ground that range from a few meters (field sensors) to 800 km (orbital sensors), with spatial resolutions varying from a few centimeters to tens of meters, respectively. The spectral configuration of passive sensors varies from multispectral (a few bands) to hyperspectral (tens to hundreds of bands) (Vane et al., 1984). Despite the perception that higher spectral information provides better detection potential, multispectral imagery has been used successfully to map soil salinity in several continents (Madani, 2005). For that purpose, not only direct indicators of salinity such as salt crust and efflorescence, but also indirect indicators such as the vegetation response to increasing salt content (stress, type, and density) were used (Metternicht and Zinck, 2003).

This chapter is based on previous studies that used ground-, air-, and satellite-borne sensors in a multisensor or individual mode to detect, map, and/or monitor salt-affected soils at surface and subsurface levels of the landscape. A variety of remotely sensed data has been used for identifying and monitoring soil salinity, including aerial photographs, video images, VIS and infrared panchromatic, multispectral, hyperspectral, and microwave images, airborne geophysics, electromagnetic (EM) induction, and ground-penetrating radar (GPR).

3.2 SURFACE SENSING OF SOIL SALINITY

3.2.1 SOIL SPECTROSCOPY

Although the soil spectrum may look somewhat monotonous, as compared with the spectra of rocks and pure minerals, it contains significant information about many soil constituents. Spectroscopy provides near-laboratory-quality reflectance or emittance data from measurements taken from far distance (meters or kilometers) (Goetz et al., 1985). Spectral information enables the identification of matter based on spectral absorption features of the material's chromophores (i.e., specific absorption of spectrally active groups), and this has been found very useful in many terrestrial and marine applications (e.g., Clark and Roush, 1984; Dekker et al., 2001). A comprehensive review of soil spectroscopy is provided by Ben-Dor et al. (1999) and Ben-Dor (2002).

Soil spectra across the VIS, NIR, and SWIR (400–2500 nm) are characterized by significant spectral absorption features that enable quantitative analysis of soil properties (Ben-Dor et al., 1999, 2002; Shepherd and Walsh, 2002; Malley and Ben Dor, 2004; Huete, 2005; Nanni and Dematté, 2006). The soil spectrum can be assessed on a point-by-point basis using a spectroradiometer, whereas a spectral-spatial overview can be obtained by plotting the same measurement points into an image using an imaging spectral scanner (also termed hyperspectral sensor). Thus, air- or satellite-borne hyperspectral sensors provide a new dimension that the traditional point or ground imaging spectroscopy, the air photography, and other multiband images cannot provide separately.

Studies examining the optimal numbers of spectral bands for soil applications using airborne and space-borne imaging spectrometers show significant variation, starting from the six spectral bands of the Landsat satellite (Ben-Dor and Banin, 1995a,b) to the 224 channels of the airborne AVIRIS (Accioly et al., 1998). Ben-Dor et al. (1999) state that, for quantitative spectral analysis of soils, optimal band width and number of channels may be strongly dependent on soil type and the properties examined. Today, it is also well established that the quality of the spectroscopy is very important for quantitative assessment of soil properties.

Irrespective of the sensor type and platform, the soil property analyzed must carry direct spectral absorption features or, at least, be correlated to a soil chromophore (indirect relationship). Farifteh (2007) reached a similar conclusion when analyzing the relationship between different salt types and contents in laboratory conditions. Using several salt contents and types (bischofite, halite, sylvite, arcanite, epsomite, and thenardite), Farifteh observed significant spectral changes such as water absorption peaks based on the hydration of salts at around 1400 and 1900 nm, and albedo changes. Spectral-based models to predict salt types and contents in the soil samples were generated using multivariate analysis. The study concluded that spectroscopy is limited in assessing directly the content and, obviously, the type of salt. Further research, including more soil properties that can be used as indirect indicators, is required to estimate the content of salt in soils using spectral means. For instance, using airborne imaging spectroscopy techniques, Ben-Dor et al. (2002) detected salt-affected areas from indirect relationships between the electrical conductivity (EC) of the water extracted from saturated soil paste and the spectral properties of the soils.

3.2.2 Passive Ground-Based Multiband Radiometers or Spectroradiometers

For surface sensing of salt-affected areas, ground-based spectroradiometers are used, such as the Field Spec Pro FR, Field Spec JR, and Field Spec VNIR (Analytical Spectral Device); the GER3700 and GER2600 (Geophysical and Environmental Research Corporation); and the PIMA SP (Integrated Spectronics). These instruments gather continuous, narrow (e.g., 3–30 nm band width) spectral measurements in the VIS and NIR to SWIR regions of the spectrum, usually within the range of 350–2500 nm, with an instantaneous field of view ranging from 1° to 22°. The spectra collected are used primarily to create spectral libraries of surface features (soils, vegetation, minerals), which are further used as endmembers to classify air- or satellite-borne data (Shepherd and Walsh, 2002), to identify the presence and spectral location of primary diagnostic spectral features of salt minerals (Howari et al., 2002), and to calibrate air or satellite imagery acquired within the same spectral range. The spectral information collected can be used in a qualitative way to identify salt-affected areas (Dehaan and Taylor, 2002; Howari et al., 2002) or in a quantitative way to relate spectral properties and abundance of salt minerals in salt-affected soils (Farifteh et al., 2004; Shuya et al., 2005).

Recently, ground-based imaging spectrometers such as the Hyperspec (Headwall Photonics), the HySpex (Norsk Elektro Optikk) and the CTHIS (Solid State Scientific Corporation) have become available for field use, although they require more attention than the ordinary point spectrometer. Despite problems such as atmospheric correction in oblique views and geometric distortion of the final image, these spectrometers can be of great utility because they provide simultaneous spatial and spectral views of the field in a real-time mode. For this reason, they may be increasingly used for soil applications in general and for soil salinity in particular.

In addition to field-based hyperspectral images covering the VIS to SWIR, thermal images can also be acquired in the field across the emitted spectral range (8–14 and 3–5 mm). Small cooled and uncooled user-friendly imaging radiometers, such as the ThermaCAMs series manufactured by FLIR systems, are used to this end. Although these spectrometers are not often used for soil salinity detection, thermal data can provide information on changes indirectly related to soil salinity, such as variations in soil moisture, vegetation stress, or albedo-based variations. Current thermal radiometers with one broad channel are small and compact and can be easily mounted on board airborne platforms. Airborne-based thermal mapping of soil salinity is an area of research that needs further development.

3.2.3 Airborne Sensors

Panchromatic, microwave, hyper- or multispectral sensors mounted on aircraft (e.g., aerial photographs, AirSAR, HyMap, DAIS-7915, AVIRIS, AISA-ES, and digital multispectral imaging [DMSI]) have the capability to sense salinized landscapes at high and very high spatial resolution (i.e., 0.25–5 m) and have a revisiting capability determined by the user. Common applications of these sensors in soil salinity studies are discussed hereafter.

3.2.3.1 Aerial Photographs, Videography, and Digital Multispectral Cameras

Scanned or hard copy aerial photographs are still a popular data source for mapping salt-affected areas, particularly because of their capability to provide historical information on landscape evolution. Some countries possess photographs from the early 1930s or 1940s that allow monitoring land degradation phenomena. The mapping of salt-affected soil units on aerial photographs usually proceeds from a combination of geomorphic features and greytones or colors. Field verification allows relating these units to variations in salt content. During the 1980s and 1990s, color-infrared photographs were used to discriminate barren saline soils (in white) and salt-stressed crops (in reddish-brown) from other soil surface and vegetation features (Rao and Venkataratnam, 1991; Wiegand et al., 1994). Color-infrared aerial photographs, with a sensitive spectral range of 380–900 nm, have also been used for mapping small areas requiring high spatial resolution and for validating satellite data (Manchanda and Iyer, 1983; Everitt et al., 1988). Color-infrared photographs have proven useful to identify variations in salinity-induced plant stress. Because the response of crops to salinity severity usually varies from season to season and from crop to crop, infrared photographs are a good means for recording multitemporal and crop-to-crop responses to salinity and changes in management practices (Ghassemi et al., 1995).

Airborne digital multispectral cameras and videography, usually with three to four bands in the VIS and NIR, were used together with color-infrared photographs to identify and assess salinity problems in agricultural areas of the United States in the 1980s and 1990s. Everitt et al. (1988) used narrow-band videography to detect and estimate the extent of salt-affected soils in Texas, while Wiegand et al. (1991, 1992, 1994) analyzed and mapped the response of cotton to soil salinity using color-infrared photographs and videography in three bands (840–850, 640–650, and 540–550 nm), with a spatial resolution of 3.4 m. By relating video and field data such as soil EC, plant height, and percent bare area these researchers determined the interrelations between plant, soil salinity, and spectral observations. Color-infrared composite and red narrow-band images were found to be better than green and NIR narrow-band images (Everitt et al., 1988; Wiegand et al., 1992, 1994). Other airborne sensors, such as the DMSI systems developed by SpecTerra Services of Perth, Western Australia, operate in the VIS–NIR regions of the spectrum but provide a wider coverage, with one extra band in the blue range, and variable pixel resolution as a function of the flight height (0.25–2 m). Due to their flexibility in frequency and time of data acquisition, rapid image availability and high spatial resolution, these systems have shown potential for qualitative monitoring and rapid detection of vegetation degradation in agricultural landscapes (Metternicht and Zinck, 2003).

3.2.3.2 Hyperspectral Sensors

Detailed spectral information can directly or indirectly pinpoint salt-affected areas (Taylor et al., 1994; Metternicht and Zinck, 1997, 2003; Ben-Dor et al., 2002). Taylor et al. (1994) were the first to show that it is possible to use airborne hyperspectral data to map salinity in soils. More specifically, they described the use of VIS–NIR–SWIR hyperspectral data collected with a 24-band airborne Geoscan for the mapping of soil salinity at Pyramid Hill, Victoria, Australia. The Geoscan imagery (Honey, 1989; Derriman and Agar, 1990) facilitated the differentiation of salt-affected soils based on the mapping of halophytic vegetation as an indirect indicator. Unfortunately, the Geoscan scanner became inoperative soon after this work was completed, and further investigation of VIS–NIR–SWIR spectral signatures of salinity had to await the development of new instruments. After new, well-calibrated instruments emerged on the market around 1997 (HyMap; Cocks et al., 1998), imaging spectroscopy was again used for soil salinity mapping. For instance, Taylor et al. (2001) characterized dryland salinity in the Dicks Creek catchment of central New South Wales, Australia, by the occurrence of spectrally distinctive smectite clays around surface salt scalds. Using the HyMap airborne sensor that acquires images over the spectral range of 450–2500 nm (i.e., VIS, NIR, and SWIR) in 128 bands, Dehaan and Taylor (2002, 2003) mapped saline areas with salt scalds, halophytic vegetation, and soils with different salinity degrees and types, at the Pyramid Hill test site (Australia). Of particular interest was the demonstration that halophytic vegetation could be mapped down to the species level, using either field or image-derived spectra. Taylor (2004) described how the shape of the hydroxyl absorption feature at 2200 nm (depth and width) changes with increasing soil salinity. This occurs both laterally, around saline discharge zones, and vertically within soil profiles where salt has accumulated at the surface as a result of evaporation. An asymmetry at shorter wavelengths, due to kaolinite, is replaced by an asymmetry at longer wavelengths, characteristic of smectites. This corresponds to the frequently observed occurrence of swelling clays around saline discharges, which often leads to the formation of dispersive soils and subsequent soil erosion. Interestingly, this spectral change could be mapped with hyperspectral imagery, and has therefore been the basis for the development of methodologies for saline soil mapping by the above workers.

Likewise, Ben-Dor et al. (2002) report good results in using the hyperspectral DAIS-7915 sensor, equipped with 70 bands across the VIS–NIR–SWIR spectral regions, together with a VIS and NIR analysis (VNIRA) approach complemented by comprehensive field and laboratory studies, to produce quantitative soil surface maps of organic matter, soil field moisture, and soil salinity. The soil selected for their study was a Vertisol, a shrink-swell clayey soil, known to be severely affected by salinity. They showed that EC was highly correlated with hygroscopic soil moisture as measured on the soil surface, and concluded that this was based on the high hygroscopic moisture content usually present along salt-affected areas (e.g., hygroscopic salts). In Figure 3.1, two processed images of the area show variations of the EC and hygroscopic moisture content at the soil surface. High EC values coincide with high soil moisture contents. In this study, multivariate analysis was used after careful data correction for atmospheric attenuation and specific selection of spectral channels.

Whiting and Ustin (1999) showed that high spatial resolution of low-altitude hyperspectral images provides important details for identifying the small, spotty morphology common to saline-sodic features in orchards and other vegetation types. They concluded that additional spectral bands would be needed to improve soil salinity mapping, especially if soil moisture content can be estimated and eliminated from the analytical process. Hyperspectral sensors covering the thermal region may adequately fulfill this need, as they provide information about soil temperature that may strongly correlate with soil moisture content (Fernandez et al., 2007). Thermal sensors can extract the soil and salt emissivity property, which holds better spectral variation than the reflected spectral region (Lane, 2002). Furthermore, emitted radiance can provide subsoil information that reflected radiation cannot. This can be further interpreted for the root zone, rather than the upper soil's crust. Figure 3.2 provides the emissivity spectra of several salt types at high and low spectral resolutions, respectively. Overall, mineral bands are distinguishable, with exception of chloride minerals that exhibit relatively flat infrared emissivity spectra. However, when chloride is contaminated with other geological materials, the emissivity spectrum behaves differently and shows the mineral features of the contaminants superposed on the emissivity spectrum of the chloride (Figure 3.3). These findings, investigated in the context of the Mars remote sensing mission (Albee et al., 1998), can be further applied to investigations on soil salinity on Earth. Operational hyperspectral airborne scanners covering the Thermal Infrared (TIR) region are important for salinity studies, although only a few are in operation (e.g., AIS mentioned by Birk and McCord, 1994) and very few studies

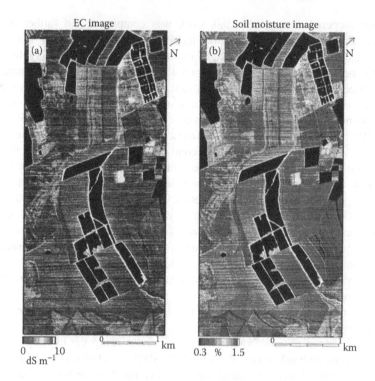

FIGURE 3.1 EC and hygroscopic moisture maps as obtained from spectral models applied on hyperspectral data.

have used them. A new hyperspectral sensor (ARES) merging the VIS–NIR–SWIR and TIR will soon be available for soil studies in general and for soil salinity in particular (Muller et al., 2004).

3.2.3.3 Microwave Sensors

Microwave sensing of salt-affected areas was undertaken in an experimental phase by the AirSAR-TOPSAR sensors of the JPL-NASA. Taylor et al. (1996), Schmullius and Evans (1997), Shao et al.

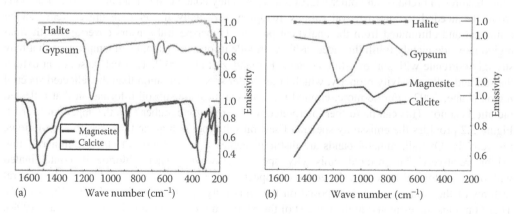

FIGURE 3.2 TIR emissivity laboratory spectra of salt minerals including chloride (halite), sulfate (gypsum), and carbonate (magnesite and calcite). (a) Full spectral resolution; (b) THEMIS spectral resolution. Spectra are offset for clarity. (From Lane, M.D., *Lunar Planet. Sci*, XXXIII, 1749, 2002.)

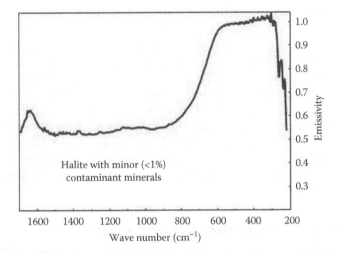

FIGURE 3.3 Spectra of halite contaminated with minor amounts (<1%) of other minerals (quartz, calcite and, possibly, and pirssonite).

(2002), and del Valle et al. (Chapter 9) highlight the adequacy of the L-band, and that of the C-band and P-band to a lesser extent, for detecting salinity features such as saline seeps, surface soil features, and halophytic vegetation associated to saline landscapes. Based on the complex dielectric properties of saline soils, Shao et al. (2002) mention that the L-band and P-band SAR can be effective in soil salinity detection. Microwave-based salinity studies, such as the ones conducted by Shao et al. (2002) and Sreenivas et al. (1995), exploit the differential behavior of the real (ε') and imaginary (ε'') parts of the dielectric constant to detect salinity. While the real part is insensitive to the presence of salts but strongly responds to soil moisture, the imagery part is highly sensitive to variations in soil EC, which varies according to the salt content of the soil. This finding enabled Taylor et al. (1996) to convert multifrequency, quad-polarized AirSAR radar data to maps of dielectric soil properties at various depths. These dielectric maps accurately delineated the extent of surface manifestations of soil salinity at the Pyramid Hill test site (Australia) and the paths of buried saline paleo-channels that are abundant in the region. The mapping of salt-affected areas also exploited the changes in surface roughness caused by macro- or microtopography (e.g., puffy versus smooth saline crusts) and the volume scattering of halophytic vegetation types that is proper to microwave data.

3.2.4 Orbital Sensors

Extensive research using satellite imagery for mapping salt-affected areas has been conducted over the last three decades, mostly with panchromatic and multispectral (VIS–NIR–SWIR and/or TIR), and with microwave to a lesser extent. Works by Epema (1990), Csillag et al. (1993), Rao et al. (1995), Metternicht and Zinck (1997), Evans and Caccetta (2000), and Fraser (Chapter 4) provide application examples from different continents and environmental settings. In general, all these works have been successful in mapping saline versus nonsaline surfaces. Some researchers have attempted to map salinity types (e.g., saline and alkaline) and severity levels (e.g., low, moderate, and high) with varying degrees of success (Kalra and Joshi, 1996; Metternicht and Zinck, 1996). To improve mapping accuracy, authors like Metternicht and Zinck (1997), Bell et al. (2001), and Castañeda and Herrero (Chapter 10) have tried approaches focusing on the fusion of data from different parts of the spectrum (e.g., VIS–NIR and microwave), whereas Lobell et al. (2007) propose the integration of multisensor satellite and multiyear yield data for mapping salinity at subsurface level.

More recently, from 2000 onward, experimental hyperspectral satellite data, from sensors such as the CHRIS on board the ESA mission PROBA-1 or the Hyperion on EO-1, have been assessed for their capability in identifying and mapping salt-affected areas (Dutkiewicz, 2006; Schmid et al., 2007). On the other hand, radar satellite-based mapping was so far limited to the use of single band, single polarization modes (e.g., ERS, JERS-1, and Radarsat). The recent launch of the ESA Envisat mission with the C-band, alternating HH or VV polarization ASAR instrument, and the ALOS mission, equipped with a full polarimetric PALSAR instrument providing L-band data, with single polarization (HH or VV) in fine-beam mode, dual polarization (HH + HV or VV + VH), or full polarimetry (HH + HV + VH + VV), will enable further research on the capability of space-borne SARs to map soil salinization. The following sections summarize application examples of multi-spectral data covering the VIS, TIR, and microwave parts of the spectrum for mapping salt-affected areas from orbital platforms.

3.2.4.1 Multispectral Visible and Thermal Infrared Sensors

Mougenot (1993) highlights that the possibility for broadband multispectral satellite images (e.g., Landsat, SPOT) to detect salt-affected areas becomes limited when salt content falls below 10%–15%. Recently, Farifteh et al. (2006) postulated that remote sensing-based thematic mapping requires identifying three types of variables, namely the measurable, the retrievable, and the hidden ones. Their view is that the potential contribution of remote sensing techniques depends on whether the number of hidden variables is larger or smaller than the other two sets of variables. As an extreme case, if the behavior of the system under study is entirely explained by hidden variables, then there is little hope that remote sensing data can provide useful information for this particular application. The spatial resolution of satellite-based sensors (usually 10–30 m) is another factor of mapping inaccuracy highlighted by several researchers.

Broadband satellite sensors that are used for the cartography of saline areas include Landsat TM, Landsat MSS, SPOT XS, Terra-ASTER, and LISS-II/III. Their spectral, spatial, and temporal configurations are presented in Table 3.1. Dwivedi et al. (2001) combined LISS-III and IRS-1C images to map salt-affected areas in the Indo-Gangetic plains of India using intensity–hue–saturation image transforms with accuracies above 85% when using multispectral LISS sensors. Ahmed and Andrianasolo (1997) compared the performance of Landsat TM and SPOT XS to map salinity at semidetailed level. They found that SPOT XS performed better than Landsat TM, although the discrimination between moderately saline areas and highly reflective eroded materials was difficult because of their similar reflectance. Verma et al. (1994) demonstrated that the incorporation of the thermal range (e.g., Landsat TM band 6) to the VIS–NIR channels helped overcome issues of spectral similarity of saline soils. In a recent study on the Aibi lake in Xinjiang, China, Shuya et al. (2005) identified saline areas dominated by sodium chlorides and sodium sulfates by using Terra ASTER imagery and the grey system theory. They report a good correlation between surface salt concentrations and band one (green) of the ASTER sensor, followed by bands two (red) and three (NIR). Likewise, Brunner et al. (2004) derived uncalibrated salinity maps from atmospherically corrected multispectral ASTER images. They found these images valuable to regionalize ground-truth activities that are essential to quantify the extent and distribution of salt-affected areas. These uncalibrated salinity maps were produced using spectral correlation techniques and comparing the results with average conductivity values derived from field point measurements.

Dwivedi et al. (Chapter 5) present a comparative study on the performance of the IKONOS multispectral imagery (4 m spatial resolution) and the LISS-III sensor on board the IRS-1D satellite, that acquires panchromatic (5.8 m spatial resolution) and multispectral (23.5 m spatial resolution) data, for mapping salt-affected soils. Using different image transformation and classification techniques, an overall accuracy of 92.4% was obtained when using IKONOS data against an overall accuracy of 78.4% and 84.3% achieved when using LISS-III multispectral data or a fusion of LISS-III panchromatic and multispectral data.

TABLE 3.1

Satellite-Borne and Airborne Sensors Utilized in Previous Soil Salinity Studies and Planned Hyperspectral Sensors *(in Italics)*

Platform	Sensor	Resolution Spatial (m)	Spectral	Temporal (Days)	Swath Width (km)
Landsat TM 4/5	Thematic Mapper	1–5 and 7: 30; 6: 120	Visible (B, G, R), NIR, SWIR, thermal IR		180
Landsat ETM+	Enhanced Thematic Mapper	1–5 and 7: 30; 8: 15; 6: 60	Visible (B, G, R), NIR, SWIR, thermal IR		
SPOT 1,2,3	HRV	XS: 20; P:10	Visible (G, R), NIR	26	60
SPOT 4	HRV; HRVIR	XS: 20; P:10	Visible (G, R), NIR; SWIR	26 (4–5)	60
SPOT 5	HRG	XS 1–3: 10; XS 4: 20; P:5	Visible (G, R), NIR; SWIR	26 (4–5)	60
IRS-1	LISS II/III PAN	1–3: 23.5; 4: 70; 5.8	Visible (G, R); NIR; SWIR	24	142
Terra	ASTER	15 m (VNIR: 3 bands), 30 m (SWIR: 6 bands), 90 m (TIR: 5 bands)	Visible (G, R), NIR, SWIR, thermal IR	16	60
PROBA	CHRIS	19 Bands: 18; or 63 bands: 36	Visible (B, G, R), NIR	7	14
EO-1	Hyperion	242 Bands: 30	Visible (B, G, R), NIR, SWIR	16	7.5
Not yet selected	*EnMAP*	*218 Bands: 30*	*Visible (B, G, R), NIR, SWIR*	*3*	*30*
IKONOS		Multispectral: 4 Pan: 1	Visible (B, G, R), NIR	16	11
Radarsat	SAR	25 (standard); 8 (fine)	C-band (5.6 cm)—HH pol	24	100
ERS	SAR	30	C-band (5.6 cm)—VV pol		100
JERS	SAR	18	L-band (23 cm) HH	44	75
Envisat	ASAR	30	C-band (HH and VV, or HH and HV, or VV and VH)		100
ALOS	PALSAR	10 (single beam); 20 (dual beam)	L-band (23 cm) (HH or VV; HH+HV or VV+VH)	46	70
NASA DC-8	AirSAR/TOP SAR	10	P-band (68 cm); L-band (23 cm); C-band (5.6 cm)		12
Space shuttle	SIR-C/X-SAR	30	L-band (23 cm); C-band (5.6 cm) X-band (3.1 cm)		
Airborne— DAIS 1915	Hyperspectral	3; 20	79 Bands: visible (B, G, R), NIR, SWIR; thermal IR		
Airborne— Hymap	Hyperspectral	2; 10	128 Bands: visible (B, G, R), NIR, SWIR		
Airborne— AVIRIS	Hyperspectral	5; 10; 20	224 Bands: visible (B, G, R), NIR, SWIR		
Airborne— CASI	Hyperspectral	2.2	288 Bands: visible (B, G, R), NIR		
Airbone— AISA-ES	*Hyperspectral*	*IFOV = 0.528 mrad*	*Up to 483 bands: visible (B, G, R), NIR, SWIR*		*FOV = 8.54°*
Airborne— AHS	*Hyperspectral*	*IFOV = 2.5 mrad*	*48 Bands, visible (B, G, R), NIR, SWIR, TIR*		*FOV = 85.92°*
Airborne— ARES	*Hyperspectral*	*IFOV = 2.5 mrad*	*192 Bands: visible (B, G, R), NIR, SWIR, TIR*		*FOV = 60°*

3.2.4.2 Satellite-Based Microwave Sensors

Data generated by past microwave sensors on board the JERS-1 (L-band) and the SIR-C/X-SAR mission of NASA and those provided by current sensors on board the Radarsat, ERS, and Envisat satellites (C-band) are applied in soil salinity studies. All these sensors provide single band and single polarization data, except for Envisat-ASAR (dual polarization) and the SIR-C/X-SAR mission (multifrequency, quad-polarized).

Applying a fuzzy classification approach on JERS-1 data, Metternicht (1998) distinguished saline from nonsaline surfaces using microtopographic variations of the terrain as indirect indicator of salinity occurrence. Changes in soil surface roughness due to different salt crust types, the presence of halophytic vegetation, and soil aggregation due to cultivation practices facilitated the mapping process. Castañeda and Herrero (Chapter 10) use the contrast of surface roughness between bare and vegetated areas, as obtained from ERS imagery (C-band), to delineate saline wetlands in Spain.

Yun et al. (2003) reported a correlation of 0.69 between the backscattering coefficients of a Radarsat-1 SAR image (C-HH) and the ε'' component of the dielectric constant measured on saline soil samples, and concluded that the C-band HH polarization SAR images can be useful for monitoring soil salinity. An innovative study by Aly et al. (2007) proposes to detect soil salinity from series of polynomial regressions of the ε'' computed on saline soil samples, simulation models, and Radarsat-1 SAR images acquired in the standard beam (Si) modes ($i = 1,3,5$). Their parametric model infers salinity (ε'') from the measurements of the Si modes of Radarsat-1 SAR without the need to apply backscattering models.

3.2.4.3 Hyperspectral Satellite Sensors

Two experimental sensors have been launched after the year 2000, namely Hyperion (30 m spatial resolution and 242 bands covering the 400–2500 nm spectral range) on board the EO-1 (NASA) and CHRIS (19 bands at a spatial resolution of 18 m or 63 bands at a spatial resolution of 36 m, within the 410–1050 nm range) on board the PROBA platform (ESA). Dutkiewicz (2006) evaluates the performance of the Hyperion imagery for mapping surface symptoms of dryland salinity using a partial unmixing technique called mixture-tuned matched filtering. She reports a kappa accuracy of 0.50 for the mapping of saltpans, small depressions with extreme soil salinity, and 0.38 for the mapping of samphire vegetation (*Crithmum maritimum*) that grows in areas of high to very high salinity. Her study demonstrates that hyperspectral satellite imagery is unable to map accurately salinity indicators related to slight or moderate soil salinity levels. It shows also that hyperspectral airborne instruments (e.g., CASI and HyMap) allow mapping saline soils with more accuracy than hyperspectral satellite instruments do (e.g., Hyperion). The performance of the CHRIS-PROBA sensor for cartography of salt-affected areas is still poorly documented and limited so far to soil degradation mapping in semiarid wetlands (Schmid et al., 2007). A German venture is to place a new orbital hyperspectral sensor (EnMAP) (Stuffler et al., 2007), which is expected to propel satellite-based soil salinity spectral mapping to a new era. The sensor, planned for launch in 2011, will consist of 123 spectral channels crossing the VIS–NIR–SWIR region, with a 30 m spatial resolution. Likewise the Canadian Space Agency is planning the launch of Hyperspectral Environment and Resource Observer, a hyperspectral sensor with more than 200 bands in the range of 400–2500 nm. With a temporal resolution of 7 days and a spatial resolution of 30 m, the satellite is planned to be in orbit in 2009 (Hollinger et al., 2006).

3.2.4.4 Multisensor, Multispectral Image Fusion

Several authors have based the mapping of salinity indicators on multisensor image fusion. For instance, Bell et al. (2001) used a fused AirSAR and Landsat TM data set together with a combined model, such as the Dubois and small perturbation model, to extract the imaginary part of the dielectric constant, with good results for salinity mapping in bare and sparsely vegetated areas.

Castañeda and Herrero (Chapter 10) combine Landsat TM and ERS data to produce an up-to-date inventory of the location and spatial extent of saline wetlands in Spain that are being degraded by intensive agricultural use.

3.2.5 CONCLUSION

Passive and active remote sensing tools, including ground- and airborne hyperspectral sensors, enable assessing salinity at the soil surface. Spectral analysis of soil salinity, both in laboratory and field, can be based on direct indicators of soil salinity, especially in the cases where salinization has reached an advanced stage. Besides, there are indirect spectral signals for soil salinity detection such as mineralogy changes, increasing moisture, color changes, varying particle size distribution, invasion of salt-tolerant plants, and increase of bare surface areas. However, these indicators can also be related to other landscape degradation processes. This one-to-many relationship between a given indicator and various degradation phenomena, together with the significant limitation of not viewing the full soil profile, impedes reaching a comprehensive understanding of the salinity problem in a 3D space, including horizontal and vertical variations in the soil mantle. The synergy of several sensors (e.g., microwave sensors with reflective and thermal sensors) and platforms (e.g., airborne and ground-based sensing) may be a key factor to better map salt-affected soils in time with sufficient reliability and accuracy.

3.3 SUBSURFACE SENSING OF SOIL SALINITY

Imaging spectroscopy also termed as hyperspectral remote sensing is a promising approach, but by itself it cannot provide the entire 3D view of the soil salinity problem. Additional tools that can explore the soil profile are required. To this end, Ben-Dor et al. (2008) have developed an approach to collect spectral data within the soil profile by using a penetrating optical fiber head (3S-HeD). Although innovative, this technology presents constraints such as limited detection depth and limited spatial coverage. Therefore, remote sensing technologies based on active EM radiation are being widely adopted. Ground-based EM methods measure EC in subsurface and substratum horizons and can thus recognize salinity anomalies in the field before salinization approaches the surface (Farifteh et al., 2007). EM induction sensors measure salinity in the soil profile by recording the soil apparent electrical conductivity (ECa), whereas GPR sensors image the subsoil layers to identify possible causes of soil salinity.

3.3.1 FREQUENCY-DOMAIN ELECTROMAGNETIC TECHNIQUES

3.3.1.1 Basic Principles and Applications in the Soil Salinity Mapping Context

A comprehensive review on the use of active geophysical methods to assess soil salinity is given by Corwin and Lesch (2005). The basic principle of operation underlying the frequency-domain electromagnetic (FDEM) techniques is described in Chapter 12. The instruments are calibrated to read in units of terrain conductivity, known as apparent conductivity. Since EC correlates strongly with other soil properties, FDEM is a powerful tool for mapping soils and detecting changes in soil types as related to salinity. Common applications of FDEM include identifying aquifers and detecting the presence of water or saline intrusions in groundwater, mapping aggregate deposits for quarry operators, detecting metal objects and mapping leaching in environmental investigations, as well as mapping permafrost and other geologic features in geotechnical engineering.

EM techniques to measure soil EC are based on a variety of devices that range from time-domain to frequency-domain. Allen (2004) reviewed all geophysics-based approaches that apply to groundwater and soil investigation. The EC imaging based on a FDEM technique was ranked as the best out of ten techniques, including passive remote sensing means. Other instruments present limitations for salinity measurement in the soil profile, as they collect very deep underground data. An EC imaging instrument is a FDEM sensor that works within a range of 30 cm to 5 m depth and

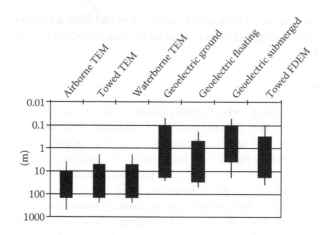

FIGURE 3.4 Ranges of detectable depths of various EM imaging devices (m). Note that airborne FDEM (not shown) has a range similar to but shallower than airborne TEM. (From Allen, D., A review of geophysical equipment applied to groundwater and soil investigation. A report to the ANVIS/sustainable irrigation travel fellowship. Available at http://downloads.lwa2.com/downloads/final_reports/David_Allen_ Fellowship_ Report_04–05.pdf [accessed November 6, 2007], 2004. With permission.)

performs best while scanning the area from about 1m above the ground. Figure 3.4 shows the position of all EM instruments in relation to sensing depths. In general, when dealing with soil salinity that occurs at 0–3 m depth, the tow FDEM is the most suitable device because of its sensitivity to depths ranging from 30 cm to 5 m, depending on the waveband (see Chapter 11). Apparently, airborne EM systems are not appropriate for soil mapping because they sense at a subsurface level deeper than most soil salinity surveys require for analyzing salinity effects on crops (root zone). Sensing with a FDEM device precludes separating the surface layer or the root zone volume that is located within the first 25 cm of the soil profile. Basically, some airborne EM systems can sense hundreds of meters into the ground, but most cannot simultaneously resolve the occurrence of shallow and deep features, as highlighted by the following equation:

$$\left\| (i\tau_0 \mu \omega) \frac{1}{2} \rho \right\| \leq \left| \frac{1}{2} \right| \tag{3.1}$$

where

i is the square root of -1

τ_0 is the conductivity

μ is the permeability of the material in the volume of exploration

ω is the angular frequency of the transmitted field

ρ is the spacing between transmitter and receiver (DuaEM)

Parameter i is set to 1 for determining the low-frequency approximation.

The FDEM device provides variable sensing depths, as a function of coil separation, and/or orientation of the height of the instrument above the ground, and/or operational frequency. For efficiency of operation, FDEM instruments transmit a time-varying sinusoidal magnetic field. The sampling volume remains solely dependent on geometry as long as the frequency of the field is consistent with the low-frequency approximation, as defined by Wait (1962). The calibration of FDEM instruments is difficult and critical to their operation. Recalibration is necessary in both the manufacturing and field domains. When the manufacturing calibration is only occasionally done, field calibration becomes important. This can be done by placing the instrument over a fixed 3D environment that does not change and is well known to the operator. Calibration and analysis of the EM data can be conducted by the following method described in Lesch et al. (1992) and in Lesch and Rhoades (1999). To calibrate the FDEM device, soil samples must be collected at different depths and then be analyzed for EC, texture, moisture, sodium adsorption ratio, as well as for other elements of interest such as Cl, SO_4, NO_3, and B. This methodology enables the production of maps of all the analyzed soil parameters at predefined sample depths. Usually, data collected using FDEM technology are spatially referenced with a d-GPS, so that they can be projected on a rectified map shortly after their reading and/or used for further GIS processing in the form of spatial interpolations to generate continuous layers of information on soil salinity.

3.3.1.2 Factors Affecting the FDEM Readings

FDEM have distinct advantages over many other active techniques. Because no contact with the ground is required, FDEM can cover large areas faster than other passive and active ground-based instruments (Rhoades et al., 1999) and can also work under a shallow canopy. Many soil minerals are poor conductors, so current flow in a saline soil is primarily through the soil moisture that fills the pore space between aggregates and, to a lesser extent, across the charged surface of clay particles. Slavich (1990) ranked the soil and environmental factors likely to have a significant effect on ECa measurements by FDEM in the following order: soil salinity > moisture content > surface charge of clay particles > bulk density. Other factors such as soil temperature and antenna position also play a role.

(a) *Soil moisture content*: For a given level of salt concentration in the soil water, ECa increases with moisture until equilibrium is reached (i.e., field capacity). Norman (1990) states that, for clay soils having more than 40% clay in the top 30 cm, the gravimetric moisture content of the soil profile should be greater than 20% to allow ECe values (as measured in laboratory) to be accurately derived from the observed ECa data (as measured in the field). This standard has been adopted for monitoring sites to allow apparent conductivity readings to be calibrated, for a later date, if required.

(b) *Soil temperature*: The temperature of the soil affects both the viscosity and the phase state (i.e., vapor or liquid) of the soil water, which in turn influences the mobility of salts. To account for the varying effect of temperature on individual conductivity readings, all readings should be normalized to 25°C.

(c) *Soil porosity*: The shape and size of the pores, as well as the number, size, and shape of the interconnecting passages, directly affect the ability of the current to flow through the soil medium. Soil porosity is reflected in soil texture and bulk density.

(d) *Amount and composition of charged clays*: Negatively charged clay particles (colloids) adsorb positively charged ions (cations) onto the particle surfaces. With the addition of water, the cations tend to dissociate themselves from the clay particles and become available for ionic conductivity. If all other soil properties are equal, a soil with higher clay content will record a higher ECa reading than a soil with lower charged clay content.

(e) *Salt concentration*: The main factor that directly influences the conduction of electrical current is salt concentration in the soil.

(f) *Antenna position*: Norman (1990) notes that all EM readings are referred to as depth-weighted estimates of ECa, since different depths contribute different amounts to the overall reading. The instrument is designed so that the current flow at a given depth is horizontal and not influenced by the current flow at any other depth. Measurements can be taken in two modes: the horizontal dipole to a depth of 2–3 m and the vertical dipole to a depth of 3–4 m. Specifications provided by the instrument maker (Geonics, Mississauga, ON, Canada) suggest that 1.5 m is the most effective depth of exploration for either dipole. In the horizontal mode, 65% of the response comes from the top 60 cm of the soil profile, while the vertical dipole receives only 36% of the signal from the top 60 cm.

3.3.1.3   EM Survey Cost

O'Leary et al. (2004) estimated the cost of using FDEM devices to be \$5.55 ha^{-1} at 15 km h^{-1} for a 30 m transect spacing. However, they did not include costs for soil sampling and specific EM calibration, because this would be a constant amount per hectare and therefore would not affect the optimal spacing. The cost of the FDEM mapping is therefore represented by the following equation:

$$EMc = 166.66/T \qquad\qquad (3.2)$$

where
T is the transect spacing (m)
EMc is the cost of EM mapping (\$ ha^{-1})

The economic evaluation of a salinized area must take into account the FDEM cost and the production loss of the land based on the current salinity status and the predicted hazard.

3.3.2   Ground-Penetrating Radar

3.3.2.1   Basic Principles

GPR is another EM-based geophysical tool that uses short radar pulses to image the subsurface. This nondestructive method uses EM radiation in the microwave band of the radio spectrum and aims at the detection of reflected signals from subsurface structures. GPR can be used in a variety of media, including rock, soil, ice, fresh water, pavements, and built structures. It can detect objects, voids, and cracks, and changes in material. GPR uses transmitting and receiving antennas. The transmitting antenna radiates short pulses of high-frequency (usually polarized) radio waves into the ground. When the wave hits a buried object or a boundary with different dielectric constants, the receiving antenna records the variations in the reflected return signal (Figure 3.5). The principles involved are similar to those of reflection seismology, except that EM instead of acoustic energy is used and reflections appear at boundaries with different dielectric constants instead of acoustic impedances. The depth range of GPR is limited by the EC of the ground and the transmitting frequency. As conductivity increases, penetration depth decreases. This is because the EM energy is more quickly dissipated into heat energy, causing a loss in signal strength at depth. Higher frequencies do not penetrate as far as lower frequencies, but give better resolution. Optimal depth penetration is achieved in dry sandy soils or massive dry materials such as granite, limestone, and concrete, where the depth of penetration is up to 15 m. In moist and/or clayey soils and in soils with high EC, penetration is sometimes only a few centimeters. Usually, standard GPR antennas are in contact with the ground to obtain the strongest signal strength, while GPR horn antennas can be used at 0.3–0.6 m above the ground. GPR images can be analyzed for deriving electrical properties and subsurface characteristics as well as for spatial mapping of water content (Basson, 1992, 2000). The resolution range is a function of the subsurface dielectric constants and the wave frequency. It may vary from a few to tens of centimeters in the relevant effective frequencies (Davis and Annan, 1986, 1989). For a certain wavelength, the penetration of the GPR waves into the subsurface is mainly a function of the conductivity of the host material; therefore, GPR waves decay significantly in conductive and saline soils. The GPR data are used to determine this decay and, at the same time, to identify subsoil layers that might affect the drainage system within salt-affected soil sites.

All electrical characteristics of the materials are determined by EC, permittivity, and permeability. In GPR technology, the permittivity (also called the dielectric constant) is the most important parameter, because any material behaves as a dielectric at higher frequencies. The GPR measures the reflected EM waves from subsurface structures. The velocity and reflectivity of the EM wave are characterized by the dielectric constant (permittivity) of the soil. When the dielectric constant of the soil is ε_r, the velocity in this material is given by

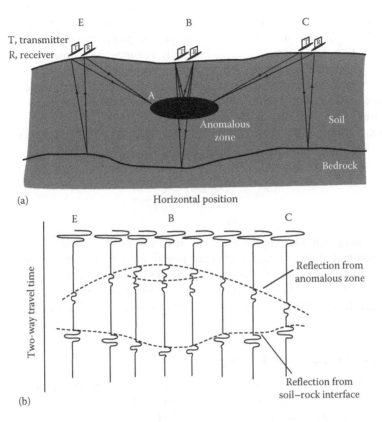

FIGURE 3.5 Schematic illustration to show how the GPR works. Representation of (a) the measurement scheme and geometry, where B is the perpendicular to the abnormal zone, and E and C are side looking positions (similar) to this zone; and (b) information recorded by the GPR at landscape positions E, B, and C. (From Davis, J.L. and Annan, A.P., *Geophys. Prospect.* 37, 531, 1989. With permission.)

$$\nu = \frac{c}{\sqrt{\varepsilon_r}} = \frac{3 \times 10^8}{\sqrt{\varepsilon_r}} \ \ (\text{m/s}) \tag{3.3}$$

The wavelength λ (m), the operating frequency f (Hz), and the velocity of the wave are related as

$$\lambda = \nu T = \frac{\nu}{f} \ \ (\text{m}) \tag{3.4}$$

Equation 3.4 shows that wavelength and frequency are in inverse proportion. An increase in EC causes a decrease in the velocity of the EM, which in turn decreases the penetrating depth (d), as shown in Equation 3.5. A GPR sensor transmits a pulsed EM wave from a transmitter located on the surface of the ground, and signals are received by a receiving antenna also on the ground surface (see Figure 3.5). The transmitted signal propagates through the subsurface material and is reflected by objects such as geological boundaries, buried objects, and groundwater. When the EM wave velocity ν is known, by measuring the travel time τ(s), the depth of the reflecting object d(m) can be calculated as follows:

$$d = \frac{\nu \tau}{2} \ \ (\text{m}) \tag{3.5}$$

The travel time is defined as the time from the timing of the transmitted signal to the timing of the received signal, which corresponds to the propagation time to the reflecting object. The reflection occurs when the EM wave encounters any electrically inhomogeneous material, especially metal. Buried metallic materials such as pipes and cables are quite easy to detect by GPR. Even insulating (dielectric) materials can be electrically detected. Any material having two different dielectric constants causes EM wave reflection. When an EM wave is incident to a flat boundary of two different materials having the dielectric constants of ε_1 and ε_2, the EM wave with amplitude of 1 is reflected by the boundary and its amplitude is Γ. Amplitude Γ is defined as the reflection coefficient of a boundary and is given by

$$\Gamma = \frac{\sqrt{\varepsilon_1} - \sqrt{\varepsilon_2}}{\sqrt{\varepsilon_1} + \sqrt{\varepsilon_2}} \tag{3.6}$$

Equation 3.6 shows that the amplitude of the reflected wave is determined by the ratio of the dielectric constant of the two materials. The range of the reflection coefficient Γ is as follows: $-1 \leq \Gamma \leq 1$. If the lower material is metal, the reflection coefficient is $\Gamma = -1$ and then the maximum amplitude is obtained. Therefore, the reflection from metallic material is always very distinct and clear. In actual GPR measurements, the radar target is not infinitively large, but it has a finite size.

3.3.2.2 Factors Affecting GPR Readings

Soil salinity increases the EC of the soil volume, which is the basic physical property that affects the radar waves, the other one being the magnetic permeability. Soils, sediments, or rocks that are dielectric will permit the passage of most of the EM energy without dissipating it. The more electrically conductive a material is, the less dielectric it is, and energy will attenuate at a much shallower depth. In a highly conductive medium, the electrical component of the propagating EM wave is conducted away in the ground, and consequently the wave as a whole dissipates. Aside from salty materials, highly electrically conductive media also include certain types of electrically conductive clays, especially if they are wet. Any soil or sediment that contains soluble salts, or electrolytes in the groundwater, creates a medium with high EC. Free ions, which allow for greater EC, act as major factors for decreasing GPR backscattering. Sulfates, carbonate minerals, iron, salts of all sorts, and any charged elemental ions (e.g., charged clay) create a highly conductive ground and readily attenuate radar energy at shallow depths. Under the very unfavorable conditions of wetness (with slightly saline water), maximum depth of GPR penetration will be less than 1 m in calcareous sediments or soils that contain clay minerals, no matter what frequency of antenna is used. Another factor affecting the GPR signal at a given frequency is the attenuation of the EM energy that increases with increasing moisture content (Daniels et al., 1995).

The lack of adequate data on soil moisture and the high spatial and temporal variations in the degree of wetness within most soils (especially the salt-affected ones) often restrict the detection to one component over another. Magnetic permeability, a measure of the ability of a medium to become magnetized when an EM field is imposed upon it, also affects the GPR ability to penetrate in the soil. Most soils and sediments are only slightly magnetic and, therefore, have usually low magnetic permeability. The higher the magnetic permeability, the more EM energy will be attenuated during its transmission, causing a destruction of the magnetic portion of the EM wave, just as the electrical component is lost with increased EC. Media that contain magnetite minerals, iron oxide cement, or iron-rich soils can all have a high magnetic permeability and therefore transmit radar energy poorly. The above-mentioned constraints are used to better manage the GPR potentials to sense at different depths. Possibly, all of the above-mentioned factors may contribute to isolate the EC factor that relates to soil salinity, as shown in the study of Shih and Myhre (1994).

The main problem associated with GPR measurements is that the observed anomalies have to be interpreted according to available information on the subsurface conditions. When characterizing soil salinity, many factors may interact and conventional methods have to be applied along with the GPR measurements. Since the backscattering energy is highly affected by the water content, the repeatability of results strongly depends upon seasonal variations in soil moisture (Daniels et al., 1995). In clayey soils with high cation exchange capacity, where most of the soil salinity problem develops, significant attenuation of the backscatter radiation may enable deeper sensing of the soil profile. To obtain strong signals in clayey terrain, the GPR has to be in contact with the soil, for which reason smooth surfaces are preferred. This makes it difficult to apply the GPR technology above vegetation canopies and over pebbled nonhomogeneous fields. In places where soluble salts and exchangeable sodium accumulate in the topsoil, high attenuation occurs and the penetration to other likely salt sources is restricted (Doolittle and Collins, 1995). In soils with high EC (ECe ≥ 4 dS m^{-1}) and/or sodium adsorption ratio above 13, GPR techniques are difficult to apply. A drawback of GPR is that the method is data- and labor-intensive for comprehensive surveys of large areas. Small-scale preliminary surveys to identify areas that require more intensive surveying would greatly enhance the usefulness of GPR for mapping and monitoring places with subsurface off-site movement of water and agrochemicals. EM induction can provide a small-scale, rapid survey prior to a more detailed, data-intensive GPR survey.

3.3.2.3 GPR Survey Cost

According to Anderson and Peltola (1996), a GPR instrument operated by two men can survey an area of more than 2 ha day^{-1} with good resolution, depending on the site characteristics. An example drawn from Finland required a GPR survey over an old gas station lot of 2 ha to detect possible petroleum leaks. The approximate cost amounted to $3500, including $2000 for GPR operation and hiring and $1500 for subsequent sampling at two observation points. In comparison, a conventional survey would have amounted to $6000, including the chemical analysis of 20 wells. The Finnish study concluded that the GPR is able to detect any kind of contamination with a reduced price as well as provide valuable information on hydrological conditions (water table and soil moisture) and other characteristics of the area (e.g., underground pipelines). For salinity mapping purposes, the cost per area might be even lower because larger areas can be covered using the same instrument and manpower.

3.4 CONCLUSIONS

Spectral passive optical-infrared technologies can be applied for salinity mapping of the upper layer of the soil mantle, using direct (e.g., salt crusts, surface sealing, deflation spots, and moist pans) or indirect (e.g., halophytic vegetation, reduction in plant cover, and decreasing crop yield) indicators of soil salinity. Passive thermal emission may provide information about the subsoil matrix, but its relation with soil salinity has not yet been studied in depth. Direct measurement of salt types is difficult to undertake in field conditions and is problematic because of similar spectral behavior of some salt types and other soil properties. Salts such as NaCl are spectrally featureless and other salts such as gypsum, with defined spectral absorption features, may be detected only in places with severe salinity and salt features apparent at the soil surface. Thermal spectra of salts and minerals may, however, add spectral information and improve soil salinity detection.

Satellite-based sensors covering the optical-infrared and microwave parts of the spectrum (e.g., Landsat and SPOT series, Terra-ASTER, LISS, IKONOS, Radarsat, ERS, ALOS, Envisat, CHRIS-PROBA, Hyperion) are cost-effective for covering large areas and for utilization on a multitemporal basis (e.g., monitoring tasks). Airborne hyperspectral sensors (e.g., AVIRIS, HyMap, AISA-ES) are generally used over relatively small areas. This is mainly due to higher cost of data acquisition, as compared to that of Landsat-like satellite imagery, and more complex image calibration and data processing.

Ground-based FDEM active tools have the ability to detect directly EC in the soil profile, at specific depths depending on the frequency used. The factors controlling the FDEM output are soil moisture, soil temperature, soil porosity, and cation exchange capacity. When all these factors are controlled and known, the ECa measurement is reliable and comparable. Only a few EM instruments, such as the EM-38 or EM-31, are suitable to operate in the soil environment (0–2 m depth). Others, usually mounted on airborne platforms, examine the earth crust at deeper depth and are not suitable for salinity detection within the soil profile. The FDEM assembly can be easily operated by a nonprofessional user to cover relatively large areas, as compared to other active means. Nevertheless, the FDEM has some constraints: (1) it is unable to cover areas as large as those covered by passive air- or satellite-borne sensors; (2) it is quite expensive to apply for monitoring tasks; and (3) it does not (yet) distinguish the topsoil (0–25 cm) from the measured soil volume. To this end, the VIS–NIR–SWIR spectral domain, which senses the upper soil surface, may add complementary information to the FDEM. Moreover, VIS–NIR–SWIR and microwave sensors can effectively monitor soil erosion and deposition, which are sometimes a direct consequence of the soil salinity problem.

Several airborne imaging spectrometry (IS) missions have demonstrated that hyperspectral information is extremely useful and adds much to the traditional field mapping capabilities. Governmental agencies increasingly recognize the IS potential to assess and monitor soil degradation and, as a result, it is expected that potential end-users will be able to implement the IS technology. New missions and sensors, including ground and orbital IS sensors as well as waveband coverage extended to the TIR region, are in progress; thus, IS remote sensing for soil salinity seems to have a promising future. As in the case of the FDEM, the GPR technique also presents constraints, since the device has to be towed over the field and the efficiency is doubtful. In general, GPR may not be a suitable tool to detect soil salinity in small-scale application areas (e.g., catchments or regional levels), but can function as an additional tool to explore salinity problem-spots.

Lastly, farmers and land managers alike appreciate information about incipient salinity problems. Thus, it is important to go beyond the level of moderate-to-high salinity, easy to identify through stressed vegetation or the presence of salt crust in places where Landsat-like sensors have proved successful, and develop approaches that allow detecting early stages of soil salinity. One relevant limitation of the multispectral tools is that they fail to provide precise information about the soil chemical status and changes induced by incipient salinity in soil properties such as aggregation, organic matter content, pH, mineral formation, and water content.

REFERENCES

Accioly, L.J.D.O., Huete, A., and Bachily, K. 1998. Using mixture analysis for soil information extraction from AVIRIS scene at the Walnut Gulch Experimental Watershed, Arizona. In *Anais IX Simpósio Brasileiro de Sensoriamento Remoto*, Santos, Brazil, pp. 1335–1344.

Ahmed, I. and Andrianasolo, H. 1997. Comparative assessment of multisensor data for suitability in study of the soil salinity using remote sensing and GIS in the Fordwah irrigation division, Pakistan. In *Proceedings of the Geoscience and Remote Sensing Symposium*, Toronto, Canada, Vol. 4, pp. 1627–1629.

Albee, A.L., Palluconi, F.D., and Arvidson, R.E. 1998. Mars global surveyor mission: Overview and status. *Science* 279: 671–672.

Allen, D. 2004. A review of geophysical equipment applied to groundwater and soil investigation. A report to the ANVIS/sustainable irrigation travel fellowship. Available at http://downloads.lwa2.com/downloads/final_reports/David_Allen_Fellowship_Report_04–05.pdf (accessed November 6, 2007).

Aly, Z., Bonn, F.J., and Magagi, R. 2007. Analysis of the backscattering coefficient of salt-affected soils using modeling and RADARSAT-1 SAR data. *IEEE Transactions on Geoscience and Remote Sensing* 45: 332–341.

Anderson, J. and Peltola, J. 1996. Ground penetrating radar as a tool for detecting contaminated areas. Available at http://www.cee.vt.edu/ewr/environmental/teach/gwprimer/gprjp/gprjp.html (accessed November 6, 2007).

Basson, U. 1992. Mapping of moisture content and structure of unsaturated sand layers with ground penetrating radar. MSc thesis (in Hebrew), Tel-Aviv University, Raymond and Beverly Sackler Faculty of Exact Sciences, Tel-Aviv, Israel.

Basson, U. 2000. Imaging of active fault zone in the Dead Sea Rift: Evrona Fault Zone as a case study. PhD dissertation, Tel-Aviv University, Raymond and Beverly Sackler Faculty of Exact Sciences, Tel-Aviv, Israel.

Bell, D., Menges, C.H., Bartolo, R.E., Ahmad, W., and VanZyl, J.J. 2001. A multistaged approach to mapping soil salinity in a tropical coastal environment using airborne SAR and Landsat TM data. In *Proceedings of the Geoscience and Remote Sensing Symposium*, Sidney, Vol. 3, pp. 1309–1311.

Ben-Dor, E. 2002. Quantitative remote sensing of soil properties. *Advances in Agronomy* 75: 173–243.

Ben-Dor, E. and Banin, A. 1995a. Near infrared analysis (NIRA) as a rapid method to simultaneously evaluate several soil properties. *Soil Science Society of America Journal* 59: 364–372.

Ben-Dor, E. and Banin, A. 1995b. Near infrared analysis (NIRA) as a method to simultaneously evaluate spectral featureless constituents in soils. *Soil Science* 159: 259–269.

Ben-Dor, E., Irons, J., and Epema, G. 1999. Soil spectroscopy. In *Manual of Remote Sensing*, 3rd edn., A. Rencz (Ed.), John Wiley & Sons, New York, pp. 111–188.

Ben-Dor, E., Patkin, K., Banin, A., and Karnieli, A. 2002. Mapping of several soil properties using DAIS-7915 hyperspectral scanner data. A case study over clayey soils in Israel. *International Journal of Remote Sensing* 23: 1043–1062.

Ben-Dor, E., Heller, D., and Chudnovsky, A. 2008. A novel method of classifying soil profiles in the field using optical means. *Soil Science Society of American Journal* 72: 1113–1123.

Birk, R.J. and McCord, T.B. 1994. Airborne hyperspectral sensor systems. *Aerospace and Electronic Systems Magazine, IEEE* 9: 26–33.

Brunner, P., Kinzelbach, W., and Li, H. 2004. Generating large scale soil salinity maps with geophysics and remote sensing. *Geophysical Research Abstracts* 6: 04745.

Clark, R.N. and Roush, T.L. 1984. Reflectance spectroscopy: Quantitative analysis techniques for remote sensing applications. *Journal of Geophysical Research* 89: 632–634.

Cocks, T., Jenssen, R., Stewart, A., Wilson, I., and Shields., T. 1998. The HyMap airborne hyperspectral sensor, the system, calibration and performance. In *Proceedings of the 1st EARSEL Workshop on Imaging Spectrometry*, Zurich, Switzerland, pp. 37–42.

Corwin, D.L. and Lesch, S.M. 2005. Apparent soil electrical conductivity measurements in agriculture. *Computer Electronics in Agriculture* 46: 11–43.

Csillag, F., Pásztor, L., and Biehl, L. 1993. Spectral band selection for the characterization of salinity status of soils. *Remote Sensing of Environment* 43: 231–242.

Daniels, J.J., Roberts, R., and Vendl, M. 1995. Ground penetrating radar for detection of liquid contaminants. *Applied Geophysics* 33: 195–207.

Davis, J.L. and Annan, A.P. 1986. High resolution sounding using ground probing radar. *Geoscience Canada* 3: 205–208.

Davis, J.L. and Annan, A.P. 1989. Ground penetrating radar for high resolution mapping of soil and rock stratigraphy. *Geophysical Prospecting* 37: 531–551.

Dehaan, R.L. and Taylor, G.R. 2002. Field-derived spectra of salinized soils and vegetation as indicators of irrigation-induced soil salinization. *Remote Sensing of Environment* 80: 406–417.

Dehaan, R. and Taylor, G.R. 2003. Image-derived spectral endmembers as indicators of salinity. *International Journal of Remote Sensing* 24: 775–794.

Dekker, A., Brando, J., Anstee, N., Pinnel, T., Kutser, H., Hoogenboom, R., Pasterkamp, S., Peters, S.,Vos, R., Olbert, C., and Malthus, T.J. 2001. Imaging spectrometry of water. In *Imaging Spectrometry: Basic Principles and Prospective Applications*, F. van der Meer and S. de Jong (Eds.), Kluwer Academic, Dordrecht, the Netherlands, pp. 307–359.

Derriman, M. and Agar, R.A. 1990. Gold and base metal exploration in the Pilbara craton, Western Australia, using the Geoscan airborne multispectral scanner. In *Proceedings of the International Geoscience and Remote Sensing Symposium*, Washington, DC, pp. 1715–1718.

Doolittle, J.A. and Collins, M.E. 1995. Use of soil information to determine application of ground penetrating radar. *Journal of Applied Geophysics* 33: 101–108.

Dutkiewicz, A. 2006. Evaluating hyperspectral imagery for mapping the surface symptoms of dryland salinity. PhD dissertation, University of Adelaide, Australia.

Dwivedi, R., Ramana, K., Thammappa, S., and Singh, A. 2001. The utility of IRS-1C, LISS-III and PAN-merged data for mapping salt-affected soils. *Photogrammetric Engineering and Remote Sensing* 67: 1167–1175.

Epema, G.F. 1990. Effect of moisture content on spectral reflectance in a playa area in southern Tunisia. In *Proceedings International Symposium, Remote Sensing and Water Resources*, Enschede, the Netherlands, pp. 301–308.

Evans, F. and Caccetta, P. 2000. Broad-scale spatial prediction of areas at risk from dryland salinity. *Cartography* 29: 33–40.

Everitt, J., Escobar, D., Gerbermann, A., and Alaniz, M. 1988. Detecting saline soils with video imagery. *Photogrammetric Engineering and Remote Sensing* 54: 1283–1287.

Farifteh, J. 2007. Imaging spectroscopy of salt-affected soils: Model-based integrated method. ITC dissertation No. 143, Enschede, the Netherlands.

Farifteh, J., Bouma, A., and van der Meijde, M. 2004. A new approach in the detection of salt affected soils: Integrating surface and subsurface measurements. In *Proceedings of 10th European Meeting of Environmental and Engineering Geophysics*, P059, Utrecht, the Netherlands.

Farifteh, J., Farshad, A., and George, R. 2006. Assessing salt-affected soils using remote sensing, solute modelling and geophysics. *Geoderma* 130: 191–206.

Farifteh, J., van der Meer, F., Atzberger, C., and Carranza, E.J.M. 2007. Quantitative analysis of salt-affected soil reflectance spectra: A comparison of two adaptive methods (PLSR and ANN). *Remote Sensing of Environment* 110: 59–78.

Fernandez, G., Palladino, M., D'Urso, G., and Moreno, J. 2007. Monitoring soil water content and soil temperature simultaneously to thermal observations from airborne data within two different experiments: *SEN2FLEX-2005 and AgriSAR-2006. Geophysical Research Abstracts* 9: 1607–7962.

Ghassemi, F., Jakeman, A.J., and Nix, H.A. 1995. *Salinisation of Land and Water Resources: Human Causes, Extent, Management and Case Studies*. The Australian National University, Canberra, and CAB International, United Kingdom.

Goetz, A.F.H., Wellman, J.B., and Barnes, W.L. 1985. Optical remote sensing of the earth, *Proceedings of the IEEE* 73: 950–969.

Hollinger, A., Bergeron, M., Maskiewicz, M., Qian, S., Othman, H., Staenz, K., Neville, R., and Goodenough, D. 2006. Recent developments in the hyperspectral environment and resource observer (HERO) mission. In *Proceedings of the Geoscience and Remote Sensing Symposium*, Denver, CO, pp. 1620–1623.

Honey, F. 1989. Technical specifications of the Geoscan Mark II AMSS. Unpublished report of Geoscan Pty. Ltd.

Howari, F.M., Goodell, P.C., and Miyamoto, S. 2002. Spectral properties of salt crusts formed on saline soils. *Journal of Environmental Quality* 31: 1453–1461.

Huete, A. 2005. Estimation of soil properties using hyperspectral VIS/IR sensors. In *Encyclopedia of Hydrological Sciences*, M.G. Anderson (Ed.), John Wiley & Sons, New York, pp. 1–15. Available at http://mrw.interscience.wiley.com/emrw/9780470848944/ehs/article/hsa064/current/abstract.

Kalra, N.K. and Joshi, D.C. 1996. Potentiality of Landsat, SPOT and IRS satellite imagery for recognition of salt affected soils in Indian arid zone. *International Journal of Remote Sensing* 17: 3001–3014.

Lane, M.D. 2002. Upcoming THEMIS investigation of salts of Mars. *Lunar and Planetary Science* XXXIII 1749.

Lesch, S.M., Rhoades, J.D., Lund, L.J., and Corwin, D.L. 1992. Mapping soil salinity using calibrated electromagnetic measurements. *Soil Science Society of America Journal* 56: 540–548.

Lesch, S.M. and Rhoades, J.D. 1999. ESAP-95 software package, version 2.01R. USDAARS, George E. Brown Jr. Salinity Laboratory, Riverside, CA.

Lobell, D., Ortiz-Monasterio, J., Gurrola, F., and Valenzuela, L. 2007. Identification of saline soils with multiyear remote sensing of crop yields. *Soil Science Society of America Journal* 71: 777–783.

Madani, A. 2005. Soil salinity detection and monitoring using Landsat data: A case study from Siwa Oasis, Egypt. *GIScience and Remote Sensing* 42: 171–181.

Malley, D. and Ben-Dor, E. 2004. Application in analysis of soils. In *Near Infrared Spectroscopy in Agriculture*, R. Craig, R. Windham, and J. Workman, (Eds.), Agronomy 44, ASA, SSSA, CSSA, Madison, WI, pp. 729–784.

Manchanda, M. and Iyer, H. 1983. Use of Landsat imagery and aerial photographs for delineation and categorization of salt-affected soils of part of north-west India. *Journal of Indian Society Soil Science* 31: 263–271.

Metternicht, G.I. 1998. Fuzzy classification of JERS-1 SAR data: An evaluation of its performance for soil salinity mapping. *Ecological Modelling* 111: 61–74.

Metternicht, G.I. 2008. Remote sensing. In *Encyclopedia of Geographic Information Science*, K. Kemp (Ed.), Sage Publications, New York, pp. 365–368. ISBN: 978-1-4129-1343-3.

Metternicht, G.I. and Zinck. J.A. 1996. Modelling salinity-alkalinity classes for mapping salt-affected topsoils in the semi-arid valleys of Cochabamba (Bolivia). *ITC-Journal*, 2: 125–135.

Metternicht, G.I. and Zinck, J.A. 1997. Spatial discrimination of salt- and sodium-affected soil surfaces. *International Journal of Remote Sensing* 18: 2571–2586.

Metternicht, G.I. and Zinck, J.A. 2003. Remote sensing of soil salinity: Potentials and constraints. *Remote Sensing of Environment* 85: 1–20.

Mougenot, B. 1993. Effets des sels sur la réflectance et télédétection des sols salés. *Cahiers ORSTOM, Série Pédologie* 28: 45–54.

Muller, A., Richter, R., Habermeyer, M., Mehl, H., Dech, S., Kaufmann, H., Segl, K., Strobl, P., and Haschberger, P. 2004. ARES: A new reflective/emissive imaging spectrometer for terrestrial applications. *Proceedings of the SPIE*, 5574: 120–127.

Nanni, M.R. and Demattê, J.A.M. 2006. Spectral reflectance methodology in comparison to traditional soil analysis. *Soil Science Society of America Journal* 70: 393–407.

Norman, C.P. 1990. Training manual on the use of the EM 38 for soil salinity appraisal. Technical Report series No. 181. Department of Agriculture and Rural Affairs, Victoria, Australia.

O'Leary, G., Grinter, V., and Mock, I. 2004. Optimal transect spacing for EM38 mapping for dryland agriculture in the Murray Mallee, Australia. In *Proceedings of the 4th International Crop Science Congress*, Brisbane, Australia (unpaginated CDrom).

Rao, B.R.M. and Venkataratnam, L. 1991. Monitoring of salt affected soils: A case study using aerial photographs, Salyut-7 space photographs, and Landsat TM data. *Geocarto International* 6: 5–11.

Rao, B., Sankar, T., Dwivedi, R., Thammappa, S., Venkataratnam, L., Sharma, R., and Das, S. 1995. Spectral behaviour of salt-affected soils. *International Journal of Remote Sensing* 16: 2125–2136.

Rhoades, J.D., Chanduvi, F., and Lesch, S.M. 1999. Soil salinity assessment: Methods and interpretation of electrical conductivity measurements. FAO Irrigation and Drainage Paper No.57. Food and Agriculture Organization of the United Nations, Rome.

Schmid, T., Gumuzzio, J., Koch, M., Mather, P., and Solana, J. 2007. Characterizing and monitoring semi-arid wetlands using multi-angle hyperspectral and multispectral data. In *Proceedings Envisat Symposium 2007*, ESA SP-636. Montreux, Switzerland, pp. 1–6.

Schmullius, C. and Evans, L. 1997. Table summary of SIR-C/X-SAR results: SAR frequency and polarization requirements for applications in ecology and hydrology. In *Proceedings of the IGARSS'97*, Singapore, pp. 1734–1736.

Shao, Y., Guo, H., Hu, Q., Lu, Y., Dong, Q., and Han, C. 2002. Study on complex dielectric properties of saline soils. In *Proceedings of the Geoscience and Remote Sensing Symposium*, Toronto, Canada, Vol. 3. pp. 1541–1541b.

Shepherd, K.D. and Walsh, M.G. 2002. Development of reflectance spectra libraries for characterization of soil properties. *Soil Science Society of America Journal* 66: 988–998.

Shih, S. and Myhre, D. 1994. Ground-penetrating radar for salt-affected soil assessment. *Journal of Irrigation and Drainage Engineering* 120: 322–333.

Shuya, H., Qinhuo, L., and Xiaowen, L. 2005. Spectral model of soil salinity in Xinjiang of China. In *Proceedings of the Geoscience and Remote Sensing Symposium*, Seoul, Korea, Vol. 6. pp. 4458–4460.

Slavich, P.G. 1990. Determining ECa depth profiles from electromagnetic induction measurements. *Australian Journal of Soil Resources* 28: 443–452.

Sreenivas, K., Venkataratnam, L., and Rao, P.V.N. 1995. Dielectric properties of salt-affected soils. *International Journal of Remote Sensing* 16: 641–649.

Stuffler, T., Kaufmann, C., Hofer, S., Förster, K., Schreier, G., Mueller, A., Eckardt, A., Bach, H., Penné, B., Benz, U., and Haydn, R. 2007. The EnMAP hyperspectral imager—An advanced optical payload for future applications in Earth observation programmes. *Acta Astronautica* 61: 115–120.

Taylor, G.R. 2004. Field and image spectrometry for soil mapping. In *Proceedings of the 12th Australian Remote Sensing Photogrametry Conference*, . Fremantle, WA, (unpaginated CDrom).

Taylor, G.R., Bennett, B.A., Mah, A.H., and Hewson, R. 1994. Spectral properties of salinised land and implications for interpretation of 24 channel imaging spectrometry. In *Proceedings of the First International Remote Sensing Conference and Exhibition*, Strasbourg, France, Vol. 3, pp. 504–513.

Taylor, G.R., Mah, A.H., Kruse, F.A., Kierein-Young, K.S., Hewson, R.D., and Bennett, B.A. 1996. Characterization of saline soils using airborne radar imagery. *Remote Sensing of Environment* 57: 127–142.

Taylor, G.R., Hemphill, P., Currie, D., Broadfoot, T., and Dehaan, R. 2001. Mapping dryland salinity with hyperspectral imagery. In *Proceedings of the International Geoscience and Remote Sensing Symposium*, Sydney, Australia, pp. 302–304.

Vane, G., Goetz, A.F.H., and Wellman, J. 1984. Airborne imaging spectrometer: A new tool for remote sensing. *IEEE Transactions on Geoscience and Remote Sensing* 22: 546–549.

Verma, K.S., Saxena, R.K., Barthwal, A.K., and Deshmukh, S.K. 1994. Remote sensing technique for mapping salt affected soils. *International Journal of Remote Sensing* 15: 1901–1914.

Wait, J.R. 1962. *Electromagnetic Waves in Stratified Media*. Pergamon Press, New York.

Wiegand, C., Richardson, A., Escobar, D., and Gerbermann, A. 1991. Vegetation indices in crop assessments. *Remote Sensing of Environment* 35: 105–119.

Wiegand, C.L., Everitt, J.H., and Richardson, A.J. 1992. Comparison of multispectral video and SPOT-1 HRV observations for cotton affected by soil salinity. *International Journal of Remote Sensing* 13: 1511–1525.

Wiegand, C.L., Rhoades, J.D., Escobar, D.E., and Everitt, J.H. 1994. Photographic and videographic observations for determining and mapping the response of cotton to soil salinity. *Remote Sensing of Environment* 49: 212–223.

Whiting, M.L. and Ustin, S.L. 1999. Use of low altitude AVIRIS data for identifying salt affected soil surfaces in Western Fresno County, CA. In *Eighth JPL Airborne Earth Science Workshop* (number 64), R.O. Green (Ed.), Jet Propulsion Laboratory, California Institute of Technology, Pasadena, CA.

Yun, S., Qingrong, H., Huadong, G., Yuan, L., Qing, D., and Chunming, H. 2003. Effect of dielectric properties of moist salinized soils on backscattering coefficients extracted from RADARSAT image. *IEEE Transactions on Geoscience and Remote Sensing* 41: 1879–1888.

Part II

*Trends in Mapping Soil Salinity
and Monitoring Salinization
with Remote and Proximal Sensing*

4 Mapping Areas Susceptible to Soil Salinity in the Irrigation Region of Southern New South Wales, Australia

David Fraser

CONTENTS

4.1 INTRODUCTION

This chapter presents the findings of a research project designed to map soil salinity in the irrigation region of southern New South Wales (NSW), Australia, using satellite imagery. The initial aims of the project were to map the spatial and temporal spread of soil salinity in dry-farming and irrigation farming regions and to evaluate and compare the rates of increase in salinity over a 10-year period. Landsat multispectral scanner (MSS) and thematic mapper (TM) imagery and in situ monitoring techniques were used for that purpose. Change detection in soil salinity would help in extrapolating the findings to the future, for better land-use planning and effective management of resources.

The Murray Valley project area extends some 400 km from Swan Hill in the west to Albury in the east. The southern boundary is the Murray River and the northern boundary lies approximately 150 km north of the river. The vast size of the project meant that it was impossible to gather data from the whole geographical area, so representative study areas were selected. Two studies were carried out to identify areas under threat from soil salinity: one in a dry-farming region and another in an irrigation region (Figure 4.1). This chapter discusses the results from the irrigation region. A predictive model was developed to monitor soil salinity by using surrogate variables. Predictive modeling based on long-term data records has assisted land managers in the Murray region for many years (Goss, 2006).

4.2 SOIL SALINITY AND REMOTE SENSING

Of all the environmental hazards facing Australia today, land degradation has been identified as being by far the worst. In recent years, soil salinity has emerged as a major land degradation issue.

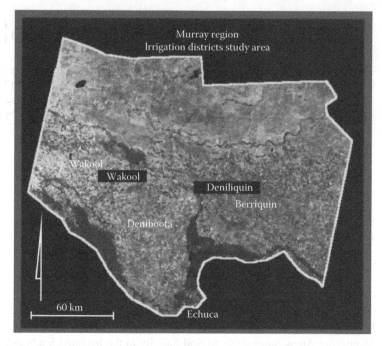

FIGURE 4.1   Map of the Murray region showing the location of the irrigation districts.

Soil salting has become increasingly evident over the past 20 years on extensive farmland areas. It has already rendered large areas sterile and unproductive, while diminishing the productivity of adjacent lands and other areas where salting is less severe. Unless effective steps are taken to halt the spread of salinity and rehabilitate damaged lands, more of its destructive blight will threaten even larger areas in the years ahead (Fraser et al., 1990). Many approaches have been and are being tested in a search for solutions; one of these includes the use of remote sensing.

4.2.1 ROLE OF REMOTE SENSING

Soil salting due to irrigation is caused by the ponding of water during flood irrigation as the excess water infiltrates into the groundwater system and causes the water level to rise, bringing salt into the root zone of crops. Soil salting is therefore a direct result of excessive groundwater recharge that, either locally or regionally, manifests itself in a rise of the water table to less than 2 m of the land surface. From this depth, capillary action in the soil structure moves water into the root zone of the uppermost soil horizons where wind and heat of the sun evaporate soil moisture, leaving a cumulative salt residue behind in the soil. This eventually kills all vegetation and leaves bare, salt-encrusted ground that is then prone to wind erosion and gullying in the worst cases.

Remote sensing, using imagery acquired from satellite-mounted multispectral scanners, is used by a number of organizations in research activities related to mapping and monitoring the spread of soil salinity. Satellite-borne multispectral scanners measure and record the intensity of the electromagnetic radiation received from the ground. The intensity of radiation emitted or reflected is recorded separately for each in a number of wavelength ranges or bands. Differences in reflectance for the various bands of the imagery can be used to distinguish land cover types. The strength of the reflectance varies from land cover to land cover and also within a given land cover type due to physical factors that may alter the health or other characteristics of the land cover. For instance, vigorously growing, healthy vegetation gives a strong reflectance in the near-infrared, but the strength of the reflectance decreases as the health of the vegetation declines. The strength of the reflectance for a particular land cover also varies among spectral bands.

The reflectance from the landscape is not recorded for the landscape as a whole, rather, it is recorded for each of many contiguous rectangular patches of land called pixels. Pixels are scanned row by row, as the scanner moves over the land. The values for each pixel are recorded digitally as brightness numbers. Image processing software converts the remotely sensed scanner data into a multicolored image of the landscape (Fraser et al., 1990). Since the data are in digital form, they can be statistically analyzed to determine spatial or spectral characteristics of the land surface features. Such characteristics can then be related back to the geographical area covered by the imagery and image maps can be created.

A specific land cover may have a unique combination of brightness numbers when compared to other land covers. The differentiation within vegetated areas is difficult, requiring more subtle analysis for such tasks as separation of vegetation types, stress, and phenological stage (Townshend and Justice, 1995).

4.2.2 IDENTIFYING AREAS SUSCEPTIBLE TO SOIL SALINITY IN THE IRRIGATED STUDY AREA

The soil salinity in the irrigated study area has not created large tracts of salt-scalded land. The effect of soil salinity is more insidious in that it slowly affects the productivity of the land due to the salt coming into contact with the plant roots through capillary rise from groundwater in specific areas. The response of the farmer is to either change the crop or turn the land over to irrigation.

The identification of areas susceptible to soil salinity was undertaken by investigating the spectral characteristics of the irrigated crops. On the imagery supplied for the project, healthy vegetation growing on irrigated land has a common spectral signature due to the very high reflectance in the near-infrared bands of the spectrum (Figure 4.2). Conversely, non-irrigated pasture associated with dry-farming tends to have low reflectance in the near-infrared.

The state government of NSW established a network of piezometers (groundwater boreholes) over the irrigation districts. The location of the piezometers is shown by the colored squares in Figure 4.3. Every month, the groundwater height and salinity readings are recorded at each piezometer site. This information has for many years provided a valuable temporal and spatial record of the changes in soil salinity and groundwater levels across the irrigation regions. Groundwater records in Western Australia have also been used over many years to indicate salinity risk in the northern region (Wilson, 2006).

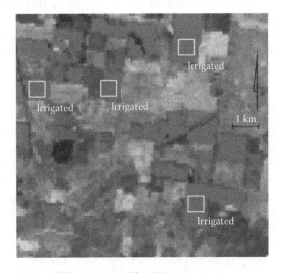

FIGURE 4.2 Polygons defining the "irrigated land" training areas.

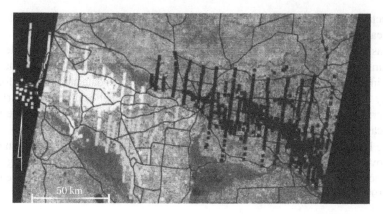

FIGURE 4.3 Network of piezometers (bores) over the irrigation districts (white, black, and yellow colors refer to different irrigation districts).

The satellite imagery provides only one source of information about the physical environment and may not contain sufficient information to allow the correct classification of detailed land cover. Spatial resolution is the major obstacle in the identification of intricate land cover (Gao and Skillcorn, 1996). Therefore, the results of any remote sensing analysis should be used in conjunction with and compared to the results from other studies. Areas affected by soil salinity, as determined from the remote sensing analysis, would be expected to coincide fairly closely with high groundwater. This association with the groundwater level provides a means of verifying any results from the remote sensing analysis using an independent data source.

According to Hill and McAllister (1990, p. 5), "the first task facing any investigator hoping to make use of remote sensing for land degradation problems is that of determining whether there is a surrogate variable or combination of surrogate variables that will portray the distribution of the problem." In the irrigation farming study, different approaches were tested and the final approach adopted was to use the concentration of irrigated land as a means of identifying areas susceptible to rising soil salinity.

The use of a surrogate variable or, in other words, an alternative indicator of the presence of soil salinity was considered to be appropriate for this study as the traditional methods tested throughout the life of the project did not provide information with an adequate level of certainty for the purpose. The lack of historical data that could be used to provide the land cover for the temporal set of imagery also meant that another approach was required. Piezometer readings were available for 2600 sites in the irrigation districts (Figure 4.3). These readings were used to determine the extent of high groundwater levels from year to year.

4.3 TECHNICAL ASPECTS OF THE APPROACH

Several data sets with different properties were used in this project. Many variables have been dealt with. The key issues in mapping soil salinity in the irrigation region will now be outlined.

This project utilized satellite imagery from 1980 through 2000. Landsat MSS imagery for the summer periods of 1980, 1982, 1986, 1988, and 1990 was used, with each image being coregistered to allow pixel-by-pixel comparison. This was achieved by image-to-image association. The 1990 MSS image was first rectified using ground-control points, and all other MSS images were registered to the 1990 image. Once both these tasks were completed, it was possible to overlay all imagery with other data sets such as the Murray region boundary. Landsat TM imagery was acquired for the summer periods of 1992, 1995, and 2000. The TM data were used to create a predictive model based on the results from the analysis of the 1980–1990 imagery.

A mask of the Murray region was placed over the imagery and all the pixels outside the region were set to null. This masking helped concentrate the statistical results on the area of concern, in this case the Murray region. Similarly, masks of the three main irrigation districts of Wakool, Deniboota, and Berriquin were placed over the imagery and all the pixels outside these districts were set to null. These irrigation districts were considered to be the primary areas within the Murray region where it was likely that soil salinity would be of most concern and would manifest itself. It was also considered that any data collected by the NSW government would be concentrated in these areas.

Piezometer readings from 1980 to 1995 for the Murray region provided a means of identifying areas of rising groundwater that may be associated with an increase in soil salinity levels. It was assumed that these groundwater readings would provide a valuable temporal link between the imagery from 1980 to 2000. The data were stored in tabulated form and in association with the coordinates of the piezometer sites. This allowed the location of each piezometer to be identified on the imagery.

The ground coordinates and groundwater height readings for each piezometer were imported into a geographic information system (GIS) and a triangulated irregular network was formed, which in turn was used to produce contouring of the data points. The polygons that enclosed the areas where the groundwater was within 3 m of the terrain surface were selected for each data set from 1980 to 1995. The 3 m level was chosen so that land susceptible to soil salinity in the future would also be identified along with the areas already impacted upon. These boundary data were superimposed over the satellite imagery. A comparison could then be made between the selected reference years to determine the nature and the spatial extent of the change in groundwater levels (Figure 4.4).

One major difficulty in dealing with multitemporal satellite imagery is the decrease in the amount and accuracy of ground-truth data collected for imagery at different times. The only ground-truth data available for this project were based on the collective knowledge gained from workshops held with local land managers and ground-truth data associated with the 2000 TM image. The ground-truth data were applied to the 1995 and 2000 TM imagery and used to produce initial land cover classifications. It was not possible to use the existing ground-truth data to train earlier imagery because of the time difference between the collection of the ground-truth data and the scanning of the images. This is due, in part, to the dynamic nature of the land use in the area. Howell and McAllister (1994) found that errors in their study were either associated with sketchy ground-truth data or occurred at the fringe of the land-use categories as in the case of irrigated annual pastures.

One land cover that could be identified clearly on all imagery was irrigated land. The irrigated land, used exclusively for rice production, was easily identified visually by the bright red appearance that these areas have on standard false-color images. These areas were used to provide

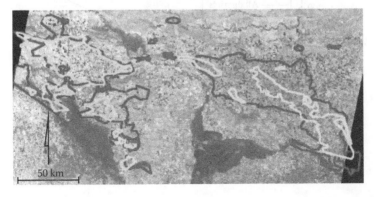

FIGURE 4.4 Groundwater within 3 m depth of the soil surface in the irrigation districts (yellow, 1980; red, 1995).

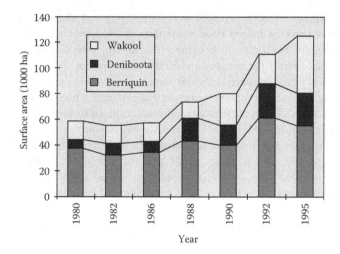

FIGURE 4.5   Contribution of each irrigation district to the total irrigated land in the period 1980–1995.

a reliable homogeneous land cover class, which was visible on the 1980 imagery through the 2000 imagery. Using a three-class image classification (irrigated land, forest, and other land), it was possible to identify the class associated with irrigated land and produce an area calculation for this class on each satellite image from 1980 to 1995 (Figure 4.5). The areas of irrigated land were then compared with the areas of rising groundwater, based on the assumption that there is a direct correlation between the application of water for irrigation and the rise of the groundwater level. Figure 4.6 shows that as the area of irrigated land increases, so does the area where the groundwater is within 3 m of the surface.

It was assumed that the irrigated land had become concentrated into specific districts over time. To test this, the area of the irrigated land within the irrigation districts of Wakool, Deniboota, and Berriquin was calculated from 1980 to 1995. This allowed the detection of trends in the area of irrigated land for these districts and also provided a means of comparing the change between districts. Some of the key findings from this study are

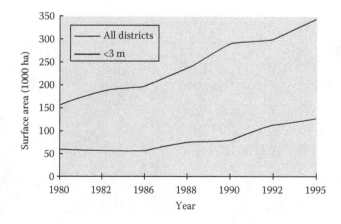

FIGURE 4.6   Increase in irrigated land (blue) and land with groundwater within 3 m depth in the period 1980–1995 (red).

1. Polygons where the groundwater is within 3 m of the surface coincide with the irrigation districts Wakool, Deniboota, and Berriquin, which also contain most of the irrigated land in the Murray region portion of the image.
2. Irrigated surface area increased in all the three irrigation districts studied from 1980 to 1995.
3. Berriquin irrigation district had the largest increase in irrigated land (18,369 ha) and also the largest increase in land with groundwater levels within 3 m of the surface (120,505 ha) from 1980 to 1995.
4. Groundwater rise over time corresponds to an increase in the area of irrigated land.
5. Areas threatened now and in the future by soil salinity are located in sectors of rising groundwater levels and coincide with high concentrations of irrigated land.

As a consequence of these observations, the groundwater level and the area of irrigated land were adopted as criteria in the belief that any trends in these variables could indirectly identify areas susceptible to soil salinity. The satellite imagery was used to identify the location of the irrigated areas and to monitor the change in areal extent from year to year. Using supervised classification and an overlay of the groundwater isolines on the classified imagery, the study revealed a spatial coincidence between the increase in the area of irrigated land and the increase in the areal extent of high groundwater.

If it is accepted that the increase in groundwater levels is directly associated with the increase in water usage due to the increase in the area of irrigated land, then it follows that the land surrounding the irrigated areas becomes increasingly exposed to rising groundwater that may bring salt with it into the soil mantle. The assumption is that water for irrigation introduced from outside the area has increased the groundwater levels not only in the irrigation districts but also in the surrounding dry-farming areas. If this trend continues, the groundwater will reach the root zone of the plants and may bring with it salt that will have a detrimental effect on the farming land surrounding the irrigated rice. At present, only small pockets of land are severely affected by soil salting but indications, from the results of this study, are that such areas will increase in number and size in the future. This increases the likelihood that such land will also be turned over to irrigation. The worst-case scenario is that the area under irrigation is going to increase to such an extent that the groundwater level will reach the soil surface and bring with it salt that will destroy the productivity of the land.

Based on this research, it was considered possible to create a predictive model to determine areas susceptible to soil salinity, using satellite imagery, piezometer readings, and water-usage records from the irrigation districts. Further research would be required to refine the method, including the identification of physical characteristics of the land such as geology and soil that may make particular areas more susceptible than others to rising groundwater. In a study at Axe Creek, Hill and McAllister (1990) used geological information supplemented by aerial photo interpretation to remove all geological formations not significant to groundwater. Once these other factors are identified, the predictive model can be refined. This approach is in line with a study by Overton et al. (1996) who created a predictive model using a GIS that incorporates groundwater depths, groundwater salinities, soil properties, and flooding frequencies to predict soil salinization risk and therefore vegetation health.

4.4 CONSTRUCTION OF THE PREDICTIVE MODEL

This section provides a brief introduction to the predictive model presented in the flowchart of Figure 4.7. The variables considered in the model are relevant to determine salinity risk in terms of the likelihood of salinization (landscape curvature) or the impact of salinization (groundwater salinity, location of roads, location of drains, creeks and rivers, and location of towns). The above variables primarily pertain to the environment, infrastructure, and downstream users.

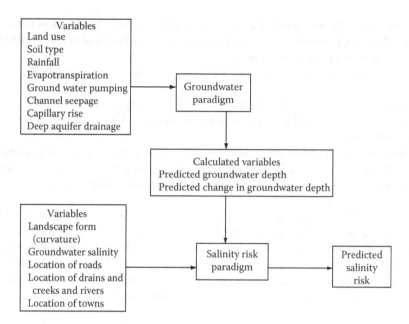

FIGURE 4.7   Flowchart of the Murray Valley model.

The model predicts the change in groundwater, which is added to the initial groundwater depth to obtain the predicted groundwater depth. These predictions are reclassified and used in risk analysis. As the groundwater paradigms have been developed specifically for each irrigation district with biennial time steps, the predictions of salinity risk from this model are specific to the study area and to this methodological approach in groundwater modeling.

Grids representing actual salinity risk values were created in a GIS from the actual values for groundwater depth and the change in groundwater. The production of grids that represent the actual salinity risk enabled the testing of the accuracy of risk predictions made by the model. Figure 4.8

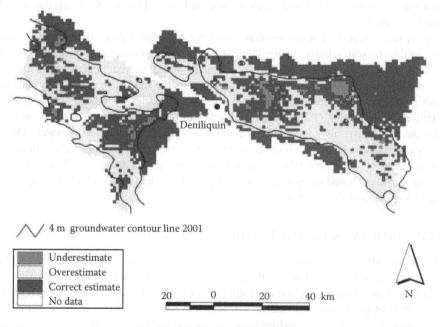

FIGURE 4.8   Predicted versus actual salinity risk for 2001 (in four time steps from 1992).

compares predicted salinity risk to actual risk. The map shows the areas where risk has been predicted correctly, overestimated, or underestimated. Due to the more generalized nature of the predictive model, a contour of 4 m depth to the groundwater was used and is also shown in Figure 4.8 (Lamble, 2000).

Field verification work revealed that the actual presence or absence of moderate-to-severe salinization on the ground does not necessarily determine an appropriate risk value, which also depends on the impact of this current salinization. Generally, the results showed that the grid cell values of the risk model were consistent with the depth to the groundwater indicators of risk. However, there were some anomalies in all districts, where piezometers with deep groundwater were in grid cells with higher risk values than warranted by the groundwater depth.

The testing of risk predictions showed that the groundwater paradigm provided the best predictions of salinity risk in each irrigation district. Areas under threat from soil salinity were located in areas of rising groundwater level. The areas under current and potential threat were determined as those areas that have high concentrations of irrigated land.

4.5 CONCLUSIONS

A method for identifying areas susceptible to soil salinity in irrigation districts was outlined, which provides a useful tool to decision makers for segregating such land in the Murray region of NSW, Australia. Characteristics of the land, which could be reliably used to detect spatial and temporal trends in soil salinity, were identified and used in the model. The groundwater level and the area of irrigated land were used as criteria in the belief that any trends in these variables could indirectly identify areas susceptible to soil salinity.

The assumption is that water introduced for irrigation from outside the region raises the groundwater level within the root zone of crops and pasture in the surrounding land. The area of land impacted upon by soil salinity is on the increase and, due to the vast area being impacted upon, it was necessary to develop a method that could be applied uniformly over the area.

This study demonstrates how remotely sensed data can be used to map and monitor over time the areal extent of areas subjected to reduced productivity due to rising groundwater and increased soil salinity. The impact of the flood irrigation used for rice production is discussed. The study establishes a link between increased groundwater levels and increased concentration of irrigated land in the managed regions. Land converted from dry-farming to irrigation farming increases the impact of soil salinity in the local region. Growing irrigated rice on less suitable soils can lead to high percolation rates, rising water tables, and increased land salinity.

This research also shows that a temporal set of remotely sensed imagery can provide the vehicle for the creation of a predictive model to determine areas susceptible to soil salinity. Further research is required to refine the method, including the identification of physical characteristics of the land such as geology and soil that may make particular areas more susceptible than others to rising groundwater.

REFERENCES

Fraser, D., Mackenzie, M., Bellamy, G., and Ellis, G. 1990. Mapping dryland salinity recharge areas using EM conductivity measurements and airborne scanner imagery. In *Proceedings of the 5th Australian Soil Conservation Conference*, Perth, Australia, Vol. 10, pp. 41–44.

Gao, J. and Skillcorn, D. 1996. Detection of land cover change from remotely sensed data: A comparative study of spatial and non-spatial methods. In *Proceedings of the 8th Australian Remote Sensing Conference*, Canberra, Australia, pp. 211–219.

Goss, K. 2006. Getting real about salt. *Focus on Salt*, (Technical Newsletter), Issue 37:1–2, CRC for plant-based management of dryland salinity.

Hill, S.R. and McAllister, A. 1990. Land classification using remote sensing and geographic information systems: A salinity case study. Land Protection Division Research Report No. 3, Department of Conservation and Environment, Melbourne, Australia.

Howell, C. and McAllister, A. 1994. Land cover classification in irrigated agriculture using multispectral satellite images. In *Proceedings of the 7th Australasian Remote Sensing Conference*, Melbourne, Australia, pp. 633–640.

Lamble, P. 2000. Creation of the Murray Valley model—a predictive model for salinity risk. PhD dissertation, RMIT University, Melbourne, Australia.

Overton, I., Lewis, M., and Walker, G. 1996. Detecting vegetation change on a semi-arid floodplain using Landsat TM and MSS. In *Proceedings of the 8th Australian Remote Sensing Conference*, Canberra, Australia.

Townshend, J.R.G. and Justice, C.O. 1995. Spatial variability of images and the monitoring of changes in the normalized difference vegetation index. *International Journal of Remote Sensing* 16(12): 2187–2195.

Wilson, G. 2006. Salinity continuing to rise in WA. *Focus on Salt,* Technical newsletter, Issue 37: 20–21, CRC for plant-based management of dryland salinity.

5 Generation of Farm-Level Information on Salt-Affected Soils Using IKONOS-II Multispectral Data

Ravi Shankar Dwivedi, Ramana Venkata Kothapalli, and Amarendra Narayana Singh

CONTENTS

5.1 INTRODUCTION

Soils with salt concentration in excess of 4 dS m^{-1} are termed saline or salt-affected, according to the U.S. Salinity Staff Laboratory (Richards, 1954). Being a severe environmental hazard, soil salinity affects crop yield and agricultural production. Globally, salt-affected soils comprise 19% of the 2.8 billion ha of arable land (Szabolcs, 1989). In India alone, estimates of 7.1 million ha (Abrol and Bhumbla, 1971) and 6.73 million ha (Anonymous, 2007) were reported to be affected by salinity and alkalinity processes. Such soils are generally not only encountered in the arid and semiarid climate but also occur extensively in subhumid and coastal zones. Some of the most unfavorable properties of salt-affected soils are high salt content, poor structure, limited microbial activity, low percolation rates, and other characteristics that restrict plant growth.

Information on the nature, spatial extent, and distribution of salt-affected soils is a prerequisite to restore their fertility and prevent further salinization and alkalization. Conventional soil surveys, apart from being labor-intensive, time-consuming, and cost-prohibitive, are impractical in inhospitable terrain. Since the late 1960s, new tools and techniques, in particular, black-and-white and color aerial photographs have been used to delineate salt-affected soils (Myers et al., 1966; Hilwig and Karale, 1973; Dale et al., 1986).

In 1972, the launch of the Earth Resources Technology Satellite-1 (ERTS-1), later renamed as Landsat-1, with a four-channel multispectral scanner (MSS), provided a synoptic view of the earth's surface every 18 days. Such data enabled researchers to generate information on the extent and spatial distribution of salt-affected soils, which allows for planning reclamation programs (Singh et al., 1977; Budd and Milton, 1982). Space-borne multispectral measurements with improved spatial, spectral, and radiometric resolutions became available from subsequent Earth observation missions such as Landsat-4 and 5 and SPOT-1, and enabled the generation of information not only on the nature but also on the magnitude of soil salinity and alkalinity (Rao et al., 1991; Toth et al., 1991; Dwivedi and Sreenivas, 1997).

With the launch of the Indian remote sensing satellites IRS-1C and IRS-1D in 1995 and 1998, respectively, multispectral data from a linear imaging self-scanning sensor (LISS-III) with 23.5 m spatial resolution and a panchromatic sensor with 5.8 m spatial resolution were used to derive detailed information on salt-affected soils (Verma et al., 1994; Dwivedi et al., 2001). Currently, space-borne spectral measurements with 4 m spatial resolution or better in multispectral mode and 1 m or better in panchromatic mode are available from IKONOS-II and Quickbird. Eldeiry and Garcia (2004) predicted soil salinity in parts of the Arkansas Valley of southern Colorado in the United States using IKONOS-II images. We report a study that was undertaken to evaluate the potential of IKONOS-II multispectral data in deriving farm-level information on soil salinity and alkalinity over an area of the Sitapur district in Uttar Pradesh, northern India.

5.2   TEST SITE AND DATA SETS

The presence of sodium-bearing minerals in the parent material and a climate of alternate wet and dry seasons with evapotranspiration exceeding precipitation, coupled with the introduction of canal irrigation, have led to the development of soil salinity and alkalinity over large areas in the Sitapur district. A test site of 10 km^2 that forms part of the administrative unit of Sidhauli tehsil (Figure 5.1) was selected. The terrain is nearly level to very gently sloping, with moderate to poor surface drainage. Soils are derived from the Indo-Gangetic alluvium of Quaternary age. The surface texture of the soils ranges from sandy loam to loam with heavier subsurface texture. The presence of CaCO$_3$ concretions in layers 30–70 cm thick known as kankar pans, occurring at about 1 m depth and acting as a hydrological barrier, is a common feature. Assured irrigation is available from the Sharda canal system. The climate is semiarid and subtropical with a pan evaporation value of 1486 mm year^{-1} (Figure 5.2) and a mean annual rainfall of about 900 mm, which is received mostly from the southwesterly monsoon.

The IKONOS-II satellite was launched from the Vandenburg Air Force Base in California by Space Imaging on September 24, 1999. Agile body-scanning configuration allows for short visit times and great flexibility in data collection over an area of user's interest with a swath of 11 km. The sensors on board IKONOS-II include a panchromatic band (450–900 nm) camera at a nominal resolution of 1 m, and a four-band MSS (450–520 nm as blue, 520–600 nm as green, 630–690 nm as red, and 760–900 nm in the near-infrared [NIR]) with a spatial resolution of 4 m. Images acquired on board the satellite are quantized to a radiometric resolution of 11 bits per pixel (0–2047 levels). Thus, IKONOS-II provides an eightfold improvement in the range of gray levels to represent target brightness, when compared to 8 bits per pixel systems such as Landsat or SPOT.

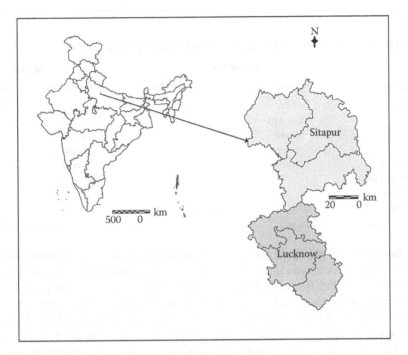

FIGURE 5.1 Location of the Sitapur test site in northern India.

The IKONOS-II multispectral image was collected for the Sidhauli site on March 28, 2003. The IKONOS-II multispectral data (Table 5.1), with nominal azimuth of 258.85°, nominal elevation of 68.02°, sun azimuth angle of 146.41°, and sun elevation angle of 61.83°, were used to derive detailed information on the salt-affected soils.

The Indian remote sensing satellite IRS-1D was launched into orbit by India's Polar Satellite Launch Vehicle (PSLV-C1) on September 29, 1997 from Sriharikota, India. IRS-1D carries a combination of three cameras: (1) a panchromatic camera (PAN) with spatial resolution of 5.8 m;

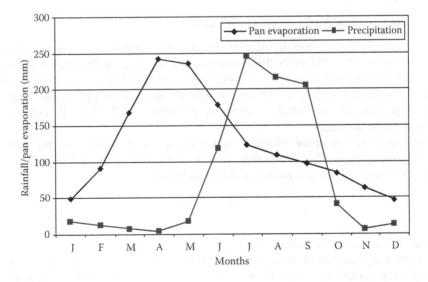

FIGURE 5.2 Mean monthly rainfall and evaporation (mm).

TABLE 5.1

Characteristic Features of IKONOS-II and IRS-1D Multispectral Data

Band Description	IKONOS-II		IRS-1D	
	Multispectral	Panchromatic	Multispectral	Panchromatic
Single band (nm)		500–730		500–750
Blue (nm)	450–520			
Green (nm)	520–600		520–590	
Red (nm)	630–690		620–680	
Near-infrared (nm)	760–900		770–860	
Shortwave infrared (nm)			1550–1750	
Ground IFOV (m)	4	1	23.5	5.8
Quantization (bits)	11	11	7	6

(2) a LISS-III operating in four spectral bands, with spatial resolution of 23.5 m in the visible and NIR bands and 70.5 m in the shortwave infrared band; and (3) a wide-field sensor with a ground resolution of 188 m. The IRS-1D LISS-III and PAN data acquired simultaneously on April 4, 2003 were used as reference, since the utility of such a data set for deriving information on salt-affected soils had already been established (Dwivedi et al., 2001). The instrumentation characteristics of IKONOS and IRS-1D satellites are given in Table 5.1. The 70.5 m spatial resolution of the fourth band (shortwave infrared) of the LISS-III sensor is not compatible with the spatial resolution of the other three spectral bands. Thus, only data of three bands, namely green, red, and NIR, were used. Additionally, topographic maps at scales of 1:25,000 and 1:50,000 were used for ground-truth collection.

5.3 METHODOLOGY

The methodology comprises (1) data preparation consisting of geometric correction, radiometric correction, and fusion of multispectral and panchromatic sensor data; and (2) digital analysis consisting of identification of spectrally homogeneous classes, ground-truth collection, spectral separability, delineation of salt-affected soils, and accuracy estimation.

5.3.1 Data Preparation

5.3.1.1 Geometric and Radiometric Corrections

Geometric and radiometric errors inherent to the raw and merged images were corrected. The geometric correction of the IKONOS-II multispectral data was done on the basis of ground control points measured using a differential global positioning system (d-GPS) with an RMS error of 1.3 m along the X-axis and 0.6 m along the Y-axis. For georeferencing of the LISS-III and PAN data, an image-to-image approach was adopted, where the georeferenced IKONOS-II multispectral data were used as the master image.

Radiometric correction of the satellite data was the next step toward data preparation. In the present study, the digital number (DN) values in each spectral band of the IKONOS-II multispectral data were converted into physical units of radiance as follows:

$$L_{ijk} = \mathrm{DN}_{ijk}/\mathrm{CalCoef} \tag{5.1}$$

where

L_{ijk} is the in-band at-sensor radiance (mW cm^{-2} sr)

DN$_{ijk}$ is the image product of the digital value of the ith image pixel in the jth spectral band

CalCoef is the in-band radiance calibration coefficient DN*mW*[cm^{2}sr]$^{-1}$ (Space Imaging, 2001)

Subsequently, the data were converted to planetary reflectance using the following equation:

$$\rho_p = \pi L\lambda d^2 / E \, sun_\lambda \cos(\theta_s) \tag{5.2}$$

where
ρ_p is the unitless planetary reflectance
$L\lambda$ is the radiance for the spectral band λ at the sensor's aperture (W/m^{-2} μm sr)
d is the earth–sun distance in astronomical units
$E \, sun_\lambda$ is the mean solar exo-atmospheric irradiance (W/m^2–μm)
θ_s is the solar zenith angle

Similarly, the DN values of the LISS-III and PAN sensor data were converted to radiance using the following equation (Anonymous, 1997):

$$L_{rad} = (DN/MaxGray)(L_{max} - L_{min}) + L_{min} \tag{5.3}$$

where
L_{rad} is the radiance for a given DN value (W/m^{-2} μm)
DN is the digital count
MaxGray is 63 for PAN and 127 for LISS-III
L_{min}/L_{max} is the minimum/maximum radiance value for a given band

The radiance values were subsequently converted to top-of-atmosphere reflectance using Equation 5.2.

5.3.1.2 Image Fusion

Several image fusion techniques have been employed to enhance the spatial resolution of a multispectral image. These techniques could be grouped into (1) merger for display, (2) merger by separate manipulation of spatial information, and (3) merger to maintain radiometric integrity (Munechika et al., 1993). Coined by Price (1987), the merger-for-display techniques are primarily concerned with enhancement of the image for an interpreter's display. The second type of image fusion technique (i.e., the merger by separate manipulation of spatial information) attempts to enhance the spatial information without disturbing the spectral information. Intensity–hue–saturation, principal component analysis (PCA), and high-pass filter procedures are examples of this type of image fusion techniques. The third type of image fusion technique addresses the issue of merging to maintain radiometric integrity. Pradines (1986), Price (1987), and Munechika et al. (1993) developed techniques for image fusion, which have been quite efficient in maintaining radiometric integrity of images. Recently, wavelet transform-based approaches have been proposed for image fusion.

In the study reported here, the PCA approach for image fusion was used. PCA transforms a multivariate data set of intercorrelated variables into a data set of new uncorrelated linear combinations of the original variables, and generates a new set of orthogonal axes. The approach for computation of the PCA comprises the calculation of a covariance matrix, eigenvalues and eigenvectors, and principal components (PCs). An inverse PCA transforms the combined data back to the original spectral space. The PCA of a multichannel image, where the first PC is replaced by a different image, aims at improving the spatial resolution of that multichannel image by introducing an image with a higher spatial resolution. Such an image that replaces PC1 is stretched to the variance and average values of PC1. The higher resolution image replaces PC1 since it contains the information common to all bands, while the spectral information is unique to each band (Chavez et al., 1991). PC1 accounts for maximum variance that can maximize the effect of high-resolution data in the fused image (Shettigara, 1992). The justification for replacing the first PC image with the stretched PAN data is that the latter are approximately equal to the first PC image. The generalized equation for PCA merging is as follows:

$$P_e = \sum_{k=1}^{n} d_k E_{ke} \qquad (5.4)$$

where
P_e is the length of PC (first, second, etc.)
k is a particular input band
E is the eigenvector matrix
n is the total number of bands
E_{ke} is the element of matrix at row k, column e
d_k is an input data file value in band k

5.3.2 GROUND-TRUTH DATA COLLECTION AND DEFINITION OF INFORMATIONAL CLASSES

Ground truth aims at establishing the relationship between spectral classes observed in the satellite data and what these classes represent on the ground. Ground truth was carried out from April 4 to April 11, 2003 following the acquisition of the IRS-1D LISS-III and PAN data synchronously to the satellite overpass. Based on the spectral classes delineated during unsupervised classification, 40 sample areas representing all the classes were selected following a stratified random sampling approach. Four areas were located in salt-affected soils for carrying out intensive field sampling in 10 m grids as well as in transects. Initially, a reconnaissance traverse of the area was made to assess trafficability and locate the sample areas. Within the sample areas, parcels of land interpreted as affected by soil salinity and alkalinity were precisely marked onto topographic maps. Observations were made with respect to terrain conditions, namely, land use and land cover, microtopography, surface drainage, presence of salt efflorescence, crop condition, density and vigor, and local relief, after recording their precise location with a d-GPS.

Soil profiles were excavated in core areas of salt-affected soil patches. Subsequently, morphological features of the profiles were recorded, and soil samples were collected from individual profiles for analysis in the laboratory. Auger bores were also studied and sampled in representative areas. Additional observations were made randomly outside the sample areas to validate the relationship already established between image elements and salt-affected soils. Soil samples were air-dried, pounded, and sieved through 2 mm sieves. Laboratory determinations included pH measurement in a 1:4 soil–water suspension using a pH-meter (Electronics Corporation of India Limited digital pH-meter 5652) and electrical conductivity of the saturation extract (ECe) measured with an electrical conductivity bridge (Systronics direct reading conductivity meter 304). The exchangeable sodium percentage (ESP) was determined following the approach of the USDA Soil Laboratory (Richards, 1954).

Salt-affected soils were grouped into three categories based on the quantity of amendment required for the reclamation of saline–sodic topsoils (0–30 cm depth). Soil category 1 requires up to 10 t ha^{-1} of gypsum, category 2 requires 10–20 t ha^{-1} of gypsum, and category 3 requires more than 20 t ha^{-1} of gypsum. The gypsum requirement (GR) was computed using the formula:

$$GR = [ESP(initial) - ESP(final) CEC]/100 \qquad (5.5)$$

where
GR is the gypsum requirement (cmol(+) kg^{-1} soil)
ESP(initial) is the ESP of the soil before reclamation
ESP(final) is the ESP of the soil to be obtained after reclamation (targeted at 15% in this study)
CEC is the cation exchange capacity of the soil (cmol(+) kg^{-1} soil)

5.3.3 Digital Analysis

Digital analysis comprises four steps, namely, preliminary digital analysis, spectral separability analysis, delineation of salt-affected soils, and accuracy assessment.

5.3.3.1 Identification of Spectrally Homogeneous Classes

Spectrally homogeneous classes were delineated using an unsupervised classification algorithm, i.e., the interactive self-organizing data clustering (ISODATA) technique. ISODATA is a clustering algorithm for delineation of spectrally homogeneous classes in an image that uses the minimum spectral distance formula to form clusters. It begins with either arbitrary cluster means or means of an existing signature set, and each time that the clustering repeats, the means of these clusters are shifted. The new cluster means are used for the next iteration.

The ISODATA utility repeats the clustering of the image until either a maximum number of iterations have been performed or a maximum percentage of unchanged pixels have been reached between two iterations. The algorithm groups the pixels with homogeneous spectral response pattern in different spectral bands using standard deviation as threshold. The process allows identifying the number of spectrally homogeneous classes available in the image. The four spectral bands of IKONOS-II were used with a definition of 25 arbitrary cluster means. On the first iteration of the algorithm, the means of the 25 clusters were arbitrarily determined. Subsequently, after each iteration, a new mean for each cluster was calculated based on the actual spectral locations of the pixels in the clusters. These new means of the clusters were used for defining clusters in the next iteration. The process continued until there was little difference between iterations. The convergence threshold was set to 95%. Efforts were subsequently made to associate these classes with the categories of salt-affected soils and other features on the ground.

5.3.3.2 Spectral Separability Analysis

Spectral signature separability plays a crucial role in improving the accuracy of the final thematic maps. Using image analysis software and identifying the training sets (a group of pixels representing a category of salt-affected soils) in the IKONOS-II multispectral data where field observations were made, the spectral response pattern of different types of salt-affected soils was generated. A total of 18 training sets was used to generate the spectral response pattern in the four spectral bands of IKONOS-II multispectral data and subsequently merged into three categories.

Though there are several techniques for assessing the spectral separability of thematic maps, transformed divergence (TD) values were used in this study. The TD values were derived as follows:

$$TD_{(i,j)} = 2[1 - \exp(-D_{(i,j)}/8)] \tag{5.6}$$

where
> $TD_{(i,j)}$ is the transformed divergence between classes i and j
> $D_{(i,j)}$ is the divergence between classes i and j

$$D_{(i,j)} = 0.5T \, [M(i) - M(j)][\text{Inv } S(i) + \text{Inv } S(j)] \, [M(i) - M(j)] + 0.5\text{Trace}[\text{Inv } S(i) \, S(j) \\ + \text{Inv } S(j) \, S(i) - 2I]$$

where
> $M(i)$ is the mean vector of class i, where the vector has N channel elements (N channel is the number of channels used)
> $S(i)$ is the covariance matrix for class i, which has N channel by N channel elements
> Inv $S(i)$ is the inverse of matrix $S(i)$
> Trace[ ] is the trace of matrix (sum of diagonal elements)
> T[ ] is the transpose of matrix
> I is the identity matrix

TABLE 5.2

Transformed Divergence Results for Saline–Sodic Soil Categories in Different Data Sets

Categories of Saline–Sodic Soils	3	2	1
IKONOS-II multispectral data			
3	0	1999.57	1999.99
2	1999.57	0	2000
1	1999.99	2000	0
IRS-1D LISS-III multispectral data			
3	0	1934.62	1999.95
2	1934.62	0	1944.13
1	1999.95	1944.13	0
IRS-1D LISS-III and PAN-merged data			
3	0	1948.77	2000
2	1948.77	0	1999.89
1	2000	1999.89	0

TD is a measure of statistical separation between category response patterns and, as it is evident from Equation 5.6, uses a covariance weighted distance between category means to determine whether category signatures are separable (Swain and Davis, 1978). Typically, TD values range from 0 to 2000. When the calculated divergence is equal to the upper bound, then the signature can be considered totally separable in the bands being analyzed. In contrast, a calculated divergence of zero means that the signatures are inseparable. In this study, any class combination with TD above 1900 was considered separable. Any class combination with TD below 1900 was merged with other classes or excluded from the classification. Wherever required, training sets were refined based on TD values. A similar exercise was carried out for IRS-1D LISS-III data, and for LISS-III and PAN-merged data. The TD values of salt-affected soils for IKONOS-II, LISS-III alone, and LISS-III and PAN-merged data are given in Table 5.2.

5.3.3.3   Delineation of Salt-Affected Soils

The final spectral response patterns for various salt-affected soil categories were generated. The digital data from the three sets, namely, IKONOS-II multispectral, IRS-1D LISS-III, and IRS-1D LISS-III and PAN-merged data, were classified using the Gaussian maximum likelihood classifier available in ERDAS Imagine Software. The maximum likelihood algorithm makes use of statistical parameters, namely, mean, standard deviation, and variance and covariance matrices. Additionally, it uses estimates of probability distribution to determine the relative likelihood that a pixel belongs to a certain class. After completing the classification of the satellite data with the Gaussian maximum likelihood classifier, it was observed that many isolated and very small patches classified as salt-affected soils were present in the thematic map. A 3×3 low-pass filter was used to eliminate these small map units in a postclassification procedure.

The Gaussian maximum likelihood per-pixel classifier was used to derive information on salt-affected soils using half of the ground-truth information collected during the periods April 4–11, 2003 and August 20–24, 2003, as well as soil chemical analysis data (Table 5.3) and ancillary information.

5.3.3.4   Accuracy Assessment

Sample areas representing different categories of salt-affected soils were selected randomly for quantitative estimation of the classification accuracy (Congalton et al., 1983). Classification results

TABLE 5.3
Classification Accuracy (%) for Various Data Sets Used in the Study

Data Set	Category-1 Soil		Category-2 Soil		Category-3 Soil		OA	Kappa Coefficient (K)
	UA	PA	UA	PA	UA	PA		
IKONOS-II multispectral	89.7	90.4	90.8	92.1	90.8	93.1	92.4	0.912
LISS-III multispectral	74.3	77.8	75.4	74.4	77.6	76.9	78.7	0.724
LISS-III and PAN-merged data	82.8	81.2	81.1	82.3	84.4	82.5	84.3	0.810

PA, producers' accuracy; UA, users' accuracy; OA, overall accuracy.

were compared with the data collected in situ. In fact, half of the data collected in situ during the field visit was used for classification of the satellite data, while the remaining half was used for the accuracy assessment. Contingency tables or error matrices are common statistical methods used to compare the results of digital analysis. An error matrix provides a natural framework for the display of the results that can be used for analysis. It presents the overall accuracy (OA) of the classification as well as the producer and user accuracy of each category (Congalton et al., 1983). With respect to the sample size for accuracy assessment, Van Genderen and Lock (1977) demonstrated that a minimum of 20 observations per class or category is required for a classification accuracy of 85%, while 30 observations per class or category are required for 90% accuracy (at 0.05 significance level).

For quantitative estimation of the classification accuracy of a color-coded soil map, sample areas representing different categories of salt-affected soils were randomly selected following the stratified random sampling approach (Congalton et al., 1983). A one-to-one comparison between the categories mapped and the ground-truth data was made. The accuracy of the classification of each data set was expressed as an error matrix from which the OA, error of omission (producer's accuracy [PA]) and error of commission (user's accuracy [UA]) (Congalton, 1991), and the kappa statistic (Cohen, 1960) were estimated. The kappa coefficient (K) was computed as follows (Bishop et al., 1975):

$$K = \frac{N \sum_{i=1}^{r} x_{ii} - \sum_{i=1}^{r} (x_i + x_{+i})}{N^2 \sum_{i=1}^{r} (x_i + x_{+i})} \tag{5.7}$$

where
  r is the number of rows in the matrix
  x_{ii} is the number of observations in row i and column i (the ith diagonal elements)
  x_{i+} and x_{+i} are the marginal totals of row r and column i, respectively
  N is the number of observations

5.4 RESULTS AND DISCUSSION

The results pertaining to the delineation of salt-affected soils, spectral separability, classification accuracy, and intrafield variability of salinity and alkalinity are discussed hereafter.

5.4.1 DELINEATION OF SALT-AFFECTED SOILS

As mentioned in the preceding sections, the IKONOS-II multispectral data were collected in the month of March, coinciding with the peak vegetative growth of Rabi crops (i.e., winter season crops). Typical Rabi crops are wheat, chickpea, mustard, potato, and sugarcane. Saline–sodic soils of category 3 remain barren with salt efflorescence on the surface, whereas saline–sodic soils of

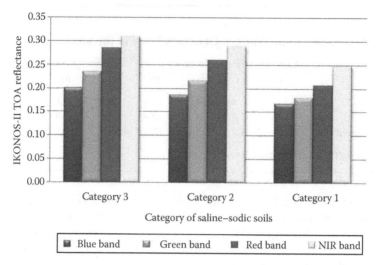

FIGURE 5.3   Mean TOA reflectance of the saline–sodic soils in the study area.

category 2 support rainy season (Kharif) crops but are left fallow in the following cropping season (October–March). Saline–sodic soils of category 1 generally have an ESP of 10–20. They support reasonably good Kharif crops, some Rabi crops like chickpea and sweat peas (*Lathyrus*) and, at places where irrigation is available, barley with poor crop stand and yield.

The mean of top-of-the-atmosphere (TOA) reflectance in IKONOS-II multispectral data for all spectral bands and the three categories of saline–sodic soils is depicted in Figure 5.3. Saline–sodic soils in category 3 have higher TOA reflectance than those in category 2, followed by those in category 1, in all the spectral bands considered in this study (i.e., blue, green, red, and NIR). However, the difference in average reflectance values for the categories mapped is more pronounced in the red and NIR regions.

5.4.2 SPECTRAL SEPARABILITY

Examination of Table 5.2 indicates a fairly good separability of the three categories of saline–sodic soils in IKONOS-II multispectral data. The IKONOS-II multispectral data covering a part of the test site are shown in Figure 5.4, while the saline–sodic soil map derived from the classification of the IKONOS image into three classes is presented in Figure 5.5. The saline–sodic soils of category 3 appear as bright white patches in the false-color composite (FCC) image of the IKONOS-II multispectral data, generated by assigning blue color to the green band (520–600 nm), green color to the red band (630–690 nm), and red color to the infrared band (760–900 nm). Saline–sodic soils are generally shown as white to bright white patches within standing crops, which appear in different shades of red color in the FCC image (Figure 5.4). The saline–sodic soils of category 2 are indicated as purple in the thematic soil map. Nonsaline and alkaline soils and other natural and cultural features like settlements, roads, and drains or streams have been assigned white color. Saline–sodic soils of category 2 appear as dull white to light brown in the standard FCC image (Figure 5.4), and are spectrally closer to category-1 saline–sodic soils. In spite of having low salt and sodium content, category-1 saline–sodic soils lie fallow during the Rabi season due to lack of assured irrigation facility. During the fallow period, these soils support grasses that could have dried up when satellite data were acquired. In the IKONOS-II image, they manifest as slightly darker than the category-3 saline–sodic soils. Soils belonging to the saline–sodic category 3 cover an area of 1173 ha and tend to be adjacent to a drain passing through the test site (Figure 5.5). Saline–sodic soils of categories 1 and 2 cover areas of 422 and 371 ha, respectively.

N

0.1 0 km

FIGURE 5.4   Standard FCC of IKONOS-II multispectral image for part of the study area.

5.4.3 CLASSIFICATION ACCURACY

The OA of delineating salt-affected soils from IKONOS-II multispectral data is 92.4%, with a kappa coefficient value of 0.912 (Table 5.3). As pointed out earlier, an attempt was also made to draw a comparison with respect to mapping salt-affected soils by using (1) IKONOS-II multispectral data with 4 m spatial resolution and LISS-III with 23.5 m spatial resolution, and (2) LISS-III and panchromatic data fused at a resolution of 6 m.

The accuracy figures for mapping various categories of salt-affected soils from IRS-1D LISS-III and from LISS-III and PAN-merged data are 78.7% (kappa value of 0.724) and 84.3% (kappa value of 0.810), respectively. The user's as well as the PA for all the categories of salt-affected soils is consistently higher in the IKONOS-II multispectral data, followed by LISS-III and PAN-merged data and LISS-III data alone. Improvements in classification accuracy of salt-affected soils derived from LISS-III, LISS-III and PAN-merged data, and IKONOS-II multispectral data could be attributed to increasing spatial resolution (23.5 m in LISS-III versus 4 m in IKONOS-II multispectral) and increasing radiometric resolution (7 bits in LISS-III versus 11 bits in IKONOS-II multispectral). Accuracy improvement as the spatial resolution of the sensor increases is in conformity with the observations made by Latty and Hoffer (1981). In contrast, Markham and Townshend (1981) observed a sharp decrease in classification accuracy with increasing spatial resolution arguably due to an increase in spectral variance, thereby reducing the spectral separability of classes in the feature space.

5.4.4 INTRAFIELD VARIABILITY

As individual fields could be identified in the IKONOS-II image due to its high spatial resolution, the feasibility of identifying intrafield variability in saline–sodic soils was evaluated. In the

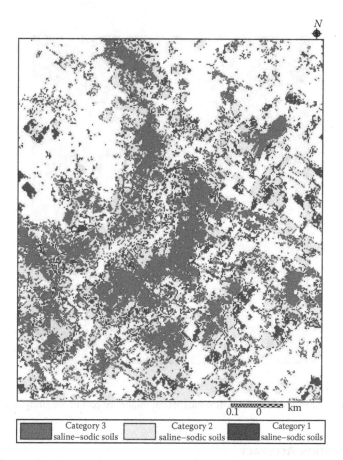

FIGURE 5.5   Saline–sodic soil map of a portion of the study area.

selected test site, there were few pockets of saline–sodic soils wherein a reclamation program had been implemented, while the rest of the saline–sodic soils were left untouched. Two sample areas were selected: one with gypsum treatment for soil reclamation and another without gypsum treatment (Figures 5.6 and 5.7, respectively). Usually, gypsum application to salt-affected soils lowers the ESP and pH, ultimately leading to their reclamation toward fertile land in due course of time.

A parcel of 0.65 ha submitted to reclamation was selected (Figure 5.6) and 65 samples were collected from the topsoil (0–30 cm depth) at a grid spacing of 10 m. Soil samples were subsequently analyzed for pH, ECe, and ESP. Table 5.4 reveals the magnitude of variation in the three soil properties analyzed for this plot. The pH ranges from 8.5 to 10.8, with mean value of 9.5 and standard deviation of 0.56. The ECe values vary from 5.6 to 41.2 dS m^{-1}, with mean value of 23.4 dS m^{-1} and standard deviation of 10.3 dS m^{-1}. ESP values range from 10.4% to 42.4%, with mean value of 24.9% and standard deviation of 9.3%. The lower values of pH, ECe, and ESP are confined to the light bluish tones of the image, while the higher values correspond to the brighter tones of the image (Figures 5.6 and 5.7). For example, the ground-truth point identification (GTP- ID) 63, 64, and 65 have lower pH and EC and also lower reflectance values, while points 27, 48, and 59 have higher pH and EC and also higher reflectance values (Figure 5.6 and Table 5.4).

In the unreclaimed pockets of saline–sodic soils, four plots were identified for appreciation of the inherent soil salinity and alkalinity and its within-field variability. Three of such plots are depicted in Figure 5.7. Plot A has a surface area of 0.26 ha, whereas plots B and C have an area of 0.28 and

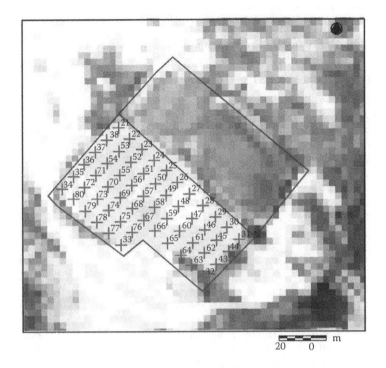

FIGURE 5.6 Intrafield variability in soil salinity in a gypsum-treated plot.

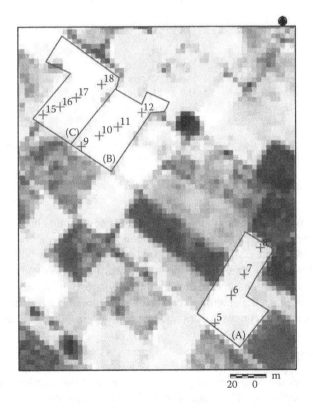

FIGURE 5.7 Intrafield variability in soil salinity in untreated plots.

TABLE 5.4
Intrafield Variability of Soil Properties in Gypsum-Treated Plots

S No	GTP-ID	pH	ECe (dS m^{-1})	ESP	S No	GTP-ID	pH	ECe (dS m^{-1})	ESP
10	21	9.9	26.9	36.2	40	55	9.2	33.3	30.4
11	22	9.7	20.8	30.2	41	56	9.8	32.8	38.4
12	23	9.2	16.4	14.9	42	57	9.8	34.5	36.8
13	24	9.3	17.6	13.8	43	58	8.9	15.5	18.9
14	25	8.9	14.5	14.1	44	59	9.2	20.2	29.6
15	26	9.0	15.6	13.4	45	60	9.1	12.6	18.4
16	27	10.3	26.4	39.1	46	61	9.2	11.1	20.2
17	28	10.2	27.1	41.5	47	62	9.8	13.2	18.8
18	29	9.7	24.1	35.4	48	63	8.5	5.6	10.8
19	30	9.9	32.6	38.4	49	64	8.6	6.2	11.0
20	31	8.7	14.8	12.1	50	65	9.1	11.5	22.1
21	32	8.5	10.6	11.7	51	66	8.9	12.2	26.5
22	33	10.8	40.8	42.2	52	67	8.9	11.6	24.2
23	34	9.5	18.2	17.4	53	68	9.0	13.6	24.7
24	35	10.7	34.9	22.8	54	69	9.1	14.2	23.6
25	36	9.9	37.9	22.0	55	70	9.5	32.1	28.5
26	37	10.0	37.6	21.8	56	71	9.6	33.2	29
27	38	10.2	36.2	21.4	57	72	9.6	30.3	40.1
28	43	8.9	10.6	14.6	58	73	9.5	31.2	42.4
29	44	8.6	11.8	10.4	59	74	9.6	38.4	18.2
30	45	8.7	14.2	20.8	60	75	9.6	31.2	18.6
31	46	9.4	17.6	21.2	61	76	9.8	41.2	20.4
32	47	10.1	35.8	28.8	62	77	9.8	16.5	31.2
33	48	10.2	34.6	31.2	63	78	8.6	13.2	16.4
34	49	9.4	16.2	21.8	64	79	9.1	17.4	16.7
35	50	9.9	35.6	28.4	65	80	10.2	22.4	38.4
36	51	10.3	32.8	35.3	Average		9.5	23.4	24.9
37	52	9.8	32.6	35.4	SD		0.56	10.29	9.28
38	53	9.4	26.5	21.7	Variance		0.32	105.78	86.10
39	54	9.4	32.8	22.5					

GTP-ID, ground-truth point identification number; pH 1:4 soil–water; ECe, electrical conductivity of saturation extract (dS m^{-1}); ESP, exchangeable sodium percentage.

0.30 ha, respectively. Within-field variability in soil salinity and alkalinity in these untreated plots was analyzed by collecting 12 soil samples, whose location is shown in Figure 5.7. A careful spatial analysis of the pH, ECe, and ESP parameters (Table 5.5) reveals that within-field variability is substantially low. This fact is supported by the low standard deviation and variance values in the untreated plots, in contrast to high standard deviation and variance of the key parameters in the treated plot. The significant within-field variability observed for the plot under reclamation (Figure 5.6 and Table 5.4) may be attributed to uneven distribution of salinity amendments in the field.

5.5　CONCLUSIONS

This study demonstrates the feasibility of generating farm-level information on saline–sodic soils from high spatial resolution satellite data. The IKONOS-II multispectral data with 4 m spatial

TABLE 5.5
Intrafield Variability of Soil Properties in Untreated Plots Shown in Figure 5.7

S No	GTP-ID	pH	ECe (dS m^{-1})	ESP
Plot A				
1	5	10.2	22.8	25.1
2	6	10.3	24.2	24.3
3	7	9.6	21.9	28.2
4	8	10.3	25.8	27.2
Average		10.1	23.7	26.2
SD		0.34	1.70	1.81
Variance		0.11	2.90	3.27
Plot B				
1	9	10.3	31.5	57.4
2	10	10.1	32.1	60.9
3	11	10.2	30.1	60.8
4	12	10.2	34.5	64.8
Average		10.2	32.1	61.0
SD		0.08	1.84	3.02
Variance		0.01	3.37	9.15
Plot C				
1	15	10.3	35.9	61.3
2	16	10.4	36.1	64.2
3	17	10.4	36.8	67.2
4	18	10.2	35.8	65.4
Average		10.3	36.2	64.5
SD		0.10	0.45	2.48
Variance		0.01	0.20	6.14

GTP-ID, ground-truth point identification number; pH 1:4 soil–water; ECe, electrical conductivity of saturation extract (dS m^{-1}); ESP, exchangeable sodium percentage.

resolution scored better than the LISS-III data and the LISS-III with PAN-merged data, with respect to accuracy of mapping of saline–sodic soils. Classification accuracy improved with increasing spatial resolution. Contrary to the general belief, within-field variability in pH, ECe, and ESP was very high in gypsum-treated plots. Such variability could be detected only with high-resolution satellite data.

The combined use of data from the LISS-IV sensor with 5.8 m spatial resolution and the Cartosat-1/2 PAN sensor with 1–2.5 m spatial resolution may also provide such information, in addition to the IKONOS data used in this study. Future earth observation missions like OrbView-I with submeter spatial resolution in multispectral mode will further enhance satellite-based mapping of soil salinity at farm level.

ACKNOWLEDGMENTS

The authors (R.S. Dwivedi and K.V. Ramana) would like to thank Dr. K. Radhakrishnan, Director, and Dr. P.S. Roy, Deputy Director, Remote Sensing and GIS, Applications Area, National Remote Sensing Agency, Department of Space, Government of India, Hyderabad, for providing necessary facilities while preparing the manuscript.

REFERENCES

Abrol, I.P. and Bhumbla, D.R. 1971. Saline and alkali soils in India, their occurrence and management. World Soil Resources, Report No. 41–42, Food and Agriculture Organization, Rome.

Anonymous. 1997. *IRS-1D Data Users Handbook.* National Remote Sensing Agency, Hyderabad, India.

Anonymous. 2007. *Nationwide Mapping of Land Degradation Using Multi-Temporal Satellite Data—Manual.* National Remote Sensing Agency, Hyderabad, India.

Bishop, Y., Fienberg, S., and Holland, P. 1975. *Discrete Multivariate Analysis: Theory and Practices.* MIT Press, Cambridge, MA.

Budd, J.T.C. and Milton, E.J. 1982. Remote sensing of salt marsh vegetation in the first four proposed Thematic Mapper bands. *International Journal of Remote Sensing* 3: 147–161.

Chavez, S.P., Sides, S.C., and Anderson, A.J. 1991. Comparison of three different methods to merge multi-resolution and multispectral data: Landsat TM and SPOT Panchromatic. *Photogrammetric Engineering and Remote Sensing* 3: 295–303.

Cohen, J. 1960. A coefficient of agreement of nominal scales. *Educational and Psychological Measurement* 20: 37–46.

Congalton, R.G. 1991. A review of assessing the accuracy of classifications of remotely sensed data. *Remote Sensing of Environment* 37: 35–46.

Congalton, R., Oderwald, R., and Mead, R. 1983. A quantitative method to test for consistency and correctness in photo interpretation. *Photogrammetric Engineering and Remote Sensing* 49: 69–74.

Dale, P.E.R., Hulsman, K., and Chandica, A.L. 1986. Classification of reflectance on colour infrared aerial photographs and sub-tropical salt-marsh vegetation types. *International Journal of Remote Sensing* 7: 1783–1788.

Dwivedi, R.S., Ramana, K.V., Thammappa, S.S., and Singh, A.N. 2001. Mapping salt-affected soils from IRS-1C LISS-III and PAN data. *Photogrammetric Engineering and Remote Sensing* 67: 1167–1175.

Dwivedi, R.S. and Sreenivas, K. 1997. Delineation of salt-affected soils and waterlogged areas in the Indo-Gangetic plains using IRS-1C LISS-III data. *International Journal of Remote Sensing* 19: 2739–2751.

Eldeiry, A. and Garcia, L. 2004. Spatial modeling using remote sensing, GIS and field data to assess crop yield and soil salinity. In *Proceedings AGU Hydrology Days 2004*, Colorado State University, Fort Collins, CO, pp. 55–66.

Hilwig, F.W. and Karale, R.L. 1973. Physiographic systems and elements of photo-interpretation as applied to soil survey in Ganges plain. *Journal of Indian Society of Soil Science* 21: 205–212.

Latty, R.S. and Hoffer, R.M. 1981. Computer based classification accuracy due to the spatial resolution using per point vs. per field classification techniques. In *Proceedings of Symposium on Machine Process of Remotely Sensed Data*, West Lafayette, IN, pp. 384–392.

Markham, B.L. and Townshend, J.R.G. 1981. Land cover classification accuracy as a function of sensor spatial resolution. In *Proceedings of 15th International Symposium on Remote Sensing of the Environment*, West Lafayette, IN, pp. 1075–1090.

Munechika, C.K., Warnick, J.S., Salvaggio, C., and Schott, J.R. 1993. Resolution enhancement of multispectral image data to improve classification accuracy. *Photogrammetric Engineering and Remote Sensing* 59: 67–72.

Myers, V.I., Ussery, L.R., and Rippert, W.J. 1966. Photogrammetry for detailed detection of drainage and salinity problems. *Transactions ASCE* 6: 332–334.

Pradines, D. 1986. Improving SPOT images size and multispectral resolution. In *Proceedings of the SPIE. Earth Remote Sensing using the Landsat Thematic Mapper and SPOT Systems* 660: 78–102.

Price, J.C. 1987. Combining panchromatic and multispectral imagery from dual resolution satellite instruments. *Remote Sensing of Environment* 21: 119–128.

Rao, B.R.M., Dwivedi, R.S., Venkataratnam, L., Ravisankar, T., Thammappa, S.S., Bhargawa, G.P, and Singh, A.N. 1991. Mapping the magnitude of alkalinity in part of the Indo-Gangetic plains of Uttar Pradesh, northern India using Landsat TM data. *International Journal of Remote Sensing* 12: 419–425.

Richards, L.A. 1954. *Diagnosis and Improvement of Saline and Alkali Soils.* Agriculture Handbook No. 60. U.S. Department of Agriculture, Washington, DC.

Shettigara, V.K. 1992. A generalized component substitution technique for spatial enhancement of multispectral images using a higher resolution data set. *Photogrammetric Engineering and Remote Sensing* 58: 561–567.

Singh, A.N., Baumgardner, M.F., and Kristof, S.T. 1977. Delineating salt-affected soils in part of the Indo-Gangetic plain by digital analysis of Landsat data. Technical report 111477, LARS, Purdue University, West Lafayette, IN.

Space Imaging. 2001. IKONOS relative spectral response and radiometric calibration coefficients. Document SE-REF-016, Rev. A. Thorton, CO.

Swain, P.H. and Davis, S.M. 1978. *Remote Sensing: The Quantitative Approach*. McGraw-Hill, New York.

Szabolcs, I. 1989. *Salt-Affected Soils*. CRC Press, Boca Raton, FL.

Toth, T., Csillag, P., and Buttner, G. 1991. Satellite remote sensing of salinity-alkalinity in the Great Hungarian plain. In *Proceedings International Symposium on Impacts of Salinization and Acidification on Terrestrial Eco-System and its Rehabilitation*, Fuchu, Tokyo.

Van Genderen, J.L. and Lock, B.F. 1977. Testing land-use map accuracy. *Photogrammetric Engineering and Remote Sensing* 43: 1135–1137.

Verma, K.S., Saxena, R.K., Barthwak, A.K., and Deshmukh, S.N. 1994. Remote sensing techniques for mapping salt-affected soils. *International Journal of Remote Sensing* 15: 1901–1914

6 The Suitability of Airborne Hyperspectral Imagery for Mapping Surface Indicators of Salinity in Dryland Farming Areas

Anna Dutkiewicz, Megan Lewis, and Bertram Ostendorf

CONTENTS

6.1 INTRODUCTION

Dryland salinity refers to human-induced salinization of the landscape in nonirrigated areas. Human-induced salinity is recognized as a major land degradation problem in rain-fed farmlands of many arid and semiarid regions of the world, including the Great Plains of North America, Africa, South America, China, India, Australia, and the Middle East (Ghassemi et al., 1995; Rengasamy, 2006).

    Catchment managers and government agencies are becoming aware of the need for improved remote sensing technologies to document the extent of salt-affected areas. In Australia, for example, the National Land and Water Resources Audit (Webb, 2000) called for improved remote sensing methods for baseline mapping and ongoing monitoring of dryland salinity. Although multispectral satellite imagery is a useful tool for mapping surface expressions of salinity within catchments and regions (Spies and Woodgate, 2004), its spatial and spectral resolutions are limited. In particular, indicator plant species and salt scalds that are characteristic of salinized environments can be difficult to discriminate with medium-resolution multispectral satellite imagery. Airborne hyperspectral imagery has the potential to overcome some of the spectral and spatial limitations of multispectral satellite imagery for monitoring salinity at both regional and farm scales. Hyperspectral sensing may allow differentiation of halophytic and nonhalophytic plant species and the detection of surface minerals associated with soil salinity, thereby providing more reliable salinity mapping.

Improved salinity maps would provide a valuable tool for land managers to target amelioration plans and evaluate remediation projects, such as plantation of saltland pasture (Barrett-Lennard et al., 2003) or revegetation with native species (Pannel, 2001). The targeted reintroduction of deep-rooted perennial vegetation is regarded as an important strategy for reducing groundwater recharge (Pannel, 2001; Clarke et al., 2002).

This chapter addresses the causes and symptoms of dryland salinity and presents a recent study that evaluated airborne hyperspectral imagery for mapping salinity indicators in dryland farming areas of southern Australia.

6.2 DRYLAND SALINITY AND REMOTE SENSING

6.2.1 Problem of Dryland Salinity

All soils contain some soluble salts, but soil salinity becomes a form of land degradation when salt concentration rises above levels that impact on agriculture (Rengasamy, 2006). Dryland salinity in Australia has much in common with many landscapes similarly affected throughout the Mediterranean climatic regions around the world. Dryland salinity is recognized as a significant land degradation problem in Australia (PMSEIC, 1999; Webb, 2000), particularly in productive dry-farming regions in southern Australia. These climatic regions experience hot, very dry summers, and cool wet winters.

Since European settlement in these regions 100–150 years ago, broad-scale clearing of original eucalypt forests, woodland, and savannah scrubland has increased the amount of water infiltrating to the groundwater system. Australian native trees and perennial shrubs are very efficient at using soil water, their roots drawing water from deep layers all year round so that very little rainwater infiltrates through to the groundwater. Over the past century, native vegetation has been replaced by shallow-rooted annual crops and pastures that use less water and for part of the year only. The excess incoming rainfall either runs off into streams and rivers or, more importantly, infiltrates below the root zone and accumulates as groundwater—a process known as groundwater recharge. As the groundwater levels rise, the benign salt stored in the soil profile dissolves and is transported close to the surface or discharged at the surface.

In Australia, salt-affected soils are classified as saline or sodic (Chhabra, 1996). Sodic soils are more typically associated with nongroundwater-related forms of soil salinity (transient salinity or dry saline land) rather than with dryland salinity (Rengasamy, 2006). Although halite ($NaCl$) is the dominant salt in many saline soils of arid to subhumid regions (Isbell et al., 1983; Fitzpatrick et al., 2003), the occurrence of soluble compounds of calcium, magnesium, potassium, iron, boron, sulfate, carbonate, and bicarbonate has been reported (Szabolcs, 1989).

The types of salt likely to form in dryland areas are dependent on rainfall patterns and the underlying regolith geochemistry. The dryland salinity mapping study presented in this chapter was conducted in the southern dryland agricultural areas of South Australia where semiarid conditions and widespread calcrete matrix give rise to saline soils dominated by halite and gypsum ($CaSo_4$).

Saline soils are defined by the concentration of salt ions, mainly Cl^-, SO_4^{2-}, Ca^{2+}, Na^+, Mg^{2+}, and to a lesser degree HCO^{3-}. The level of soil salinity or salt ion concentration can be determined by measuring the electrical conductivity (EC) of a solution of soil and water (e.g., $EC_{1:5}$) or that of the saturated soil extract (ECe), or with ground-based geophysics EC instruments. Salinity at the land surface can be broadly divided into seven classes: nonsaline (<2 dS m^{-1}); moderately low (2–4 dS m^{-1}); moderate (4–8 dS m^{-1}); moderately high (8–16 dS m^{-1}); high (16–32 dS m^{-1}); very high; and extreme salinity (>32 dS m^{-1}) (Maschmedt, 2000).

Dryland salinity discharge sites can be recognized through a variety of surface indicators relating to vegetation, soils, and position in the landscape, rather than from direct soil measurements. Three stages of saline discharge have been defined (Chaturvedi et al., 1983). The primary stage involves surfacing water or waterlogging (varying levels of salinity); the intermediary stage corresponds to declining plant condition and the invasion of salt-tolerant plant species or halophytes (moderate to high salinity); and the final process is the annihilation of all vegetation and the development of permanent salt water or dry salt deposits on the surface (very high to extreme salinity). Dryland salinity is most often associated with low-lying areas such as flats, catchment drainage lines, lake fringes, and land depressions, or with the base of hills where there is a change in the surface gradient to a gentler slope (break in slope) (Department of Primary Industries, 2004).

Salt accumulation in soils impairs plant growth by reducing water availability to the plants. Mild salinity creates stressful conditions for many plants and induces health symptoms such as (1) yellowing of leaves of crops and pasture (chlorosis) and drying before the growing season has ended; (2) slow growth rate and stunting of leaves; (3) whole plant growth rate decline; (4) reduced number of plants per area; (5) greater susceptibility to disease; and (6) inhibition of seed germination (Larcher, 1980). These indicators become quite noticeable as salinity spreads and patchy areas of discoloration or dieback appear in croplands or pastures. Deteriorating crops and pastures are often replaced by salt-tolerant plants or halophytes, which are conspicuous indicators of salinity. Halophytic plants have adapted to saline soils by developing distinctive leaf, stem, and root structures that regulate salt ion uptake, eliminate and dilute saline water, or allow the plant to tolerate salt accumulation (Larcher, 1980).

Halophytic plants associated with mild salinity include sea barley grass (*Critesion marinum*) and Wimmera ryegrass (*Lolium rigidum*) (Department of Primary Industries, 2004). As salinity increases to moderate levels (8–16 dS m^{-1}), these halophytes are progressively replaced by more tolerant species. Ultimately, when severe levels of salting are attained (>16 dS m^{-1}), the only species remaining are highly tolerant plants such as samphire (*Halosarcia pergranulata*) and rounded noon flower (*Disphyma crassifolium*) (Department of Primary Industries, 2004). These easily recognizable halophytes are stunted bushes, often reddish in color with semisucculent leaves. Many of these halophytic species grow in distinct zones associated with particular soil salinity levels and topographic features (Barrett-Lennard et al., 2003).

6.2.2 Mapping Dryland Salinity with Conventional Remote Sensing Methods

Inventory and monitoring programs have generally adopted traditional approaches to mapping expressions of salinity at the land surface, including the inference of salinization from maps of other soil properties and visual interpretation of aerial photographs. The mapping of salinity from aerial photo interpretation is limited to the severe symptoms that are visually detectable in the imagery. These photo interpretations are commonly compiled into soil landscape maps that infer the occurrence of salinity from combinations of soil properties, position in the landscape, and some field sampling. Each map unit may contain diverse soil and salinity characteristics, but the exact location and extent of salinization within map units are not documented.

Over the last two decades, several studies have assessed multispectral satellite imagery for mapping and monitoring surface salinity, with mixed results (Hick and Russell, 1988; Wheaton et al., 1992; Mougenot et al., 1993; Furby et al., 1995; 1998; Metternicht and Zinck, 1997; Kiiveri and Caccetta, 1998; Caccetta et al., 2000; Howari, 2003; Metternicht and Zinck, 2003). The major problems of direct mapping of salt-affected soils with broadband imagery occur where saline soils are covered with salt-tolerant vegetation (Furby et al., 1995; Howari, 2003) and where there is no evidence of salt crust (Howari, 2003). Furby et al. (1995) also noted that nonsaline sandy soil surfaces were sometimes confused with bare, severely salt-affected areas. Most studies have attempted to overcome the spectral and spatial limitations of the imagery by combining multitemporal images with ancillary information such as soil and terrain data (Furby et al., 1995, 1998; Evans et al., 1996; Kiiveri and Caccetta, 1998; Caccetta et al., 2000; Evans and Caccetta, 2000; Thomas, 2001).

6.2.3 Capabilities of High Spectral Resolution Imagery

The high spectral and spatial resolutions of airborne hyperspectral imagery may improve salinity mapping because of its potential to detect spectrally distinct salt-tolerant plants and saline soils. Therefore, understanding the spectral characteristics of soil and vegetation is fundamental to the mapping of salinity indicators with hyperspectral imagery.

The complexity of soil properties makes spectral identification of salt minerals in soils problematic (Csillag et al., 1993), compounded by the fact that one of the most common salts, halite, has an essentially featureless spectrum in the optical wavelengths (Crowley, 1993). Nevertheless, the investigation of the spectral properties of salt crusts has identified distinctive spectral features relating to numerous evaporite minerals (Crowley, 1993; Drake, 1995; Ben-Dor et al., 2003). Saline crusts are combinations of evaporite minerals and have been shown to exhibit narrow diagnostic absorption features, particularly in the shortwave infrared (SWIR) region (1300–2500 nm), corresponding to overtones of fundamental vibrations of hydrogen-bonded structural water molecules and other ions such as carbonate and nitrate (Hunt et al., 1972; Crowley, 1993; Drake, 1995). Some soil spectral characteristics pertaining to irrigation-induced salinity have also been identified by Dehaan and Taylor (2002b), who reported that increasing salinity could be inferred from the reduced intensity of the 2200 nm hydroxyl feature and high reflection at 800 nm.

Unlike the distinctive, narrow mineral absorption features, most plant pigments, biochemical and structural components have relatively broad and often overlapping spectral absorption features (Elvidge, 1990). Spectral differences between plant types are difficult to detect because most plants comprise the same biochemical constituents. For plant species discrimination, spectral variation between species must be greater than variation within species (Schmidt and Skidmore, 2001). However, some spectral variation between plants can be attributed to specific adaptations. Plants adapted to arid environments, for example, may have distinct spectral features resulting from their scleromorphic characteristics, such as reduced leaf area, thickening leaf tissue, and waxy coatings (Lewis et al., 2000). In particular, Lewis found that the foliage of semisucculent salt-tolerant chenopods display multiple SWIR absorption features, which were attributed to waxy cuticle leaf coatings. Dehaan and Taylor (2002b) found the spectral signature of a salt-tolerant succulent shrub (*H. pergranulata*) to be characterized by two peaks at 555 and 624 nm, due to accessory pigments, and a distinctive slope to the infrared reflectance plateau between 1250 and 1400 nm.

By exploiting these distinctive spectral characteristics, hyperspectral image analysis has been able to discriminate vegetation types (Lewis et al., 2000; McGwire et al., 2000; Lewis, 2002; Underwood et al., 2003) and has proven to be particularly valuable in geological mapping. Mineral mapping has used full spectral mixture analysis, which identifies all the scene components (Resmini et al., 1997; Chabrillat et al., 2000; Kruse and Boardman, 2000), as well as partial unmixing methods, which target specific minerals of interest. Frequently used partial unmixing methods include matched filtering (MF) (Boardman et al.,1995; Stocker et al., 1990; Harsanyi and Chang, 1994), mixture-tuned MF (MTMF) (Boardman, 1998), and least-squares spectral band fitting (Clark et al., 1990). Various studies have shown that partial unmixing methods are suitable for geological mineral mapping (Kruse et al., 1999; Kruse and Boardman, 2000; Kruse et al., 2000), mapping clay soils (Chabrillat et al., 2002), and mapping evaporite minerals (Crowley, 1993).

More specifically, studies by Dehaan and Taylor (2002a, 2006) and Taylor et al. (2001) investigated the potential for mapping soil and vegetation indicators of irrigation-induced and dryland salinity using airborne hyperspectral imagery over a limited area, comparing the output maps to local descriptions of vegetation and ground-based geophysical data. The most successful feature extraction methods were partial unmixing techniques using MF, the related MTMF, and spectral feature fitting.

Dryland salinity affects broader areas and a wider range of landscapes than irrigation-induced salinity, presenting a challenge to inventory with remote sensing. Many soil and vegetation indicators associated with dryland saline discharge sites have distinguishing spectral features and cover significant landscape portions, making them amenable to survey with airborne hyperspectral imagery. The following section presents a recent evaluation of airborne hyperspectral imagery for mapping salinity indicators in a dryland agricultural area of southern Australia.

6.3 MAPPING DRYLAND SALINITY INDICATORS IN SOUTHERN AUSTRALIA WITH HYMAP IMAGERY

Preliminary field investigations characterized soil and vegetation expressions of salinity to determine the feasibility of hyperspectral discrimination, and collected reference spectra to support image analysis. Airborne imagery was acquired during the dry season, with the objective of assessing whether this time of the year was suitable for mapping dryland salinity indicators.

Since dryland salinity affects broad areas, it was important to identify and test hyperspectral feature extraction methods that could be applied consistently to the multiple swaths of airborne imagery required to provide regional coverage. The transferability of analyses across several images was essential to demonstrate the potential of airborne hyperspectral imagery for farm-scale mapping over broad areas. Following the approach used by Dehaan and Taylor (2002a), commercially available image analysis was employed, so that successful techniques could be applied readily to baseline salinity mapping and monitoring by regional land managers. The analysis aimed to identify image spectra related to salinity indicators and subsequently produce distribution maps thereof. These salinity indicator maps were validated with extensive field sampling.

6.3.1 STUDY SITE

The study area covered 140 km^2 of Point Sturt Peninsula on the eastern side of the southern Mt Lofty Ranges in South Australia (Figure 6.1). The Peninsula juts into Lake Alexandrina near the mouth of the Murray River and is part of the Murray–Darling Basin catchment. The Murray–Darling Basin depression is filled with Quaternary and Tertiary sediments, which contain vast salt deposits that have accumulated and concentrated over tens of thousands of years from rainfall, aeolian deposition of airborne marine salts, and evaporation (Herzeg et al., 2001).

At present, the area has a Mediterranean climate, with an average annual rainfall of 425 mm, (Figure 6.2) and is used predominately for dryland cropping and grazing. The peninsula is low-lying, with sandy, loamy, and calcareous soils overlying rocky calcrete and clayey sediments. The lowest points in the area experience evaporative discharge of the saline groundwater and are characterized by symptoms such as waterlogging, exposed saline soils due to plant dieback, saltpans, and hypersaline lakes. Some of the saline areas are thought to be naturally occurring, while salinity in other areas has been exacerbated by extensive land clearing that has contributed to aquifer recharge and rising saline groundwaters (Thomas, 2001). Of recent concern are the increasing salt loads in swales and flats on the peninsula that directly feed into Lake Alexandrina (Dooley and Henschke, 1999).

Nowadays, annual grasses and pastures cover most of the study site, with mixed agriculture of grazing and cropping in the west and predominately grazing in the east. Some riparian wetlands

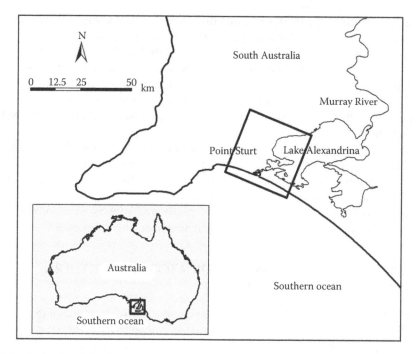

FIGURE 6.1   Study site: Point Sturt Peninsula bordering Lake Alexandrina near the mouth of the Murray River, South Australia.

thrive in coastal areas and small patches of remnant native vegetation still persist (Michalski, 2000). The dominant halophytic plants growing in saline areas are samphire (*H. pergranulata*) and sea barley grass (*C. marinum*). Samphire is a low, succulent, perennial, native shrub with numerous woody stems. The shrub can appear green or change to purple-red in highly saline areas, tolerating soils with salinity levels greater than 14 dS m^{-1} (Department of Primary Industries, 2004). Sea barley grass is an introduced annual grass species found growing in a variety of soils and is widespread in low to moderately saline soils (3–14 dS m^{-1}). Both halophytic species exhibit zonation over sizable areas, making them ideal candidates for mapping with remote sensing imagery.

6.3.2   FIELD CHARACTERIZATION OF DRYLAND SALINITY

Surface expressions of salinity vary throughout the year in this highly seasonal climate. They are most acute at the end of the hot, dry summer season, when moist saline discharge areas contrast with

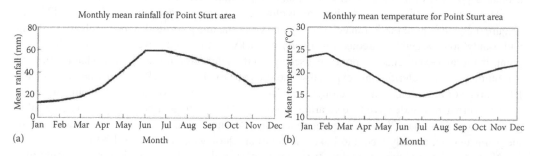

FIGURE 6.2   Monthly mean rainfall (a) and temperature (b) at Hindmarsh Island, immediately south of Point Sturt (1989–2003). (Data from the Bureau of Meteorology (http://www.bom.gov.au/).)

surrounding dry pastures and crops. Consequently, this study aimed to evaluate late-summer dry season hyperspectral imagery for mapping dryland salinity.

Soils, salinity, and vegetation were assessed at representative sites in the Point Sturt area to identify the major indicators and levels of salinity present in late summer (March). Soil sampling and field spectral collection focused on two main saline discharge sites (sites A and B in Figure 6.7), which contained a variety of representative salinity indicators including saltpans and samphire. Site A is centrally located in a low-lying inland area that contains the largest salt-affected sector on the peninsula. This area is underlain by a regional clay aquifer and is characterized by large saltpans over 20 ha in extent where the aquifer flow terminates. Site B, located near the southern coast, contains smaller saltpans.

The surface soil and salt crust of the saltpans were generally dry during field visits in March 2003 and the following year 2004. Surface soil samples were collected from saltpans and adjacent areas at sites A and B. Site A had moist grayish sandy-clay soils, with a fine, thin, white crust at the pan edges. The saltpan at site B was covered with a coarse, thick, white salt crust, underlain by very dark gray wet mud.

Salinity of soil samples was determined using standard $EC_{1:5}$ measurements. $EC_{1:5}$ values within the bare saltpans were greater than 20 dS m^{-1}, confirming the extreme salinity of the site. $EC_{1:5}$ values of 10 dS m^{-1} were recorded at the pan edge. Just 10 m away from the pan edge, EC values decreased to 5 dS m^{-1}. This sharp EC drop signals the vegetation transition zone from samphire to sea barley grass.

A selection of samples was subjected to qualitative x-ray diffraction (XRD) analysis at CSIRO, Australia, in order to determine soil mineralogy. Dominant minerals of the salt crusts at Site A were halite and quartz, followed by gypsum. Similarly, Site B saltpan was dominated by halite with traces of gypsum. Calcite and clay minerals were found to be minor constituents in most of the soil samples. The presence of gypsum is unsurprising because large Holocene gypsum deposits are common along the South Australian coastline (Warren, 1982).

6.3.3 Spectral Characterization of Dryland Salinity

Field and laboratory reflectance spectra were measured for a representative variety of soils, vegetation types, and salinity indicators at sites A and B and other accessible salt-affected sites on the peninsula. The field spectra were collected for comparison with image-derived spectra, to corroborate the landscape components mapped with the airborne imagery. Field spectra were acquired with an ASD Fieldspec Pro Spectrometer (Analytical Spectral Devices, Boulder, CO), which provided 1 nm spectral sampling over the spectral range of 350–2500 nm. In-situ reflectance spectra were collected from the dominant landcover types including red- and green-phase samphire, background soil surrounding the samphire canopies and saltpans. The location of the ground-sampling sites was determined with a differential GPS with 1–2 m accuracy. To minimize spectral noise, spectra were smoothed in Spectra Solve (Lastek Pty. Ltd., Adelaide, Australia) using the Savitzky–Golay least-squares algorithm (Savitsky and Golay, 1964), which preserved the depth and shape of the absorption features.

Spectra of oven-dried saltpan soil samples from Site A were characterized by deep broad absorption features at 1450 and 1950 nm (Figure 6.3) due to undissociated water (Hunt and Salisbury, 1970). These spectra also contained a steep visible ramp, a broad rounded shoulder near 800 nm, a broad shallow absorption feature centered at 2200 nm, a narrow feature at 1750 nm, and a triple feature near 1450 nm. Inferring soil mineralogy from the 2200 nm absorption feature is problematic, because both gypsum and clay spectra contain features in this region. However, the conspicuous feature at 1750 nm and the triple feature around 1450 nm are consistent with gypsum spectra from the USGS spectral library (Figure 6.5). The significant gypsum content was confirmed by the XRD results. Halite and quartz, the other main saltpan minerals identified by XRD, are essentially featureless with high albedo at all wavelengths and are therefore difficult to detect spectrally.

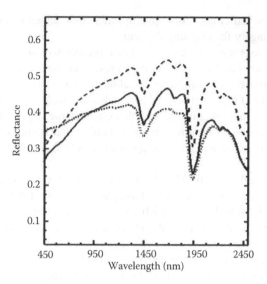

FIGURE 6.3 Comparison of salt-crust spectra from site A (solid and dashed line) and site B (dotted line).

The spectra of the coarse, thick, white salt crust collected at Site B were characterized by a bright and flat visible/near-infrared (NIR) region, and shallow absorption features at 2200 and 1755 nm (Figures 6.3 and 6.4). The very shallow features suggested the presence of small quantities of gypsum or clay. XRD analysis found that the salt crust at site B contained dominant halite and traces of gypsum. The lack of well-defined absorption features, together with the bright, flat visible/NIR slope, is consistent with the XRD analysis finding that crystalline halite is the dominant salt-crust component. The soil spectra from sites A and B do not show any significant iron oxide feature at 870 nm, consistent with the lack of red hues in the gray soil.

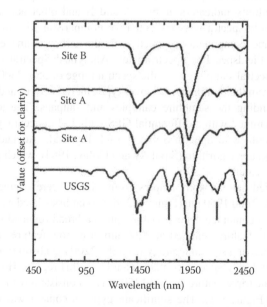

FIGURE 6.4 Continuum removed salt-crust spectra from site B (top), site A (middle), and USGS spectral library gypsum spectrum (bottom) with 1450 nm triple, 1750 nm, and broad 2200 nm features.

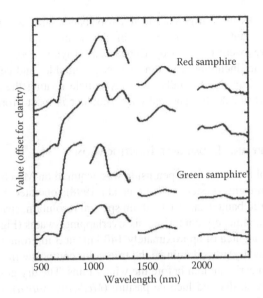

FIGURE 6.5   Field spectra of red (top) and green (bottom) phase samphire. Noisy wavebands are removed.

The surface soils were very dry and most saltpans were completely devoid of surface water during March. The lack of surface moisture meant that spectral features pertaining to soil properties could be maximized (Lobell and Asner, 2002). Therefore, an ideal time to exploit the soil and saltpan spectral characteristics in hyperspectral imagery was late in the dry season.

The individual spectra of green and red samphire (Figures 6.5 and 6.6) were dominated by typical photosynthetic vegetation characteristics such as chlorophyll absorption, a red edge, and a NIR plateau containing deep water absorption features. The spectra of green samphire, often seen

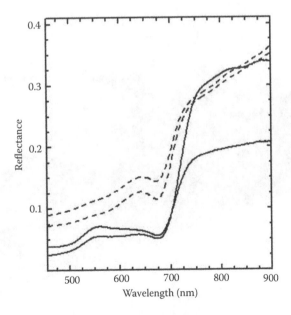

FIGURE 6.6   Detail of the NIR spectral region of green (solid line spectra) and red (dashed line spectra) samphire spectra.

with red tips, exhibit peaks in both green and red wavebands, and not surprisingly red samphire spectra contain a significant reflectance peak in the visible red wavelength region. Similar samphire spectral features were reported by Dehaan and Taylor (2002b). Samphire spectra have additional features of note, including a sloping red edge shoulder and cellulose-lignin features at 2100 and 2300 nm, possibly due to stalks or waxy cuticle coating the succulent beaded leaves (Elvidge, 1990; Lewis et al., 2000). The red edge shoulder is most prominent in the samphire's green-phase spectra.

6.3.4   HyMap Imagery and Endmember Identification

Hyperspectral imagery of the Point Sturt peninsula was acquired on March 14, 2001 by the HyMap airborne hyperspectral imaging system (Cocks et al., 1998), operated by HyVista Corporation, Sydney, Australia. The 126 band data, at 10–20 nm spectral resolution, spanned a wavelength range from 430 to 2486 nm. The imagery comprised six overlapping swaths (Figure 6.7) obtained during parallel flights, covering an area of approximately 140 km^2 at 3 m ground resolution.

Large fields of yellow dry grass dominate the true-color composite image (Figure 6.7), because the data were acquired in early autumn before the winter rains. The dry grass fields are interspersed with greener irrigated areas, dryland lucerne pasture (*Medicago sativa*), lake-fringe marshes, and remnant native mallee (*Eucalyptus* sp.) stands. Bright areas of exposed terrain surface, such as dry saltpans, small quarries, and sand dunes, are visible in the centre of the image.

The calibrated radiance-at-sensor images were corrected for solar irradiance and absorption by atmospheric water vapor and gases using the ACORN 4.0 (Imspec LLC, Palmdale, CA) software. This correction converted the imagery to apparent surface reflectance, thus enabling comparisons between image spectra and field or laboratory-derived spectra. After noisy broad water absorption spectral regions were removed, reducing the data to 117 bands, and cross-track illumination correction was applied to four of the six image swaths. Lake-water areas were masked out in these swaths to reduce the effect of low water reflectance on the image swath spectral statistics. The cross-track

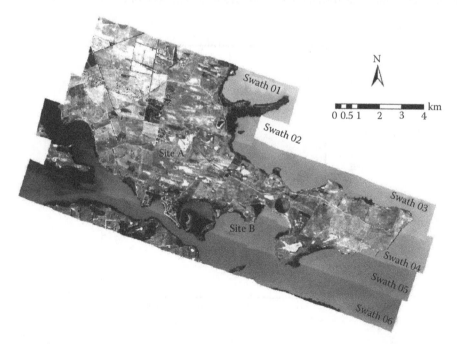

FIGURE 6.7   Mosaic of six parallel hyperspectral image swaths in true color, covering approximately 140 km^2. The locations of study sites A and B are displayed.

correction was not applied to the two water-dominated swaths because the process was found to exaggerate across-track effects.

Image analysis aimed to distinguish the saline deposits from nonsaline sandhills and quarries, and also identify perennial halophytic vegetation, while excluding other perennial vegetation (dryland lucerne and native species) in a background matrix of dry grass. Standard hyperspectral analyses were used to extract spectrally pure endmembers related to salinity indicators and subsequently map their distribution over the landscape. As described by Kruse et al. (1999), this involved data noise reduction, spectrally pure pixel location, endmember extraction and interpretation, followed by partial unmixing. The minimum noise fraction (MNF) transformation (Green et al., 1988; Lee et al., 1990) was employed to remove noise and reduce the hyperspectral data volume. The most spectrally pure pixels were then extracted using the pixel purity index (PPI) (Boardman et al., 1995) function in ENVI (ITT Visual Information Solutions, Boulder, CO). Spectral endmembers relating to relevant soil and vegetation types were isolated through a combination of automated and manual endmember selection and culling. The MNF transform and endmember extraction were applied to all image wavebands and also to spectral and spatial subsets in order to assess their effectiveness in generating endmembers related to salinity.

Two image swaths encompassing the major recharge areas produced the widest range of meaningful soil and vegetation endmembers (Figure 6.8). The endmembers were identified and interpreted through comparison with field spectra and USGS library spectra (Clark et al., 1993). While the USGS spectral library was suitable for interpreting soil mineral spectra, it did not contain

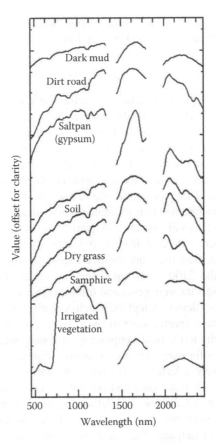

FIGURE 6.8   Selection of endmembers extracted from one image swath.

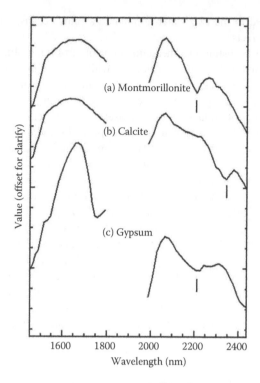

FIGURE 6.9 Detail of image-derived soil endmember spectra. The absorption features correspond to USGS mineral spectra of (a) montmorillonite or kaosmectite, (b) calcite, and (c) gypsum.

any pertinent Australian or halophytic vegetation spectra; these were provided by the ASD field spectra from known locations at Point Sturt (described in Section 6.3.1).

In general, spectral characteristics identified were associated with kaolinite/smectite mixtures (kaosmectites), calcite (absorption feature centered at 2343 nm) and gypsum (Figure 6.9a through c), dry grass, irrigated pasture, and water. Clay and gypsum spectra were primarily associated with saltpans, while calcite spectra were associated with unsealed roads characterized by exposed calcrete rubble. XRD analysis of saltpan soils at site A confirmed the presence of dominant halite, subdominant gypsum, minor calcite, and minor kaolin group minerals with possible smectites. At site B, halite also dominated with traces of gypsum and quartz.

The samphire endmembers extracted from two image swaths contained the same features seen in the field spectra: collapsed red edge shoulder, shallow chlorophyll absorption, and finer cellulose-lignin features in the 2100–2200 nm range. To enable mapping samphire across all images, additional reference spectra were generated from image regions of interest (ROI) contained within known homogeneous, dense, samphire patches with minimal soil exposure in each image swath. Even though these spectra were likely to represent mixtures of samphire and some background soil, their similarity to the field samphire spectra suggested that ROI spectra from each swath were suitable references for image feature extraction (Figure 6.10).

Samphire and saltpans were selected as the best candidates for salinity mapping in the dry season, because they were widely distributed and generally occurred in homogeneous patches over sizeable areas. Furthermore, both indicators had defining spectral characteristics. In the case of the saltpans, the conspicuous gypsum characteristics were selected as the best features to exploit in hyperspectral discrimination of saltpans. The image and field saltpan spectra did not contain any distinct absorption features relating to saline minerals other than gypsum.

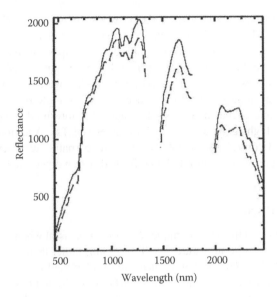

FIGURE 6.10 Comparison of samphire endmember (dashed) and samphire mean spectrum (solid). The ROI spectrum was generated from a region that encompassed a known samphire patch.

6.3.5 MAPPING SALINITY INDICATORS

6.3.5.1 Methods

Consistent detection and mapping of endmembers were required across the entire study area, while minimizing repetitious processing of separate image swaths. Ideally, reference spectra from a single image swath or an external source should be used to map target indicators across multiple image swaths, thereby reducing the need for repeated endmember extraction. Because this study targeted specific salinity indicators, partial unmixing was considered more appropriate than complete linear unmixing because the former is most suited to mapping selected scene components (Boardman et al., 1995). Furthermore, partial unmixing methods were considered suitable for mapping soils with well-defined spectral absorption features pertaining to salt minerals.

MF was used to map saline soils, applying a single reference spectrum to multiple swaths. The analysis was performed using the full wavelength range of the imagery (455–2455 nm) and also using spectral subsets focused on characteristic gypsum absorption features at 1750 and 2200 nm. MF maximizes the response of each reference spectrum, while minimizing the response of the unknown, unwanted background signatures pixel by pixel (Harsanyi and Chang, 1994). Outputs from the MF analysis were grayscale images ("score" image band) with high values indicating that the image spectra closely matched the reference spectrum.

Preliminary tests suggested that MF was not suitable for discriminating and mapping the salt-tolerant vegetation in the study area. By contrast, MTMF has been used successfully to map nonhalophytic vegetation (Williams and Hunt, 2002) as well as some halophytic species (Taylor et al., 2001; Dehaan and Taylor, 2002a). The partial unmixing method of MTMF was described by Boardman (1998) as a superior MF method, having the added advantage of highlighting pixels significantly different from background spectra but which do not match the reference spectra. MTMF was implemented as the method most likely to successfully map vegetation indicators of salinity. MTMF outputs MF "score" image bands and an "infeasibility" image band. The best match to a reference spectrum is obtained when the MF score is high (near 1) and the infeasibility score is low (near 0). A ratio of the MTMF score to infeasibility values was used to produce the final samphire maps.

Reference spectra for vegetation discrimination were mean spectra of known samphire areas in each of the images. The partial unmixing was performed using the full spectral range of 455–2455 nm, as well as reduced wavelength subsets encompassing the chlorophyll absorption region (455–789 nm), the red edge (638–789 nm), or the entire visible/NIR region (455–1050 nm).

The grayscale images resulting from the partial unmixing were thresholded, based on expert knowledge of the main saline areas, to produce salinity symptom maps. All unmixing was performed on nongeoreferenced image swaths and the threshold maps were subsequently georeferenced using in-flight geographical lookup tables (GLTs). Additional georeferencing was conducted using ground control points (GCPs) to improve spatial accuracy of the swaths and to allow for more accurate ground-truthing to be performed within a geographical information system (GIS) framework.

6.3.5.2 Results and Discussion

Gypsum discrimination using the MF technique was most successful when the analysis was confined to the 1750 nm gypsum absorption feature (1690–1790 nm) rather than the larger 2200 nm feature. MF discrimination with the 1750 nm bands successfully identified most saltpans, while excluding nonsaline soil surfaces on vegetated dunes, rubble-surfaced roads, and roadside quarries. The highly saline soils with some samphire cover, were also successfully mapped, adjacent to the large bare saltpans. Mapping based on the broad 2200 nm gypsum feature was unreliable, because it coincided with the 2200 nm clay hydroxyl feature, corroborating findings of Crowley (1993). Furthermore, subsetting the data to include both spectral features (1690–2310 nm) also incorrectly mapped rubble-surfaced roads and sand dunes.

MF unmixing with an endmember derived from a single swath effectively mapped saltpans in four other image swaths. The georeferenced mosaic combining all image swaths is shown in Figure 6.11. Areas indicating high gypsum abundance coincide with the location of known saltpans throughout the peninsula. The most significant area mapped is the central low-lying plain where the clay aquifer flow terminates (site A). Here, numerous bright areas coincide with the known location of saltpans and highly saline soil surfaces. Other gypsum-rich areas correspond to a small coastal

FIGURE 6.11 Gypsum map produced by partial unmixing, where bright areas with abundant gypsum coincide with the location of known saltpans and surrounding highly saline soils.

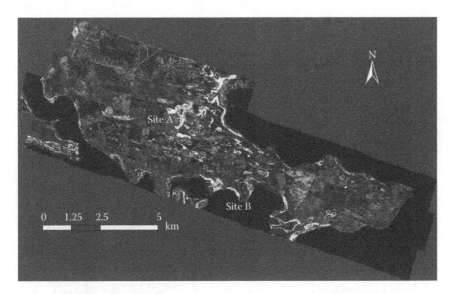

FIGURE 6.12 Samphire map produced by partial unmixing, where bright pixels indicate areas of high abundance.

saltpan at site B. Unmapped dark areas within saltpans were due to some remnant surface water that masked the mineral absorption features in the image spectra.

The best MTMF mapping of samphire was achieved with the entire visible to SWIR range of the HyMap imagery (455–2455 nm). The georeferenced mosaic of samphire distribution produced from four image swaths is shown in Figure 6.12. The northern-most image swath was not analyzed because it contained no patches of samphire. The maps reveal high abundance of samphire surrounding the numerous saltpans and fringing some coastal areas. MTMF proved capable of discriminating areas of samphire from irrigated pasture, perennial dryland alfalfa, and native vegetation stands.

Figure 6.13 shows the hyperspectral mosaic overlaid with the samphire and saltpan threshold maps within a GIS software. In general, saltpans and samphire areas were successfully mapped in the main groundwater discharge region (site A) and along the lake shore (including site B).

6.3.5.3 Accuracy Assessment

To assess the accuracy of the salinity maps, over 100 ground reference sites were selected through random and stratified sampling using GIS software. The sampling sites were all located within 500 m of the road network to allow for visual inspection of those sites that fell inside inaccessible paddocks. Remote sampling sites were located from the direction and distance indicated by the differential GPS. To take into account GPS and image georegistration errors, sampling points comprised circles of 7.5 m radius to cover areas approximately equivalent to 3×3 m^2 pixels of the airborne imagery (i.e., 9×9 m^2 on the ground). The presence or absence of each salinity symptom was recorded at sample locations. Samphire was scored as present if the cover was greater than 10%. Obvious saltpans and exposed soil sites containing highly salt-tolerant plant species, such as samphire, were labeled as saltpans.

Standard error matrices (Congalton, 1991) were constructed for the samphire and saltpan maps (Tables 6.1 and 6.2) using field observations from the ground reference sites. Separate matrices were required for the samphire and saltpans, because the two indicators were mapped independently over a different number of image strips and hence utilized a different number of sampling sites. Furthermore, both soil and vegetation indicators were often coincident in the field,

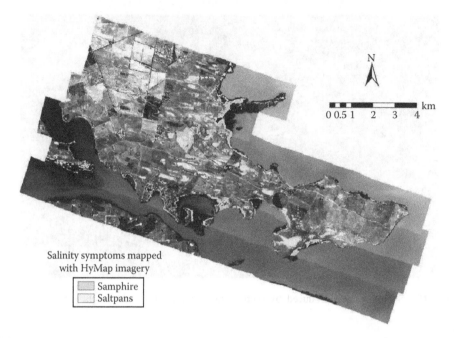

FIGURE 6.13 Combined salinity symptom maps produced from HyMap image analysis. Samphire and saltpans are mainly mapped in the central groundwater discharge region (including site A) and along the southern coast (including site B).

so their assessment as mutually exclusive categories in a single error matrix was not appropriate. Each error matrix records two categories: the presence or absence of a particular indicator at each field sampling site. The overall accuracy of the samphire and saltpan maps equals 89% and 88%, respectively.

In addition, KHAT statistic values were calculated through kappa analysis (Congalton, 1991). The KHAT correlation value, which ranges from 0 to 1 in this case, is a measure of how well the salinity symptom map agrees with the field data. KHAT for both samphire and saltpan maps was 0.6, indicating moderate agreement between the image mapping and field samples (Congalton, 1991).

Pixels within native eucalyptus stands were sometimes mistakenly mapped as saltpans. This misclassification is most likely due to the native vegetation spectra containing shallow absorption features between 1720 and 1760 nm that coincide with the gypsum absorption feature centered at 1750 nm. Absorption features of the mallee vegetation in this spectral region are most likely due to

TABLE 6.1
Error Matrix for the Samphire Map (129 Sites over Four Image Swaths)

	Samphire	Field		Total
		Present	Absent	
Image	Present	17	4	21
	Absent	12	96	108
	Total	29	100	129

Notes: Total accuracy, 88%; KHAT, 0.6.

TABLE 6.2
Error Matrix for the Saltpan Map (137 Sites over Five Image Swaths)

	Saltpan	Field		Total
		Present	Absent	
Image	Present	17	2	19
	Absent	16	102	118
	Total	33	104	137

Notes: There are more saltpan sampling sites, because saltpans were mapped across more image swaths than samphire. Total accuracy, 88%; KHAT, 0.6.

nongreen plant components like dry leaves and bark (Elvidge, 1990), and characteristics more typical of species adapted to arid environments (Lewis et al., 2000). The other main inaccuracy in the highly saline (gypsum) soil areas was the omission of pixels around saltpan fringes resulting from the thresholding of the saltpan map. Overall, subjective fine-tuning of threshold values may improve map accuracy, although it would be preferable to develop an objective and consistent method for optimizing threshold values to produce more accurate maps. Currently, on-the-ground expert knowledge in one or more areas where salinity indicators appear is essential to enable the selection of reasonable threshold values.

6.4 CONCLUSIONS

The expression of salinity at the land surface can be recognized through a variety of soil and vegetation indicators that are symptomatic of varying levels of salinity. Accurate delineation of salt-affected land and monitoring its change over time are urgently required. The enhanced spectral resolution of hyperspectral imagery offers potential for improved discrimination of soil and vegetation surface indicators of salinity. The study presented here illustrates some of the requirements and the potential for identification and delineation of key indicators of salinized land with airborne hyperspectral imagery.

6.4.1 POTENTIAL OF HYPERSPECTRAL IMAGERY FOR MAPPING SALINITY INDICATORS

In contrasting seasonal environments, expressions of salinity vary markedly throughout the year. In our case study, characterized by Southern Australian Mediterranean climate, several key indicators of dryland salinity were most evident in late summer–early autumn. At this time, soil moisture and saline seepage in dryland farming areas is at the annual minimum, enhancing the potential for discrimination of mineral absorption features in exposed soils. Furthermore, this appeared to be a suitable time of the year to distinguish homogeneous patches of perennial samphire from native mallee vegetation lucerne and senesced annual grasses that dominate the landscape. However, fewer halophytic indicators were available to map because annual grasses had senesced and were indistinguishable on the ground. Further work is required to determine the best season for mapping vegetation indicators of lower salinity levels in emerging saline seeps. For example, Dutkiewicz and Lewis (2006) found that halophytic annual grasses, such as sea barley grass (*C. marinum*), may be spectrally distinguishable during late spring flush because they tend to occupy low-lying wetter areas in the landscape and therefore senesce later than other grasses.

Partial unmixing analysis of the HyMap hyperspectral imagery enabled effective documentation of the distribution of samphire and gypsum-dominated saline soils across the study region. Samphire, an important salinity indicator, was discriminated from nonhalophytic native vegetation and pasture plants including summer-growing dryland lucerne. Field and image spectra of samphire displayed a gently sloping red edge, shallow chlorophyll absorption, and a subtle double reflectance peak in the visible green and red wavelengths, which are all distinguishing features consistent with high-resolution samphire spectra measured in the field (Dehaan and Taylor, 2002b). Discrimination was most successful when using the entire image spectral range (450–2455 nm), suggesting the importance of the SWIR region for mapping halophytic chenopods with scleromorphic characteristics.

Highly saline soils and saltpans were discriminated from other soils with high albedo on the basis of the spectral characteristics of the salt minerals. A single image-derived endmember was used as reference spectrum for mapping highly saline soils with gypsum using a reduced set of data spanning the 1750 nm absorption feature. Nonsaline soils associated with quarries, unpaved roads, and sand dunes were successfully excluded. These encouraging results suggest that other calcium and magnesium salts associated with dryland salinity (Crowley, 1993) may also be mapped with hyperspectral imagery through the judicious isolation of absorption features or combinations of absorption features.

This study demonstrates some of the potential for hyperspectral analysis to overcome the limitations of multispectral imagery in salinity mapping. Multispectral imagery, captured in appropriate seasons, has been and must be interpreted with other information such as position in the landscape to locate highly saline areas. Previous Landsat-based mapping of saline areas at Point Sturt (Thomas, 2001) required multitemporal images and terrain data. In contrast, the hyperspectral analysis directly mapped samphire, a key salinity indicator species, and soils containing gypsum, a common saline mineral in many semiarid landscapes, based on their spectral characteristics. In addition, this study identified saline soil minerals in areas with substantial vegetation cover, i.e., areas omitted in the previous multispectral mapping.

6.4.2 Benefits and Constraints of Hyperspectral Technology

Both soil and vegetation salinity indicator maps were produced using readily available software and standard hyperspectral analyses that involved partial unmixing. This makes the benefits of hyperspectral remote sensing potentially accessible to operational salinity assessment and management programs. Furthermore, consistent mapping was achieved across the multiple swaths of the airborne survey of the study area. This also enhances the scope for airborne hyperspectral imagery to be used in the inventory of larger catchments or agricultural landscapes. The airborne maps also have the advantage of higher spatial resolution that is appropriate for farm-scale study. In addition, the successful discrimination of samphire from dryland lucerne and native vegetation suggests that high-resolution imagery may have the potential for monitoring recharge reduction strategies, such as the planting of dryland lucerne and native vegetation (Pannel, 2001; Clarke et al., 2002) and saltland pastures being promoted for the productive use of saline land (Barrett-Lennard et al., 2003).

Even though airborne hyperspectral imagery and available processing methods are suitable for salinity mapping over small to medium-sized catchments, uptake of the technology may continue to be hindered by the perceived high cost of airborne image acquisition and the requirement for sophisticated image analysis. At this time, catchment managers may be most likely to accept airborne hyperspectral salinity mapping for targeted high-value agricultural areas.

New hyperspectral satellite sensors are being developed to replace the current Hyperion system that is condemned, including EnMAP in Europe (Kaufmann et al., 2006) and the Canadian Space Agency's HERO (Hollinger et al., 2006), with launch dates proposed for 2011 and 2009, respectively. These systems have a spatial footprint similar to that of multispectral satellite imagery, but with the high spectral resolution of airborne hyperspectral imagery. Hyperspectral satellite imagery would be most readily suited to mapping spectrally distinct indicators of dryland salinity across broader landscapes, including saline soil mineralogy and more subtle spectral properties of halophytic vegetation (Dutkiewicz et al., In Press), decline in crop and pasture vigor, and delayed senescence in salinity discharge areas.

ACKNOWLEDGMENTS

The authors gratefully acknowledge the partial funding for the research presented in this chapter by the Cooperative Research Centre for Plant-Based Management of Dryland Salinity (now Future Farm Industries CRC). The hyperspectral imagery was supplied by PIRSA (Primary Industries and Resources South Australia). The authors would like to thank individuals at the University of Adelaide, Rural Solutions South Australia (RSSA), the Department of Water, Land, and Biodiversity Conservation (South Australia), and CSIRO (Commonwealth Scientific and Industrial Research Organization) for their assistance. We would also like to thank the Point Sturt Landcare Group and local landowners for their support.

REFERENCES

Barrett-Lennard, E.G., Malcolm, C.V., and Bathgate, A. 2003. *Saltland Pastures in Australia: A Practical Guide*. (Land, Water and Wool Sustainable Grazing on Saline Lands Sub-program), Department of Agriculture, South Perth, Western Australia.

Ben-Dor, E., Goldshleger, N., Benyamini, Y., Agassi, M., and Blumberg, D.G. 2003. The spectral reflectance properties of soil structural crusts in the 1.2- to 2.5-μm spectral region. *Soil Science Society of America Journal* 67: 289–299.

Boardman, J.W., Kruse, F.A., and Green, R.O. 1995. Mapping target signatures via partial unmixing of AVIRIS data. In *Summaries of the 5th JPL Airborne Earth Science Workshop*, JPL Publication 95-1, Pasedena, CA, vol. 1, pp. 23–26.

Boardman, J.W. 1998. Leveraging the high dimensionality of AVIRIS data for improved sub-pixel target unmixing and rejection of false positives: Mixture tuned matched filtering. In *Summaries of the 9th Annual JPL Airborne Earth Science Workshop*, JPL Publication 97-21, Pasadena, CA, vol. 1, p. 55.

Caccetta, P.A., Allen, A., and Watson, I. 2000. The land monitor project. In *Proceedings of the 10th Australasian Remote Sensing and Photogrammetry Conference*, Adelaide, Australia. Available at http://www.cmis.csiro.au/rsm/research/pdf/CaccettaP_lmpaper2000.pdf.

Chhabra, R. 1996. *Soil Salinity and Water Quality*. Balkema Publishers, Rotterdam, the Netherlands.

Chabrillat, S., Goetz, A.F.H., Krosley, L., and Olsen, H.W. 2002. Use of hyperspectral images in the identification and mapping of expansive clay soils and the role of spatial resolution. *Remote Sensing of Environment* 82: 431–445.

Chabrillat, S., Pinet, P.C., Ceuleneer, G., Johnson, P.E., and Mustard, J.F. 2000. Ronda peridotite massif: Methodology for its geological mapping and lithological discrimination from airborne hyperspectral data. *International Journal of Remote Sensing* 21: 2363–2388.

Chaturvedi, L., Carver, K.R., Harlan, J.C., Hancock, G.D., Small, F.V., and Dalstead, K.J. 1983. Multispectral remote sensing of saline seeps. *IEEE Transactions on Geoscience and Remote Sensing*, GE-21: 239–251.

Clark, R.N., Gallagher, A.J., and Swayze, G.A. 1990. Material absorption band depth mapping of imaging spectrometer data using the complete band shape least-squares algorithm simultaneously fit to multiple spectral features from multiple materials. In *Proceedings of the 3rd Airborne Visible = Infrared Imaging Spectrometer (AVIRIS) Workshop*, JPL Publication 90-54, Pasadena, CA, pp. 176–186.

Clark, R.N., Swayze, G.A., Gallagher, A.J., King, T.V.V., and Calvin, W.M. 1993. The US geological survey digital spectral library: Version 1 (0.2 to 3.0 μm). US Geological Survey, Open File Report 93-592. Available at http://speclab.cr.usgs.gov/spectral.lib04/clark1993/spectral_lib.html.

Clarke, C.J., George, R.J., Bell, R.W., Hatton, T.J. 2002. Dryland Salinity in South-Western Australia: its origins, remedies and future research directions. Australian Journal of Soil Research 40, 93–113.

Cocks, T., Jenssen, R., Stewart, A., Wilson, I., and Shields, T. 1998. The HyMap airborne hyperspectral sensor: The system, calibration and performance. In *Proceedings of the First EARSeL Workshop on Imaging Spectroscopy*, Zurich, Switzerland, pp. 37–42.

Congalton, R.G. 1991. A review of assessing the accuracy of classifications of remotely sensed data. *Remote Sensing of Environment* 37: 35–46.

Crowley, J.K. 1993. Mapping playa evaporite minerals with AVIRIS data: A first report from Death Valley. *Remote Sensing of Environment* 44: 337–356.

Csillag, F., Pasztor, L., and Biehl, L.L. 1993. Spectral band selection for the characterization of salinity status in soils. *Remote Sensing of Environment* 43: 231–242.

Dehaan, R. and Taylor, G.R. 2002a. Image-derived spectral endmembers as indicators of salinization. *International Journal of Remote Sensing* 24: 775–794.

Dehaan, R.L. and Taylor, G.R. 2002b. Field-derived spectra of salinized soils and vegetation as indicators of irrigation-induced soil salinization. *Remote Sensing of Environment* 80: 406–417.

Dehaan, R.L. and Taylor, G.R. 2006. Mapping soil indicators of dryland salinity using HyMap hyperspectral imagery. *International Journal of Geoinformatics* 2: 61–78.

Department of Primary Industries. 2004. Salinity indicator plants. A guide to spotting soil salting. Department of Primary Industries, Victoria, Australia. Available at http://www.dpi.vic.gov.au/dpi/vro/vrosite. nsf/pages/water_spotting_soil_salting.

Dooley, T. and Henschke, C. 1999. Salinity management within the Goolwa to Wellington LAP area. A technical report produced for the Goolwa to Wellington Local Action Planning Board, PIRSA Rural Solutions, Nuriootpa, South Australia.

Drake, N.A. 1995. Reflectance spectra of evaporite minerals (400–2500 nm): Applications for remote sensing. *International Journal of Remote Sensing* 16: 2555–2571.

Dutkiewicz, A. and Lewis, M. 2006. Spectral discrimination of sea barley grass and the implications for mapping salinity with hyperspectral imagery. *Journal of Spatial Sciences* 51(2): 143–149.

Dutkiewicz, A., Lewis, M., and Ostendorf, B.O. 2006. Mapping surface symptoms of dryland salinity with hyperspectral imagery. In *The International Archives of the Photogrammetry, Remote Sensing and Spatial Information Sciences*, Vol. 34, Part 30 CDrom.

Dutkiewicz, A., Lewis, M., and Ostendorf, B.O. In Press. Evaluation and comparison of hyperspectral imagery for mapping surface symptoms of dryland salinity. *International Journal of Remote Sensing* (Accepted date 03 Aug 2007).

Evans, F.H. and Caccetta, P.A. 2000. Salinity risk prediction using Landsat TM and DEM-derived data. In *Proceedings of the 10th Australasian Remote Sensing and Photogrammetry Conference*, Adelaide, South Australia.

Evans, F.H., Caccetta, P.A., and Ferdowsian, R. 1996. Integrating remotely sensed data with other spatial data sets to predict areas at risk from salinity. In *Proceedings of the 8th Australasian Remote Sensing Conference*, Canberra, Australia.

Elvidge, C.D. 1990. Visible and near infrared reflectance characteristics of dry plant materials. *International Journal of Remote Sensing* 11: 1775–1795.

Fitzpatrick, R.W., Merry, R.H., Cox, W., Rengasamy, P., and Davies, P.J. 2003. Assessment of physio-chemical changes in dryland saline soils when drained or disturbed for developing management options. Technical Report 2/03, CSIRO Land and Water and the National Dryland Salinity Program, Adelaide, South Australia.

Furby, S.L., Wallace, J.F., Caccetta, P., and Wheaton, G.A. 1995. Detecting and monitoring salt-affected land: A report from the LWRRDC project detecting and monitoring changes in land condition through time using remotely sensed data. Remote Sensing and Image Integration Group, CSIRO Division of Mathematics and Statistics, Perth, WA, Australia. Available at http://www.cmis.csiro.au/rsm/research/salmapmon/salmapmon_full.htm.

Furby, S.L., Flavel, R., Sherrah, M., and McFarlane, J. 1998. Mapping salinity in the Upper South East Catchment in South Australia: A report from the LWWRDC project mapping dryland salinity (CDM2). Remote Sensing and Image Integration Group, CSIRO Mathematical and Information Sciences, Western Australia.

Ghassemi, F., Jakeman, A.J., and Nix, H.A. 1995. *Salinisation of Land and Water Resources: Human Causes, Extent, Management and Case Studies*. University of NSW Press, Canberra, Australia.

Green, A.A., Berman, M., Switzer, P., and Craig, M.D. 1988. A transformation for ordering multispectral data in terms of image quality with implications for noise removal. *IEEE Transactions on Geoscience and Remote Sensing* 26: 65–74.

Harsanyi, J.C. and Chang, C.I. 1994. Hyperspectral image classification and dimensionality reduction: An orthogonal subspace projection approach. *IEEE Transactions on Geoscience and Remote Sensing* 32: 779–785.

Herzeg, A.L., Dogramaci, S.S., and Leaney, F.W.J. 2001. Origin of dissolved salts in a large, semi-arid groundwater system: Murray Basin, Australia. *Marine and Freshwater Research* 52: 41–52.

Hick, P.T. and Russell, W.G.R. 1988. *Remote Sensing of Agricultural Salinity*, WA. Remote Sensing Group, CSIRO Division of Exploration Geoscience, Western Australia.

Hollinger, A., Bergeron, M., Maskiewicz, M., Qian, S., Othman, H., Staenz, K., Neville, R., and Goodenough, D. 2006. Recent developments in the hyperspectral environment and resource observer (HERO) mission. Geoscience and Remote Sensing Symposium, IGARSS 2006. IEEE International Conference, Denver, CO, pp. 1620–1623.

Howari, F.M. 2003. The use of remote sensing data to extract information from agricultural land with emphasis on soil salinity. *Australian Journal of Soil Research* 41: 1243–1253.

Hunt, G.R. and Salisbury, J.W. 1970. Visible and near-infrared spectra of minerals and rocks: I. Silicate minerals. *Modern Geology* 1: 283–300.

Hunt, G.R., Salisbury, J.W., and Lenhoff, C.J. 1972. Visible and near-infrared spectra of minerals and rocks: V. Halides, phosphates, arsenates, vanadates and borates. *Modern Geology* 3: 121–132.

Isbell, R.F., Reeve, R., and Hutton, J.T. 1983. Salt and sodicity, Chapter 9. In *Soils: An Australian Viewpoint*. CSIRO Division of Soils, Melbourne, Australia, pp. 107–117.

Kaufmann, H., Segl, K., Chabrillat, S., Hofer, S., Stuffler, T., Mueller, A., Richter, R., Schreier, G., Haydn, R., and Bach, H. 2006. EnMAP: A hyperspectral sensor for environmental mapping and analysis. Geoscience and Remote Sensing Symposium, IGARSS 2006. IEEE International Conference, Denver, CO, pp. 1617–1619.

Kiiveri, H.T. and Caccetta, P.A. 1998. Mapping salinity using decision trees and conditional probability networks. *Digital Signal Processing* 8: 225–230.

Kruse, F.A., Boardman, J.W., and Huntington, J.F. 1999. Fifteen years of hyperspectral data: Northern Grapevine Mountains, Nevada. In *Proceedings of the 8th JPL Airborne Earth Science Workshop*, Pasadena, CA, vol. 99-17, pp. 247–258.

Kruse, F.A. and Boardman, J.W. 2000. Characterization and mapping of kimberlites and related diatremes using hyperspectral remote sensing. In *Aerospace Conference Proceedings, 2000 IEEE*, Big Sky, MT, vol. 3, pp. 299–304.

Kruse, F.A., Boardman, J.W., Lefkoff, A.B., Young, J.M., Kieren-Young, K.S., Cocks, T.D., Jenssen, R., and Cocks, P.A. 2000. HyMap: An Australian hyperspectral sensor for solving global problems-results from USA HyMap data acquisitions. In *Proceedings of the 10th Australasian Remote Sensing and Photogrammetry Conference*, Adelaide, Australia, pp. 296–311.

Larcher, W. 1980. *Physiological Plant Ecology*. Springer-Verlag, Berlin-Heidelberg, Germany.

Lee, J.B., Woodyatt, A.S., and Berman, M. 1990. Enhancement of high spectral resolution remote-sensing data by a noise-adjusted principal components transform. *IEEE Transactions on Geoscience and Remote Sensing* 28: 295–304.

Lewis, M. 2002. Spectral characterization of Australian arid zone plants. *Canadian Journal of Remote Sensing* 28: 1–12.

Lewis, M.M., Jooste, V., and Degasparis, A.A. 2000. Discrimination of arid vegetation with hyperspectral imagery. In *Proceedings of the 10th Australasian Remote Sensing and Photogrammetry Conference*, Adelaide, Australia, pp. 949–960.

Lobell, D.B. and Asner, G.P. 2002. Moisture effects on soil reflectance. *Soil Science Society of America Journal* 66: 722–727.

Maschmedt, D. 2000. *Assessing Agricultural Land: Agricultural Land Classification Standards Used in South Australia's Land Resource Mapping Program*. PIRSA, Adelaide, South Australia, Australia.

McGwire, K., Minor, T., and Fenstermaker, L. 2000. Hyperspectral mixture modeling for quantifying sparse vegetation cover in arid environments. *Remote Sensing of Environment* 72: 360–374.

Metternicht, G. and Zinck, J.A. 2003. Remote sensing of soil salinity: Potentials and constraints. *Remote Sensing of Environment* 85: 1–20.

Metternicht, G. and Zinck, J.A. 1997. Spatial discrimination of salt- and sodium-affected soil surfaces. *International Journal of Remote Sensing* 18: 2571–2586.

Michalski, C. 2000. Point Sturt and Districts Landcare Group, Natural Heritage Trust Point Sturt and Districts Natural Resource Strategy. 1999 Project Officer's Report, Natural Heritage Trust funding for the Point Sturt and Districts Landcare Group, South Australia.

Mougenot, B., Pouget, M., and Epema, G.F. 1993. Remote sensing of salt-affected soils. *Remote Sensing Reviews* 7: 241–259.

Pannel, D.J. 2001. Dryland salinity: Economic, scientific, social and policy dimensions. *Australian Journal of Agriculture and Resource Economics* 45: 1–22.

PMSEIC. 1999. *Dryland Salinity and Its Impact on Rural Industries and the Landscape.* Prime Minister's Science, Engineering and Innovation Council. Occasional Paper Number 1, Department of Industry, Science and Resources, Canberra, Australia.

Rengasamy, P. 2006. World salinization with emphasis on Australia. *Journal of Experimental Botany, Plants and Salinity,* (special issue) 57: 1017–1023.

Resmini, R.G., Kappus, M.E., Aldrich, W.S., Harsanyi, J.C., and Anderson, M. 1997. Mineral mapping with hyperspectral digital imagery collection experiment (HYDICE) sensor data at Cuprite, Nevada, USA. *International Journal of Remote Sensing* 18: 1553–1570.

Savitsky, A. and Golay, M.J.E. 1964. Smoothing and differentiation of data by simplified least squares procedures. *Analytical Chemistry* 36: 1627–1639.

Schmidt, K.S. and Skidmore, A.K. 2001. Exploring spectral discrimination of grass species in African range-lands. *International Journal of Remote Sensing* 22: 3421–3434.

Spies, B. and Woodgate, P. 2004. Technical report: Salinity mapping methods in the Australian context. Prepared for the Programs Committee of the Natural Resource Management, Ministerial Council through Land and Water Australia and the National Dryland Salinity Program.

Stocker, A.D., Reed, I.S., and Yu, X. 1990. Multi-dimensional signal processing for electro-optical target detection. *Proceedings of the SPIE International Society of Optical Engineering* 1305: 218–231.

Szabolcs, I. 1989. *Salt-Affected Soils.* CRC Press, Boca Raton, FL.

Taylor, G.R., Hempill, P., Currie, D., Broadfoot, T., and Dehaan, R.L. 2001. Mapping dryland salinity with hyperspectral imagery. In *IEEE 2001 International Geoscience and Remote Sensing Symposium,* University of NSW, Sydney, Australia, vol. 1, pp. 302–304.

Thomas, M. 2001. Remote sensing in South Australia's land condition monitoring project. Unpublished report, PIRSA Sustainable Resources, Land Information Group, Adelaide, Australia.

Underwood, E., Ustin, S., and Dipietro, D. 2003. Mapping non-native plants using hyperspectral imagery. *Remote Sensing of Environment* 86: 150–161.

Warren, J.K. 1982. The hydrological setting, occurrence and significance of gypsum in late Quaternary salt lakes in South Australia. *Sedimentology* 29: 609–637.

Webb, A. 2000. Australian dryland salinity assessment 2000. Technical report, National Land and Water Resources Audit, Brookfield, Queensland, Australia.

Wheaton, G.A., Wallace, J.F., McFarlane, D.J., and Campbell, N.A. 1992. Mapping salt-affected land in Western Australia. In *Proceedings of the 6th Australasian Remote Sensing Conference,* Wellington, New Zealand, pp. 369–377.

Williams, A.P. and Hunt, J.R. 2002. Estimation of leafy spurge cover from hyperspectral tuned matched filtering. *Remote Sensing of Environment* 82: 446–456.

7 Applications of Hyperspectral Imagery to Soil Salinity Mapping

Thomas Schmid, Magaly Koch, and José Gumuzzio

CONTENTS

7.1 INTRODUCTION

Soil salinity is an important component of soil degradation in arid and semiarid regions where there is insufficient water to leach away soluble salts. Salinization is a complex process with severe consequences for the soil environment and involves hydrological, climatic, geochemical, agricultural, social, and economic aspects. Soluble salts accumulate in the soil and water compartments, where salt concentration increases due to natural or, frequently, human-induced processes.

They precipitate in a wide range of soluble minerals or stay in the soil solution. Salinity problems can develop and spread rapidly within the landscape. The causes and processes of salinization are reasonably well understood, but finding a solution to this problem can be a difficult and challenging task (Tanji, 1990). The assessment and monitoring of soil salinity with conventional methods are time-consuming and expensive. Experience shows that remote sensing technology can contribute efficiently to the study of soil salinity. However, there are some constraints that affect the detection of salinity from remote sensing.

Saline soils contain soluble salts that influence the growth of most plants that are not adapted to such conditions, and reduce crop yield. Therefore, salinity in soils is related to salt tolerance of plants and crops (Maas, 1990). Soils strongly affected by salinity have usually a salic horizon characterized by salt accumulation (Soil Survey Staff, 2006). The amount of salt required to recognize a salic horizon represents only a small percentage of the soluble minerals present in the total soil mass, but has nevertheless an important effect on plants and crops. The dilution of the soluble minerals with respect to the soil mass is an additional difficulty when it comes to identifying salinity by remote sensing. Field and laboratory spectroradiometry helps study the composition of saline soils, but remote sensing is often limited in identifying types of soluble minerals.

Often, slightly or moderately saline soils significantly affect crops that are moderately tolerant to salinity. This occurs mainly within the root zone and does not necessarily manifest through surface salinity features. Frequently, plants and crops respond to salinity long after it has reached a critical level. In these cases, it is difficult to map soil salinity with sufficient precision using remote sensing technology.

The main constraints to the use of remote sensing technology for detecting salt-affected soils are:

- Relatively low ratio of soluble minerals with respect to the soil mass (mass dilution effect), together with temporal and spatial changes in the chemical and mineralogical composition
- Variable effect of salinity on plant species and crops according to their salt-tolerance levels
- Soil salinity refers commonly to the root zone, and there is a time lag before its surface manifestation becomes apparent (e.g., vegetation stress, salt crust/efflorescence)
- Soil salinity gradients and temporal changes may affect plants and crops, as well as saline accumulations near or on the soil surface, in specific periods and locations

Numerous studies have successfully assessed soil salinity using a wide range of remote sensors (Metternicht and Zinck, 2003; Farifteh et al., 2006). The spectral and spatial resolutions of the sensors play an important role in detecting soil salinity and monitoring soil salinization. In the context of optical remote sensing, high-resolution hyperspectral sensors provide a large number of spectral bands that allow identifying surface features in more detail than the multispectral sensors do. Furthermore, airborne hyperspectral sensors provide high spatial resolution imagery that contributes to improved identification of soil salinity (Ben-Dor et al., 2002; Dehaan and Taylor, 2003; Dutkiewicz, 2006). However, the complex nature of soil salinity calls for an integrated approach where data obtained from multisensors are combined with field measurements and laboratory analyses, and this greatly improves the assessment and prediction of soil salinity (Metternicht and Zinck, 2003; Schmid, 2005; Farifteh, 2007).

The aim of this chapter is to map soil salinity using hyperspectral data within semiarid and arid areas of Spain, where human activities often take place in saline soil conditions. Hyperspectral data were collected to obtain high-resolution information on main characteristics of salt-affected areas, and this information was then adapted to broadband sensors in order to carry out change detection analysis. We review the use of remote sensing to determine soil salinity, present a methodological framework, and analyze two case studies where soil salinity affects sensitive wetland ecosystems.

7.2 SOIL SALINITY AND REMOTE SENSING

7.2.1 SURFACE FEATURES AND INDICATORS OF SOIL SALINITY

Salinity causes soil properties to change and therefore influences the reflectance of features that occur at the soil surface, including salt crusts and efflorescence as well as variations in surface structure and texture. Salinity also causes changes in plant communities and crop performance that influence the spatial variation of the vegetation cover (Mougenot, 1993; Metternicht and Zinck, 2003; Farifteh et al., 2006).

Areas affected by salinity are often identified by the presence of white patches of precipitated salt that are closely associated with the occurrence of small depressions where evaporation and capillary processes are favored. This is an evidence of secondary salt enrichment of the soil mass in the form of covering or filling of pores and channels. Visible salt efflorescence and salt crusts obviously indicate salt accumulation, but this does not necessarily mean that there is high salinity within the root zone. In certain cases, salt-affected soils are associated with saline seeps where salt crusts form at the surface through evaporation and salt sedimentation.

The problems associated with salinity are not always evident. In the case of irrigation-induced salinization, salt accumulation becomes apparent when it significantly affects cropping. Sometimes, high salt concentration in the soil solution can be inferred from the health status of the crops, especially when there is a spatial salinity gradient. This is reflected by irregular spatial distribution of plants and crops, with strong variations in height and density, and the occurrence of barren soil patches. However, in the case of moderate salinity or uniform salt distribution, crop damage is more homogeneously spread over the fields and the effect of salinity is usually more difficult to detect through vegetation alone (Rhoades, 1990). In addition, the time lag between the onset of soil salinization and its first sign of surface effect on plants and cropping practices often prevents timely detection. Moreover, plants growing in saline conditions suffer from osmotic stress and this gives them an aspect similar to plants affected by water shortage.

Halophytic plants are often found to invade abandoned fields and are good indicators of the degradation caused by increasing salinization. That means that changes in the spatial distribution and surface extent of vegetation species with different thresholds in salinity tolerance may provide useful information on soil salinity gradients in an area.

7.2.2 POTENTIAL OF HYPERSPECTRAL IMAGERY

Hyperspectral remote sensing instruments offer new remote sensing capabilities. The number of spectral bands in hyperspectral sensors can vary from several tens to hundreds of channels, compared to multispectral sensors with only a few spectral bands (Van de Meer, 2000).

Hyperspectral imagery from different imaging spectrometers has been successfully applied to a variety of studies in arid and semiarid regions that include the identification and mapping of geological materials (Chabrillat et al., 2000; Crowley et al., 2003), soil characteristics (Chabrillat et al., 2002; Ben-Dor et al., 2006), vegetation cover (Okin et al., 2001; Asner and Heidebrecht, 2002), and wetland areas (Koch et al., 2001; Schmid et al., 2004). Salinity studies have been carried out with AVIRIS (Crowley, 1993; Whiting and Ustin, 1999), DAIS 7915 (Ben-Dor et al., 2002; Schmid, 2005; Schmid et al., 2005a), and HyMap (Dehaan and Taylor, 2003; Koch et al., 2006; Farifteh et al., 2007). Hyperspectral airborne sensors serve as a basis for the implementation of future satellite-borne sensors. Further advantages of hyperspectral sensors are that the acquisition time and orientation of the flight path can be determined by the user. However, processing of hyperspectral images is a complex procedure when several flight lines are involved and large areas are to be surveyed.

The added advantage of satellite sensors is that they provide a global coverage of the earth's surface condition at different temporal resolutions. This is an important point when studying soil salinity, as salt concentration is closely related to seasonal changes as well as to time cycles of several years or decades. Studies on salinity have been carried out with hyperspectral satellite-borne sensors such as Hyperion (Dutkiewicz, 2006) and Proba-1/CHRIS (Barducci et al., 2007; Schmid et al., 2007).

7.2.3 ADVANTAGES OF AN INTEGRATED APPROACH

To overcome the constraints that affect the mapping of soil salinity, we suggest applying an integrated approach where different data sources are combined and the techniques are adapted to the specificity of each case study.

Ground observations and radiometric measurements indicate that the main factors affecting the reflectance are the quantity and mineralogy of salts, moisture content, color, and surface roughness (Mougenot, 1993; Drake, 1995). Field spectroradiometer data are often used to develop spectral libraries and catalogs of selected spectra, which can be implemented as endmembers in classification or aid in the identification of spectral features obtained with sensors of lower spatial and spectral resolutions. Field spectrometry, field observations, and ancillary data improve dramatically the results obtained by hyperspectral sensors (Dehaan and Taylor, 2002; Schmid et al., 2005a).

The spatial and spectral resolutions of the sensors may hamper the detection of salt-affected soils in places where the saline areas are smaller than the pixel size. The resulting spectral characteristics of an area covered by a pixel will therefore be related to a mixture of the different components within that pixel. Thus, the individual components must be known to identify the spectral features of salinity. Field and laboratory studies in test sites must be very detailed, while the field and laboratory radiometers are able to determine the spectral characteristics of individual minerals, soil and plant properties as well as their conditions (Crowley, 1991; Mougenot, 1993; Okin et al., 2001; Farifteh and Farshad, 2002; Howari et al., 2002; Ben-Dor et al., 2003). The comparison of these detailed spectral characteristics with those obtained from hyperspectral sensor pixels is an important step to identify the corresponding conditions and characteristics of surface features within any area of interest. Again, the advantages of satellite-borne sensors over airborne sensors are improved temporal resolution, optional high temporal resolution with off-nadir capabilities (e.g., SPOT, ALOS), and worldwide coverage. However, only few hyperspectral satellite-borne sensors are currently operational (i.e., Hyperion and Proba-1/CHRIS), and the high demand for this type of data also limits the possibility of data acquisition. Future missions such as the Environmental Monitoring and Analysis Program (EnMAP), a German satellite mission scheduled for launch in 2010 (Kaufmann et al., 2006), can help overcome this situation.

Synergistic use of data from a variety of sensors with different spatial and spectral resolutions is important for discriminating and detecting changes in surface features. Field spectroscopy, together with hyperspectral and possibly multispectral data, allows determining soil salinity characteristics and carrying out change detection studies. Synergistic mapping approaches that integrate multi-sensor and multitemporal data, supported by field spectrometry and ancillary data, and use advanced processing methods, were successfully applied to change detection and monitoring studies of saline wetland areas in Central Spain (Schmid et al., 2005a).

Section 7.3 presents a methodological framework for synergistic use of multisensor, multiscale, and multisource data sets in soil salinity studies. Subsequently, this approach is illustrated with two case studies dealing with soils affected by variable degrees of salinization.

7.3 METHODOLOGICAL FRAMEWORK

In an initial step, site-specific abiotic and biotic surface features were identified during spring and summer months when soil moisture was maximum and minimum, respectively. The moisture content varies with the time of year and influences the dynamic characteristics of saline soils with dramatic changes from one season to another as well as interannual variations. Ancillary data included cartographic and thematic maps as well as aerial photographs.

The application of remote sensing techniques is time-specific, which means that data are obtained at specific moments in time and reflect thus the conditions at these moments. Our field studies were carried out during the summer season, when salinity symptoms were most likely to appear at the surface. Moisture, soil surface features, and vegetation types are the principal factors to be determined in areas affected by salinity.

7.3.1 DATA SOURCES

Data were acquired from various sources at different spatial scales. Hyperspectral data were collected with a field spectroradiometer (ASD FieldSpec Pro) covering the visible, near-, and shortwave infrared (VNIR-SWIR) regions of the spectrum. During field campaigns, georeferenced spectral data were obtained for selected surface cover types (e.g., soil, vegetation, and rock outcrops) and within test plots representing areas with saline as well as nonsaline soils. Surface cover samples (i.e., soil and vegetation) were collected at specific observation points, after spectral measurement was completed. Field spectral information was complemented with field observations, physical, chemical, and mineralogical characterization of the soil surface samples, and classification of soils and vegetation.

Hyperspectral data were acquired during different flight campaigns depending on the study site and the type of sensor. They included images from hyperspectral airborne sensors such as digital airborne imaging spectrometer (DAIS 7915) and HyMap. Airborne data acquisition was jointly organized by the German Aerospace Centre (DLR) and the HYVISTA Corporation within the EU Framework Programs of Hyperspectral Sensors (HySens) and Hyperspectral Europe (HyEurope) in the years 2000, and 2004, respectively.

Multispectral data were acquired from a number of broadband sensors such as Landsat TM, ETM+, and Terra-ASTER. The Landsat TM and ETM+ data were obtained commercially from the U.S. Geological Survey, National Center for Earth Resources Observation and Science (EROS), Sioux Falls, South Dakota. The ASTER data were acquired as part of a Research Data Agreement Program (AP-0072: Land degradation monitoring in environmentally sensitive areas of NE and Central Spain) with the contribution of the Japan Aerospace Exploration Agency (JAXA).

Preprocessing of the hyperspectral data included radiometric, atmospheric, and geometric corrections carried out through a DLR and HYVISTA cooperation (Koch et al., 2005; Schmid, 2005; Schmid et al., 2005b). The radiometric and preliminary geometric corrections of the multispectral Landsat and ASTER data were also carried out by the aforementioned organizations. The Landsat and ASTER data were acquired as level 1G and level 1B products, respectively. The quantized pixel values were converted to surface reflectance data applying calibration parameters and atmospheric correction procedure as described in Schmid (2005). Ancillary data were used in different phases of the work and were obtained in numerical and digital format from bibliographical and cartographic sources in several institutions.

7.3.2 METHODOLOGICAL STEPS

The proposed integrated methodological approach is composed of six main parts (Figure 7.1) that correspond to the different data sources (field, ancillary, hyperspectral, multispectral, and cartographic data) and the type of mapping methods applied to determine surface characteristics related to soil salinity in arid and semiarid areas.

Part I consists in the processing of various field data sets with the aim of creating a site-specific spectral library (Schmid et al., 2004). Field data include surface features with different abiotic and biotic characteristics, which were obtained by making measurements with the field spectroradiometer, carrying out basic soil and mineralogical laboratory analyses, and classifying the lithology, geomorphology, and vegetation types. The final information contained in the spectral library was twofold: (1) spectral characteristics of landscape components describing the soils, salt crusts, and

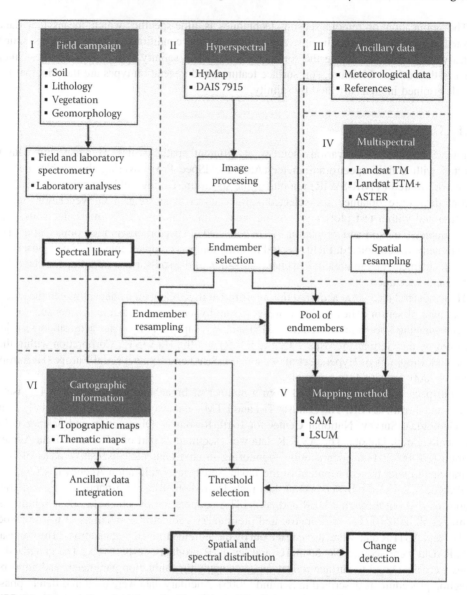

FIGURE 7.1 Integrated methodology for identifying soil salinity.

vegetation types of the area, obtained by collecting spectrally homogeneous, representative samples of these components at the field scale; and (2) assessment of the spectral heterogeneity in selected areas, on the basis of multiple components present within one pixel of the hyperspectral sensors used. Furthermore, laboratory spectra were obtained with the same spectroradiometer, adapting the acquisition technique to include an accessory known as contact reflectance probe that was added to the ASD FieldSpec Pro. A final site-specific spectral library containing different field data and key reference spectra was created and used for further identification and verification of results obtained in work related to soil salinity.

Part II focuses on integrating hyperspectral data from airborne and satellite-borne sensors. These data have high spectral and/or spatial resolutions and provide the basis for the determination of surface characteristics related to soil salinity. The hyperspectral data were processed to obtain image-derived endmembers using the ENVI image-processing program (Research Systems

Inc., 2006). A minimum noise fraction (MNF) procedure (Green et al., 1988) was used to reduce interband correlation and data redundancy, and the inherent dimensionality of the data was determined using normalized eigenvalues with magnitudes exceeding 1. The corresponding MNF components were input to the pixel purity index (PPI) procedure (Boardman et al., 1995) to select the purest pixels. The MNF and PPI results were projected on a n-dimensional visualizer (Boardman, 1993) to determine the image-derived endmembers. The identification and labeling of the endmembers were carried out by comparing key field spectra from the spectral library and verifying their location in the field.

Ancillary data in Part III were valuable sources of information for determining the meteorological conditions prior to and at the time of data acquisition, and included references to studies that may assist in the identification of individual endmembers. Finally, a collection of endmembers was ready for use with the selected mapping methods. Endmembers were resampled when the information about the image-derived endmembers was adapted to the spectral resolution of the multispectral data, as explained in the following part. These endmembers were also added to the pool of endmembers for further use with the mapping methods.

The integrated methodology contemplated using different types of data to enable determining the spatial and spectral characteristics of soils affected by salinity, and included studies based on time series to carry out change detection analysis. This is possible using either hyperspectral or multispectral data acquired at different dates. When applying hyperspectral data, the image-derived endmembers were introduced directly into Part V, namely the mapping methods. However, when using multispectral data in Part IV, the spatial resolution of the data was altered by the resampling to equate the resolution of the hyperspectral data used to derive endmembers. This resampling was carried out to determine the endmember location in the multispectral data as well as for comparison when obtaining the distribution maps of the individual endmembers. It is worth noting that the latter type of data opens up a much greater possibility of using time-series data, which have been gathered regularly over decades. As mentioned earlier, soil salinity varies over different seasons and longer time periods. Therefore, one of the aims of the integrated methodology was to apply data from different periods acquired by different sensors.

The mapping methods of Part IV included the spectral angle mapper (SAM) and linear spectral unmixing (LSU) techniques. The SAM technique is based on the well-known coefficient of proportional similarity, which measures the difference in the shapes of the spectral curves (Kruse et al., 1993). The LSU expresses the fact that if there are n land cover types present within the surface area covered by a single pixel and if each photon reflected from the pixel area interacts with only one of these n cover types, then the integrated signal received at the sensor in a given band will be the linear sum of the n individual interactions (Mather, 2004). In this study, an unconstrained LSU was used in order to estimate how well the linear mixture model described the data. Thresholds can be set for both mapping techniques, thus improving the final results obtained for the individual endmembers. These thresholds are based on field knowledge and the spectral library database.

The cartographic information available in Part VI supplied ancillary data that were integrated into the final spatial and spectral distribution results. The final step was to detect changes in the spatial and spectral distributions obtained for different dates.

7.4 CHARACTERIZATION AND MAPPING OF SALINE WETLANDS: LA MANCHA CASE STUDY

Two study areas were selected to implement the integrated approach for identifying salt-affected areas in arid and semiarid zones. The study areas (Figure 7.2) are located inland and include wetlands and steppe plains as well as areas used for agriculture. The La Mancha study area is used as a case study in this section and the Los Monegros area is used in Section 7.5.

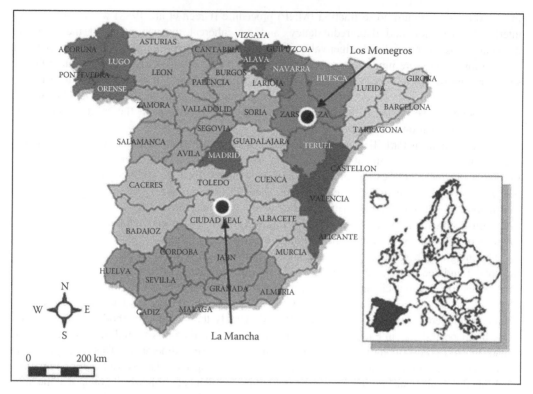

FIGURE 7.2 Location of the study areas.

7.4.1 STUDY AREA CHARACTERIZATION

Small wetlands are scattered over the semiarid landscape of La Mancha Alta (Central Spain). These are areas of great importance regarding environmental quality and biodiversity, especially because of the migratory movements and wintering of waterfowl in the south of Europe. The current situation of the wetlands is dramatic, as it has been estimated that 60% of their surface area has disappeared or is in the process of disappearing, and only about 3% remains relatively well preserved (Casado de Otaola and Montes del Olmo, 1995; Oliver and Florin, 1995; Cirujano, 2000).

Wetland systems are very complex and subject to natural (seasonal) as well as human-induced changes, causing important ecosystem degradation. Reduction or disappearance of the wetlands is linked to the development of large irrigation systems in the region with significant reduction of the groundwater resources, land use changes, wetland drainage, waste disposal, and wastewater effluence. As a consequence, wetlands dry up and are exposed to progressive salinization.

The monitoring and control of the wetlands often face practical difficulties due to factors such as their small size and changes in soil moisture, saline surface composition (salt efflorescence), vegetation cover, or land use that modify over short time periods the ecosystem dynamics. The challenge in this study area is to detect, with adequate accuracy and at different temporal and spatial scales, any rapid changes in the composition of the land cover (soil, vegetation, and water).

There are a variety of wetland types in the study area of La Mancha (Oliver and Florin, 1995) (Figure 7.3). One type was selected for the purpose of studying soil salinity: an anthropogenic-affected floodplain with an area of about 500 ha located in the alluvial plain of the river Cigüela and belonging to the Pastrana property where land use changes caused the wetland to dry up and, by the same token, increased soil salinity with subsequent invasion of halophytes. The river Cigüela

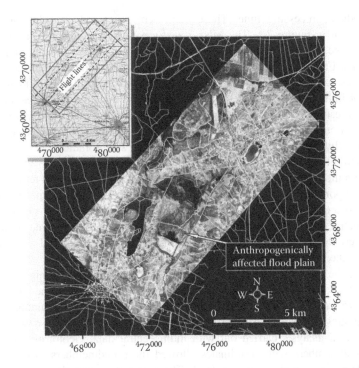

FIGURE 7.3 Study area with a mosaic of DAIS 7915 data (false-color composite at wavelengths of 0.889, 1.757, and 2.180 μm in the RGB channels) and superimposed cartographic information (roads and urban area) at scale 1:25,000.

meanders in north-south direction, forming a floodplain with quaternary sediments rich in gypsum and calcium carbonates. Tertiary sandstone and limestone formations occur in the surrounding uplands. Salt-affected soils including Salic Fluvisols, Gypsic Solonchaks, Gypsic Kastanozems, and Calcic Gypsisols (FAO, ISRIC, and ISSS, 1998) are present in this test site. The upland soils are not salt-affected (Haplic and Petric Calcisols; Calcaric, Chromic and Rhodic Cambisols; and Calcaric Leptosols). The natural vegetation of the wetlands is hygrophytic (*Phragmites australis* and *Typha domingensis*). Areas degraded by salinization have halophytic vegetation (*Salicornia europea*, *Suaeda vera*, *Puccinellia fasciculata*, *Limonium carpetanicum*, *Lygeum spartum*, and *Tamarix canariensis*).

7.4.2 REMOTE SENSING DATA AND PROCESSING SPECIFICATIONS

Multispectral Landsat ETM+ and hyperspectral DAIS 7915 data were acquired for the summer season, 1 day apart on June 28 and June 29, 2000, respectively. Additional multispectral TM and ASTER data were obtained on June 17, 1987 and June 2, 2002. Spectral field data were obtained during the hyperspectral field campaign and in June 2001 and 2003.

The reflective bands of the DAIS 7915 sensor were used to determine surface features of wetland components. Image-derived endmembers were identified in the floodplain of the river Cigüela. Reference endmembers, derived from field measurements, represent pure endmember spectra. However, image-derived endmembers were obtained at the spatial scaling of the corresponding sensor, and this determined the type of wetland component that would be eventually selected. The DAIS 7915 imaging spectrometer contains 72 narrow-band channels in the V-SWIR spectrum (0.45–2.45 μm). Three flight lines were obtained extending over an area of 105 km². Processing was carried out in specific areas of individual flight lines, in order to better manage the data set and reduce processing time. These data were processed according to the integrated methodology shown in Figure 7.1. The MNF transformation reduced the data to a maximum of 11 components. The PPI procedure used 50,000 iterations to select the purest pixels from the hyperspectral data. Endmembers were selected using the n-dimensional visualizer.

The spatial resolution of the DAIS 7915 sensor often makes it difficult to identify the spectral characteristics of individual cover types when complex and dynamic processes are involved. For instance, a spatial resolution of 5×5 m or greater is seldom dominated by one type of surface cover. Therefore, spectral curves were established for the cover types present in selected areas using the different sensors (hyperspectral, multispectral, and field spectroradiometer) in order to relate the information to different spatial scales (Schmid, 2005).

Depending on the spatial resolution of the sensor, the final spectral signal reflects the mixture of the different components. Field spectra collected with the ASD field spectroradiometer help understand the spectral curves obtained with airborne hyperspectral and satellite multispectral sensors. The spectral curves generated by the ASD instrument can also provide additional information in cases where the image pixels are located in highly dynamic areas and where the spatial and spectral scales of the different land covers vary at short distance.

Results from the hyperspectral data focused on the determination of high-resolution image-derived endmembers. These were obtained by applying the second part of the integrated approach. A final pool of 75 image-derived endmembers formed the basis for determining the spectral and spatial characteristics of the salt-affected soils in the study area. They also served as the basis for change detection analysis carried out with multispectral data. Geolocation of the individual image-derived endmembers with coordinates allowed these endmembers to be identified in the hyperspectral data set as well as in the field. The selected endmembers were compared with key spectral curves compiled from a spectral library (Schmid et al., 2005a) using the spectral feature-fitting (SFF) algorithm (Clark et al., 1991).

7.4.3 Characterization of Surface Features in Salt-Affected Areas

Detailed field survey provided spectral curves for a range of surface covers that determined the abiotic and biotic characteristics of the region (Figure 7.4). Field spectral curves of saline and nonsaline features show important differences throughout the spectral range (Figure 7.4a). Electric conductivity of the samples with salinity is 37.8 and 5.5 dS m^{-1} for a strongly saline salt crust and a slightly saline soil sample, respectively (Schoeneberger et al., 2002). The soils in the upland areas surrounding the wetlands are nonsaline, and this feature often coincides with the boundary of salt-affected and wetland areas where soil moisture and water content increase near to the soil surface. In general, surface reflectance increases with salt precipitation at the terrain surface. However, salt-affected soils have often a relatively low percentage of soluble salts in relation to the bulk soil mass. In the case of soils with low or moderate salinity, it is difficult to discriminate salinity levels on the soil surface due to the minor presence of soluble minerals as compared to other soil mineral constituents. In contrast, in soils with high salinity and percentages of soluble salts in the range of 3%–5%, it is possible to distinguish salinity levels if the soil surface shows enough evidence of salt efflorescence. Ground validation is therefore essential to relate surface features, salt content, and reflectance.

The reflectance characteristics of surface samples with high gypsum content include absorption features throughout the spectral range (Figure 7.4b). X-ray diffraction (XRD) showed that the sample is rich in gypsum but contains also some calcite and smectite. The main absorption features of gypsum at 1.196, 1.440, 1.530, 1.740, 1.936, 2.208, and 2.262 μm (according to Crowley, 1991) were determined with the spectroradiometer, both in laboratory and field conditions.

Salinity is very closely related to the hydrological conditions in the soil (Figure 7.4c), and therefore different moisture conditions have an important effect on the reflectance spectra of soils (Schoeneberger et al., 2002). A change in the moisture state from moderately moist to wet,

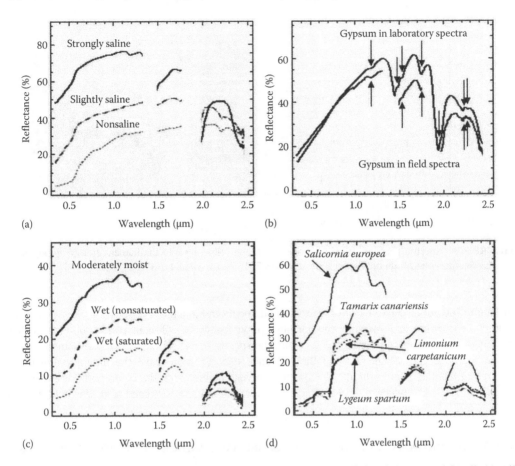

FIGURE 7.4   Reflectance spectra of (a) different levels of soil surface salinity, (b) gypsum-rich soil, (c) soil moisture influence, and (d) dominant halophytic vegetation.

nonsaturated to saturated, lowers the overall reflectance and causes the spectral absorption features to be less pronounced.

Natural halophytic vegetation (Figure 7.4d) covers a large extent of the wetlands, and could be a good indicator of the salinity levels found in different soils. Certain vegetation species, such as *S. europea*, are able to survive in strongly saline soils and cover these areas without the competition of other vegetation types. As soils become less saline, other plants start competing. Steppe vegetation with *L. spartum* indicates a well-preserved wetland, whereas *L. carpetanicum* tends to grow in degraded areas. The *T. canariensis* is a tree well adapted to saline conditions and is usually found at the edges of wetland areas. These plants are specifically adapted to the prevailing environmental conditions. In comparison, annual vegetation dries out in summer (our measurement period) as high temperatures increase the evaporation rate. This effect cancels water availability for plants and causes the accumulation of soluble salts at the soil surface.

A typical natural saline area contains bare soil with patches of salt efflorescence, and is covered by sparse steppe vegetation (albardinal) characterized by *L. spartum*. However, the shape of the curve tends to describe the spectral characteristics of the soil. Details of the spectral curve obtained with the ASD field spectrometer show a slight influence of vegetation at around 0.7 μm (Figure 7.5). The soil is very slightly saline with a carbonate content of 14.5% and an organic matter content of 4.9%. XRD shows that gypsum and clay minerals are abundant. In areas with salt efflorescence, the soil contains very abundant gypsum, common hexahydrite, calcite and phyllosilicates, and traces

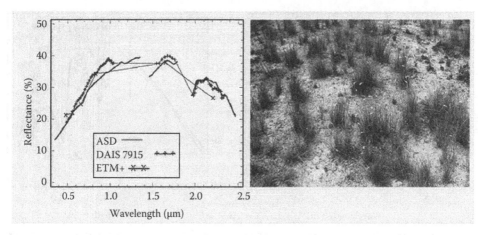

FIGURE 7.5 Spectral characteristics of steppe vegetation (albardinal) in the human-affected floodplain of La Mancha, generated from field spectrometer, DAIS 7915 (+) and Landsat ETM+ data (×).

of halite. The curve derived from the ASD field spectroradiometer data shows a clear absorption feature for gypsum at 1.74 μm. Another absorption feature at 2.3–2.34 μm is more difficult to associate with a specific mineral, but is within the spectral range of calcite. The combined spectral signal of the different minerals and the influence of vegetation make the interpretation of the absorption features more difficult. The spectral curves obtained with the DAIS and ETM+ sensors (Figure 7.5) clearly match this characteristic shape of the curve obtained with the field spectroradiometer.

7.4.4 MULTITEMPORAL CLASSIFICATION OF SALINE AREAS AND CHANGE DETECTION ANALYSIS

Change detection analysis was carried out using a selection of high-resolution image-derived endmembers (Figure 7.6) and applying a LSU model (LSUM) to the multispectral data from different sensors. Landsat ETM+ data were acquired one day earlier than the hyperspectral DAIS data. Meteorological records show that the atmospheric conditions were very similar on both days.

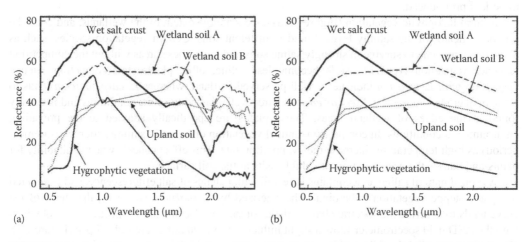

FIGURE 7.6 Spectral curve transformation of image-derived endmembers from (a) DAIS 7915 data to (b) Landsat ETM+ data. (From Schmid, T., Koch, M., and Gumuzzio, J., *IEEE Trans. Geosci. Rem. Sens.*, 43, 2516, 2005a. With permission.)

Therefore, the Landsat ETM+ and DAIS data were used as reference, and change detection was carried out between Landsat TM and ETM+ data (1987–2000) and between Landsat ETM+ and Terra-ASTER data (2000–2002).

7.4.4.1   Spectral Characteristics and Constraints on Endmember Selection

The endmembers obtained from DAIS 7915 (Figure 7.6a) show that the hygrophytic vegetation is vigorous with a characteristic spectral reflectance in the red/near-infrared region of the spectrum. The vegetation includes *P. australis* and *T. domingensis* as well as tree species like *T. canariensis*. This latter species is associated with the halophytes and, depending on soil and hydrological conditions, often indicates salinity. Wetland soil A presents high reflectance in the visible and NIR range, which is influenced by seasonal flooding with the formation of a smooth, slightly saline depositional surface and abundance of gypsum and calcite. The spectral characteristics of wetland soil B, moderately saline, are associated with higher gypsum content than that of soil A, together with abundant calcite. In general, soil B shows a lower reflectance than soil A because of soil surface roughness resulting from plowing. The wet salt crust shows high reflectance in the visible and NIR wavebands, while reflectance decreases in the mid-infrared due to the moisture conditions. The salt crust is strongly saline with 10% carbonate content. Upland soils, forming the boundary of the wetland area, are nonsaline with 11% carbonates, 0.9% Fe_2O_3, and very abundant clay minerals. The key absorption features identified in this endmember are located at 0.87 and 2.2 μm for iron oxides and clay minerals, respectively.

The transformation of these five endmembers from the hyperspectral space to the multispectral ETM+ resolution (Figure 7.6b) shows that important spectral characteristics are maintained at the corresponding wavelengths. The general form of the curves for the Landsat TM spectral data is almost identical to that of the hyperspectral data. There is a slight variation when the endmembers are transformed to the Terra-ASTER waveband configuration. The reason for this variation is that the ASTER sensor lacks a band in the visible blue spectral region corresponding to the Landsat TM or ETM+ band 1, but has additional bands in the mid-infrared region.

The LSUM and the change detection procedure have to account for the sensors with a lower number of spectral bands such as Landsat TM and ETM+. Therefore, a maximum number of five endmembers was used (i.e., LSUM requires a maximum number of endmembers equal to $n - 1$ the number of spectral bands used). For the ASTER data, the six spectral bands whose band centers and spectral range are comparable with the Landsat bands were used. Keeping the same number of bands between different multispectral sensors is essential to compare the performance of the sensors and carry out the change detection analysis. An unconstrained LSUM was implemented that allowed estimating how well the linear mixture model described the data.

Endmembers (Table 7.1) were selected after several tests were carried out using image-derived endmembers from the pool of endmembers initially chosen for the LSUM. This implies that different combinations of endmembers were used to evaluate the relationship between endmembers derived from the hyperspectral image and their implementation into the multispectral data. This phase of the research also involved evaluating the goodness of fit of the unmixing models created using different combinations of endmembers.

7.4.4.2   Change Analysis Using Abundance Images of Endmembers

Statistical distribution of the abundance values for individual endmembers (Table 7.2) was made for the three image dates (Landsat TM of June 1987, Landsat ETM+ of June 2000, and Terra-ASTER of June 2002). The resulting abundance values or proportions lying within 0–1 were highest for the hygrophytic vegetation with percentages higher than 98.7% for all dates. Wetland soil A had the highest performance with the ASTER data (93.8%) and the lowest with ETM+ (75.2%). Wetland soil B had the highest performance with ETM+ (76.6%) and the lowest with TM (39.8%). The wet

TABLE 7.1

Selection of Image-Derived Endmembers for the Spectral Unmixing Analysis

Endmember	Description	Characteristics
Hygrophytic vegetation	*P. australis* and *T. domingensis* enclosed by *T. canariensis* (halophytic vegetation)	Thick vigorous vegetation and tree crowns along the river and in water ponding areas
Wetland soil A	Desiccated soil surface area exposed to seasonal flooding	Soil with abundant gypsum and 40% carbonate
Wetland soil B	Cultivation practices in fallow and cropland	Soil with very abundant gypsum and 20% carbonate; irregular surface due to plowing
Wet salt crust	Salt crust in depressions along drainage ditches where the soil surface is exposed with moist condition	Composed of starkeyite, hexahydrite, gypsum, and calcite
Upland soil	Exposed cultivated soils in uplands surrounding the floodplain	Very abundant phyllosilicates with an average of 10% carbonate and 0.9% Fe_2O_3

Source: From Schmid, T., Koch, M., and Gumuzzio, J., *IEEE Trans. Geosci. Rem. Sens.*, 43, 2516, 2005a. With permission.

TABLE 7.2

Descriptive Statistics of the Endmembers Representing Abundance Values

	Hygrophytic Vegetation	Wetland Soil A	Wetland Soil B	Wet Salt Crust	Upland Soil	RMS Error
TM 1987						
Minimum	0.01	−0.55	−0.98	−0.74	−0.13	0.000
Maximum	0.52	1.32	0.86	0.63	1.12	0.015
Mean	0.13	0.51	−0.10	−0.27	0.42	0.007
SD	0.05	0.28	0.32	0.16	0.24	0.002
Percent < 0	0.00	4.00	60.18	94.60	0.64	0.00
Percent 0–1	100.00	93.05	39.82	5.40	99.21	100.00
Percent > 0	0.00	2.95	0.00	0.00	0.15	0.00
ETM 2000						
Minimum	−0.20	−1.73	−1.17	−0.81	−1.21	0.000
Maximum	1.04	2.18	3.20	1.18	0.93	0.030
Mean	0.18	0.31	0.34	−0.07	0.00	0.003
SD	0.15	0.39	0.37	0.21	0.29	0.002
Percent < 0	0.88	20.75	22.14	62.44	50.02	0.00
Percent 0–1	99.09	75.21	76.63	37.55	49.98	100.00
ASTER 2002						
Minimum	−0.23	−0.94	−0.91	−1.18	−1.31	0.000
Maximum	0.99	2.61	1.09	0.66	0.80	0.019
Mean	0.17	0.72	−0.03	−0.35	0.07	0.003
SD	0.09	0.20	0.23	0.12	0.20	0.002
Percent < 0	1.34	0.20	55.91	98.49	43.62	0.00
Percent 0–1	98.66	93.78	44.09	1.51	56.38	100.00
Percent > 0	0.00	6.02	0.00	0.00	0.00	0.00

Source: From Schmid, T., Koch, M., and Gumuzzio, J., *IEEE Trans. Geosci. Rem. Sens.*, 43, 2516, 2005a. With permission.

salt crust in general had a low performance with the highest value for ETM+ (37.6%). In the case of TM and ASTER, practically all the abundance values have low mean (−0.27 and −0.35) and standard deviation (0.16 and 0.12). The values have shifted into the negative scale and, therefore, the results for this endmember were treated with caution. The performance of the upland soils was best for TM (99.2%) and lowest for ETM+ (50.0%).

When the unconstrained model is used, proportions may lie outside the range of 0–1 due to one or more of the following reasons: (1) the linear mixture model does not adequately fit the data; (2) the endmembers are poorly chosen and do not represent the extremes of the distribution of the reflectance values of each endmember; (3) the number of endmembers is not sufficient to describe the spectral properties of the data set (Mather, 2004).

7.4.4.3 Evaluation of Errors in the Change Detection Process

The root-mean-square (RMS) error image indicates the goodness of fit of the model. The higher the maximum RMS value, the worse the model fit in terms of determining the distribution of the abundance values. The maximum RMS values attained in the error images resulting from the linear unmixing of the Landsat TM, ETM+, and ASTER data sets were 0.015, 0.030, and 0.019, respectively. These values are low and lie within the range of acceptance (Bryant, 1996; Koch, 2000). The RMS image is an important indicator of surface features that were not included in the LSUM. As mentioned earlier, only five endmembers could be used at any one time to carry out the LSUM, making it difficult to fully represent the highly dynamic environment proper to salinized landscapes. Furthermore, an endmember that is present in one year is not necessarily present at another time. So the temporal aspect of the analysis complicates the selection of an ideal endmember combination. A threshold value had to be determined to obtain the pixels with high abundance value. For the pixels to contain an abundance value for an endmember with no less than 95% purity, the threshold was set at 1.64 standard deviation. Field verifications showed this threshold to be the best value for comparison of the final abundances. Therefore, the difference between corresponding endmember abundances for the three dates was used to indicate the spatial change in a particular endmember (Figure 7.7).

7.4.4.4 Analyzing the Nature of Changes

The changes observed in the surface characteristics are closely related to the anthropogenic influences in the area. Hygrophytic vegetation (Figure 7.7a and e) covered a surface area of 81 ha (5.4% of the total area) in 1987, 117 ha (7.8%) in 2000, and 67.5 ha (4.5%) in 2002. These changes are due to construction works aimed at channeling the river and coincide with low annual rainfall in 1987. Furthermore, natural vegetation was cleared for cropping in the eastern part of the study area in 2002. Wetland soil A (Figure 7.7b and f) is a slightly saline soil confined to a desiccated area in the southern part of the floodplain. The soil covered a surface area of 61.5 ha (4.1% of the total area), 72 ha (4.8%), and 63 ha (4.2%) in 1987, 2000, and 2002, respectively. The total surface area is fairly constant. However, a spatial redistribution is observed between the different image dates. This soil type was also identified on the outer fringes of the area, close to the uplands with little or nonsaline characteristics. An explanation is that the carbonate content in the soil of the desiccated area is as high as that of the upland soils where the spectral properties of carbonate dominate over those of other salt minerals and therefore influence the unmixing result. Changes in the moderately saline wetland soil B (Figure 7.7c and g) were largely influenced by agricultural activities in the east and west of the floodplain. The surface area was 72 ha (4.8%), 30 ha (2.0%), and 82.5 ha (5.5%) in the three selected years, respectively. The higher figures indicate larger areas left fallow and the lower figure reflects cropping expansion. Wet salt crusts (Figure 7.7d and h) occur along drainage ditches and in depressions. The area occupied in the corresponding years are 87 ha (5.8%), 72 ha (4.8%), and 78 ha (5.2%). The spatial distribution in 1987 is again related to construction works carried out on the river. Improved drainage and the exposure of new soil surface areas allowed the

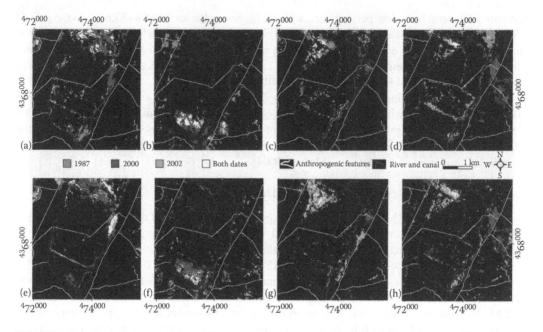

FIGURE 7.7 Change detection analysis between June 17, 1987 and June 28, 2000 for (a) hygrophytic vegetation, (b) wetland soil A, (c) wetland soil B, (d) wet salt crust; and between June 28, 2000 and June 2, 2002 for (e) hygrophytic vegetation, (f) wetland soil A, (g) wetland soil B, and (h) wet salt crust. (From Schmid, T., Koch, M., and Gumuzzio, J., *IEEE Trans. Geosci. Rem. Sens.*, 43, 2516, 2005a. With permission.)

formation of salt crusts. In 2000, a large wet salt crust area appeared to the east of the floodplain, extending into the uplands. Areas where a cereal crop had been recently harvested and the stubble and straw were left covering the field, were identified with the spectral characteristics of wet salt crust, causing spectral confusion. Field verification and the use of lithological maps showed that the distribution was accurately identified in salt-affected areas, but not so in salt-free areas.

The land surface conditions of the study area show important changes between 1987 and 2002. Human activities have introduced major changes to the hydrology of the region. Proper management of the floodplain is therefore important, especially as this area is also used for both agricultural and recreational purposes.

7.4.5 DISCUSSION

The methodological approach described in this section improves the identification and characterization of soils in saline wetlands. Image-derived endmembers from high spatial and spectral resolution data were able to determine surface features associated with soil salinity. Identification and verification of the endmembers were based on data obtained at field scale.

Application of this information across scales, including resampling of the hyperspectral endmembers to the multispectral resolution, affects the level of detail regarding the spectral absorption features of the different land cover types. However, the general spectral characteristics are maintained and, therefore, this information can be applied to an archive of multispectral data such as Landsat and similar satellite missions.

The change detection carried out with the selected combination of endmembers performed reasonably well. The LSUM was limited by the number of bands of the Landsat TM/ETM+ image, which has six reflective bands that restricted the maximum number of endmembers to five. Therefore, careful selection of the endmembers is very important since it has a decisive effect on the final result. Furthermore, a first change detection was applied to two multispectral (TM and

ETM+) sensors with very similar characteristics regarding the number and width of the reflective bands. A second change detection was undertaken between two sensors (ETM+ and ASTER) with different band characteristics. This type of temporal comparison will become more common as remote sensing programs such as Landsat and SPOT come to an end, and data from new sensors within the framework of the Global Monitoring for Environment and Security (GMES) and the Global Earth Observation System of Systems (GEOSS) become available.

7.5 HYPERSPECTRAL MAPPING OF PLAYA LAKES: LOS MONEGROS CASE STUDY

Significant land use changes, mainly in the form of changing agricultural practices, are transforming the semiarid karstic landscape of Los Monegros in NE Spain that encloses numerous saline lakes (playa lakes), forming a unique habitat (Samper-Calvete and García-Vera, 1998; Koch, 2000; Castañeda and Herrero, 2005, 2008). The introduction of irrigation systems is changing the water balance and soil properties of the playa lakes and surrounding areas (Samper-Calvete and García-Vera, 1998; Sánchez et al., 1998). As a result, significant changes in the distribution and extent of salt-affected soils are occurring (Isidoro et al., 2006). Thus, the main objective of this study is to map the temporal and spatial distribution of salt-affected soils in a semiarid karstic wetland area, using hyperspectral images of two dates representing the wet and dry seasons. A two-step approach was designed to this end. First, the spatial distribution of salt-affected soil groups was mapped for each image date by using supervised classification. Second, changes in the soil distribution and status were analyzed as related to seasonal changes affecting the playa lakes and their environment. The soil groups show different contents of soluble salts and their mineralogy is closely related to the underlying geological units and the water cycle (due to climate and irrigation).

Wetlands of semiarid areas are highly dynamic ecosystems undergoing short-term seasonal changes that often preclude the detection and identification of long-term human-induced changes. Monitoring and assessing the effects of landscape transformation (e.g., from rain-fed to irrigated agriculture) on soil and water quality require a good understanding of the nature and magnitude of the changes, and the possibility to separate natural (seasonal) from human-induced changes. This is important for establishing a baseline or reference for future measurement of noncyclical changes due to human-induced landscape transformations.

7.5.1 MATERIALS AND PREPROCESSING

Two pairs of hyperspectral images from the airborne HyMap sensor were obtained in 2004 as part of the HyEurope flight campaign carried out by the German Aerospace Centre (DLR) and the HYVISTA Corporation. The dates of the hyperspectral images were selected so that they would match as close as possible the wet and dry seasons of the playa lake environment (i.e., May 17 and August 12, 2004).

The HyMap image flightlines were oriented so that they crossed the plateau area where the playa lakes are located from N to S and NNE to SSW (Figure 7.8). In this way, several landscape components were covered by the strips, including irrigated and nonirrigated fields, playa lakes and small karstic depressions, and the edge of the Ebro River escarpment in the south (Figure 7.9).

Fieldwork was conducted on June 30, 2006 and July 7–8, 2007, and consisted of collecting ground information (field observations, photographs) and soil samples that were subsequently analyzed in the laboratory and used for validating the image processing results.

Atmospheric and geometric corrections of the HyMap data were performed by the DLR, and the data sets were preprocessed to ground reflectance data with a spatial resolution of 5 m. Further refinement of the geocoding was performed by the authors using high-resolution orthophotos (M.A.P.A., 2007) of the area as well as digital topographic maps at scale 1:25,000.

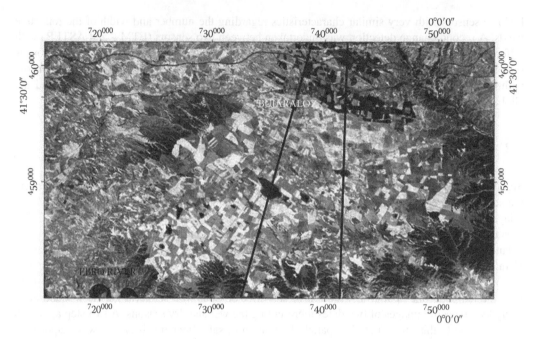

FIGURE 7.8   Landsat TM band 5 image of Los Monegros study area with HyMap flightlines.

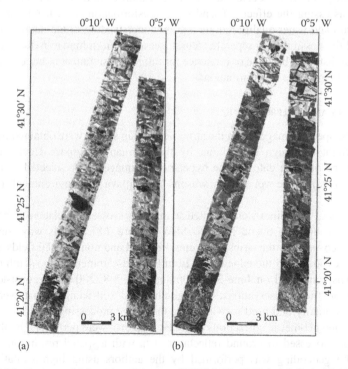

FIGURE 7.9   HyMap images of May (a) and August (b) 2004. The NIR band distinguishes clearly between vegetation (bright), soil (gray), and water/shadow areas (dark).

7.5.2 IMAGE ENDMEMBER EXTRACTION

Image endmembers of representative wetland features (salt crust, soil, vegetation, and water) were extracted from the HyMap data using the following standard processing steps: (1) MNF rotation of all 126 HyMap bands and selection of the first 15 MNF bands containing most of the nonredundant information; (2) application of the PPI to the 15 MNF bands to identify pure pixels; and (3) display of the resulting data cloud with the n-dimensional visualizer to select the endmembers.

The resulting pool of endmembers was further inspected in terms of their location within the HyMap image area, their proximity to field survey and soil sampling points, and their spectral characteristics. The latter was done by comparing the image endmember spectra with the field spectral library obtained from the wetland area in La Mancha (Section 7.4). This integrated procedure enabled the identification and labeling of the endmembers. A total of 32 and 27 endmembers was obtained, respectively, for the HyMap images of May and August 2004. These endmembers represent mainly soils and their conditions. Vegetation endmembers were only marginally represented in the final pool of endmembers, because the image area used for endmember extraction was limited by application of a mask to the nonvegetated area, with no rain-fed or irrigated fields. Nevertheless, some soil endmembers show the influence of sparse vegetation, which is unavoidable in an agricultural area.

7.5.3 CLASSIFICATION AND MAPPING OF SALT-AFFECTED SOILS

A supervised classification method, namely the SAM, was adopted since this is a physically based classification method where image spectra are compared and matched to reference spectra (endmembers). SAM compares the angle between an endmember spectrum and each image pixel spectrum. Smaller angles represent closer matches to the endmember spectrum. The advantage of using this supervised method is that it is relatively insensitive to cross-illumination and albedo effects which are present in the HyMap images (Mather, 2004).

The two aforementioned sets of endmembers were used separately in the SAM classification of the HyMap images. The 32 endmembers extracted from the wet-season image were applied to that same May image, and the 27 dry-season endmembers were applied to the August image. In both cases, a small similarity angle of 0.1 rad was used as threshold and a vegetation mask was applied to ensure that only non- or mostly unvegetated areas are considered by the classifier.

A class distribution report was generated for each SAM classification output to assess the number and percentage of pixels included in each SAM class. Based on the reports, a further reduction of the initial set of endmembers (reference spectra) could be achieved by excluding those endmembers that were associated with classes having very small percentages of classified pixels. In other words, the classes that were insignificantly represented in the SAM classification outputs were excluded, and the whole classification procedure was carried out once again with a reduced but representative set of reference spectra.

It turned out that a set of 15 endmembers was sufficient to accurately map the distribution of the main soils in the study area using the wet-season May image, while 16 endmembers were required for the same task when using the dry-season August image. Inspection of the individual SAM classes (for each date) with respect to the field survey points, the soil sample analysis data, and the mean spectral class characteristics allowed merging several subclasses into main classes based on spectral and physicochemical similarities. This reduced the total output classes from 15 to 8 for the wet-season image, and from 16 to 9 for the dry-season image. The final classification results are shown in Figure 7.10a and b. They form the basis for the change detection analysis that follows.

7.5.4 CHANGE DETECTION ANALYSIS OF SALT-AFFECTED SOILS

The SAM classification results in Figure 7.10 show the distribution of pixels that match the soil reference spectra used in the classification with a high degree of similarity (very small angle

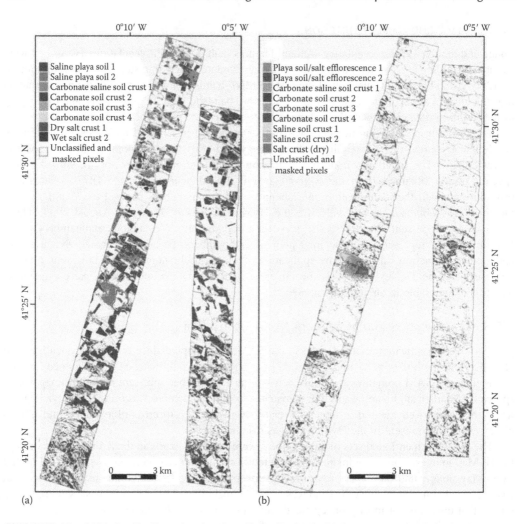

FIGURE 7.10   SAM classifications showing the soil distribution in (a) the wet season and (b) the dry season.

between pixel and reference vectors). Therefore, many pixels appear in both classification data as unclassified or masked areas (both shown in white). Unclassified pixels are either mixed soil pixels that do not satisfy the small angle threshold criterion or represent a soil type for which no reference spectra was used. Masked pixels, on the other hand, represent vegetated areas (crops) or areas with some dry vegetation that is of no interest in the change detection analysis.

The resulting SAM classes for both image dates can be associated to the following main surface soil groups: carbonate-rich soils, playa soils, saline soils, and salt crusts, each with different levels of salinity and surface crust development. Their mean class spectra for the wet- and dry-season images are, respectively, shown in Figure 7.11a and b.

In order to determine how individual soil classes changed from the wet to the dry season, a change detection analysis of the SAM classification outputs in Figure 7.10 was conducted. The method used is based on categorical (spectral class) comparison rather than on radiometric (reflectance values) comparison of equivalent image bands, and does not require radiometric normalization of the two images as for example in the single-band differencing method (Jensen, 1996).

A cross-tabulation of the two classifications (Table 7.3) shows the changes among mapped classes (i.e., from-to), as well as areas that remained unchanged. The SAM classes of the May image represent the initial state image (columns) and the SAM classes of the August image (i.e., dry

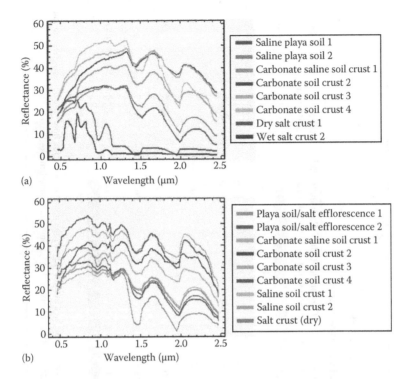

FIGURE 7.11 Mean spectra of SAM classes representing soils in (a) the wet season and (b) the dry season.

season) represent the final state image (rows). The table further lists the percentages of pixels from the initial state classes that fall into the final state classes including unclassified (class not present) and masked (vegetated) pixels.

From Table 7.3 and Figure 7.10, it is obvious that the nature and spatial distribution of the initial and final classes have changed significantly between the May and August images. Some plowed fields with relatively large areas of homogeneous soil cover in May became covered with dry vegetation (stubble/straw) in August, right after the harvest season, or are occupied by irrigated vegetation (both vegetation types being represented by the masked pixel class in Table 7.3). The former is probably the reason why the soil classes in the August image (Figure 7.10b) show a more fragmented distribution with more scattered and smaller soil patches than in the May image (Figure 7.10a). Furthermore, the cross-tabulation of the initial and final classes highlights a significant change in the status of individual soil classes (i.e., initial carbonate-rich soils becoming more saline), while the nature of the soils remains the same (i.e., initial carbonate-rich soil crusts correlating well with final carbonate-rich soil crusts, and so on, but excluding unclassified or masked pixel areas). This is best expressed by looking at the mean spectra of individual classes of the initial and final state images.

In Figure 7.12, the same mean class spectra of Figure 7.11 are shown again, but this time they are grouped into three main categories of surface features: (1) salt crusts (mainly in playa lakes), (2) playa soils with salt efflorescence or saline playa soils, and (3) carbonate-rich soil crust with different salinity levels. The left-hand side of this figure shows the class spectra of the initial state image (May) and the right-hand side shows the corresponding class spectra of the final state image (August). This figure illustrates in a graphical manner the change detection trends observed in Table 7.3. While soil groups in the initial state image tend in general to correspond to the equivalent soil groups in the final state image, their mean spectra however have undergone some changes. These changes can be attributed to increased soil crust formation and precipitation of salts at the terrain surface in the dry season.

TABLE 7.3

Change Detection Statistics Showing the Percentage of Pixels Falling in Each SAM Class

Class Description	Initial State (Wet Season)								Class Total (%)
	Dry Salt Crust 1	Wet Salt Crust 2	Saline Playa Soil 1	Saline Playa Soil 2	Carbonate-Rich Saline Soil Crust 1	Carbonate-Rich Soil Crust 2	Carbonate-Rich Soil Crust 3	Carbonate-Rich Soil Crust 4	
Unclassified	2.4	52.0	1.0	4.7	40.6	72.4	50.5	5.0	100
Salt crust (dry)	11.5	66.4	0.0	0.0	0.0	0.0	0.0	0.0	100
Saline soil crust 1	2.6	0.0	1.9	11.7	0.1	0.0	0.0	0.1	100
Saline soil crust 2	9.3	0.0	26.1	16.2	0.0	0.0	0.0	1.3	100
Playa soil with salt efflorescence 1	38.6	0.0	16.2	22.2	0.0	0.0	0.0	10.0	100
Playa soil with salt efflorescence 2	35.6	1.6	38.5	4.4	0.0	0.0	0.0	0.7	100
Carbonate-rich saline soil crust 1	0.1	0.0	9.0	14.7	6.8	0.5	1.7	13.1	100
Carbonate-rich soil crust 2	0.0	0.0	0.0	4.8	18.8	4.2	6.2	3.7	100
Carbonate-rich soil crust 3	0.0	0.0	0.0	0.8	3.1	3.4	8.2	2.3	100
Carbonate-rich soil crust 4	0.1	0.0	7.3	7.2	1.7	1.6	5.0	63.0	100
Masked pixels (vegetation)	0.0	0.0	0.04	13.2	29.0	18.0	28.5	9.8	100
Class total (%)	100	100	100	100	100	100	100	100	100

(Final state (dry season))

Notes: Column classes are initial state classes (wet season) and row classes are final state classes (dry season). Highlighted in light gray are the main classes of the final state into which the initial classes are broken down. Highlighted in dark gray are the initial classes with the highest percentage (>10%) of pixels that fall into unclassified/masked pixels.

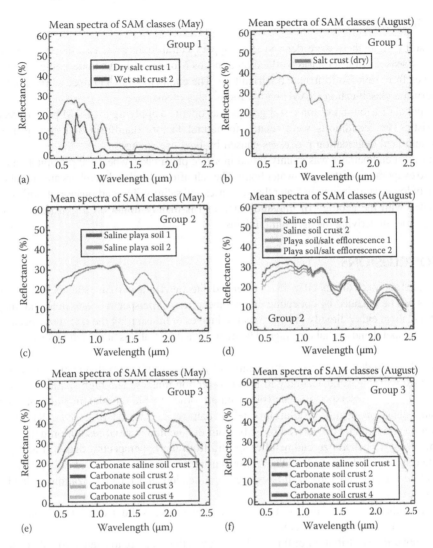

FIGURE 7.12 Comparison of mean SAM spectra of the main surface soil groups between the wet (May) and dry (August) seasons.

During the wet season (May), the salt crusts in group 1 (Figure 7.12a) are strongly affected by shallow flooding and salts are diluted within the sediments as well as in the water. The spectral characteristics change toward the dry season (Figure 7.12b), with the formation of salt crusts at the soil surface. In fact, some of the areas with salt crust spectra undergo a change toward group 2 and form saline soil crusts in August (Figure 7.12d). In group 2, the spectral characteristics of saline soils in the playa area (Figure 7.12c) are strongly influenced by salinity, and indications of vegetation are present. At the peak of the dry season in August (Figure 7.12d), the soil is in a dry state and the influence of vegetation has disappeared. The carbonate-rich soil crusts in group 3, with carbonate content of 45% and slight to moderate salinity, have in general high reflectance. The spectral characteristics in May (Figure 7.12e) show the presence of some vegetation that disappears in August (Figure 7.12f) due to dry conditions, and more intense soil crusts are formed. The relatively higher abundance of carbonates over other salt minerals hinders the identification of the saline spectral features. An increase in surface salinity starts to become noticeable within the visible and mid-infrared spectral range, although absolute confirmation requires soil surface analysis.

7.5.5 DISCUSSION

Selected soil endmembers were derived from hyperspectral images for two dates representing the wet and dry seasons of a semiarid wetland area in Los Monegros that is undergoing important land use changes from rain-fed to irrigated agriculture. The endmembers were used as reference vectors in a supervised classification (SAM) with various rates of success.

The spatial distribution of three soil groups was obtained applying the SAM procedure, and soil characteristics and conditions were related to natural factors (landform and geology), seasonal changes, and land degradation processes caused by land use changes.

A salt-crust endmember was mainly found in saline playa lakes where the precipitation of salts is high due to evaporation of the open-water body. The salt-affected soil endmembers are probably related to the use of low-quality irrigation water that increases soil salinity. The carbonate-rich soil endmember is associated with the geology of the study area and is mainly found on a gypsiferous/calcareous mudstone facies underlying the agricultural fields.

7.6 CONCLUSIONS

The integrated approach discussed in this chapter and illustrated with two case studies shows that several advantages can be obtained by collecting endmembers from hyperspectral data (field and image data) and applying them either directly to hyperspectral images or transferring them to multispectral images with better temporal and spatial coverage. The main advantages of this approach include:

1. The ability to detect surface components in hyperspectral images that cannot be otherwise mapped with the same accuracy in lower-resolution images. The high spatial and spectral resolutions of hyperspectral data (field and airborne) enable the identification of subpixel components in medium-resolution satellite data (e.g., TM, ETM+, and ASTER). This identification cannot be derived directly from the medium spatial resolution data.
2. The ability to generalize endmember spectra from the hyperspectral data to multispectral data (e.g., TM/ETM+ and ASTER) allows using sequences of new and old multitemporal satellite images and thus facilitates the identification and characterization of landscape changes.
3. Results obtained from small test sites, which have been validated on the ground, can be extended to wider areas in semiarid environments.
4. A noteworthy contribution of this methodology is the capacity to use endmembers derived from hyperspectral information in the analysis of existing multispectral data from different sensors.
5. The same approach can be used with other and future hyperspectral and multispectral sensors to enhance results and carry out monitoring tasks. The methodology forms the basis to continue integrating data obtained from current and future sensors in order to delineate and monitor salinity components in semiarid environments.
6. The approach could therefore be applied to ecosystems throughout semiarid environments in different regions of the world.

ACKNOWLEDGMENTS

The authors would like to thank and gratefully acknowledge the support of the following institutions: the UK Natural Environment Research Council Equipment Pool for Field Spectroscopy (NERC EPFS) for the loan of the ASD FieldSpec Pro spectrometer and the German Aerospace Centre (DLR) for providing the DAIS 7915 data, collected during the EU Framework HySens Programme 2000–2003 (HS-2000ES2), and the corresponding preprocessing of the data. Furthermore, we would like to thank DLR and HYVISTA Corporation with respect to the HyMap data and the HyEurope program that was carried out during 2004, as well as the Japan Aerospace Exploration Agency (JAXA) for providing the ASTER image as part of the user proposal AP-0072. The authors would also like to give special thanks to P.M. Mather (Nottingham University) and the editors G. Metternicht and A. Zinck for their valuable comments and revision of the manuscript.

REFERENCES

Asner, G.P. and Heidebrecht, K.B. 2002. Spectral unmixing of vegetation, soil and dry carbon cover in arid regions: Comparing multispectral and hyperspectral observation. *International Journal of Remote Sensing* 23: 3939–3958.

Barducci, A., Guzzi, D., Marcoionni, P., Pippi, I., and Raddi, S. 2007. Proba contribution to wetland monitoring in the coastal zone of San Rossore natural park. In *Proceedings of Envisat Symposium 2007*, ESA SP-636, Montreux, Switzerland, pp. 1–6.

Ben-Dor, E., Patkin, K., Banin, A., and Karnieli, A. 2002. Mapping of several soil properties using DAIS-7915 hyperspectral scanner data—a case study over clayey soils in Israel. *International Journal of Remote Sensing* 23: 1043–1062.

Ben-Dor, E., Benyamini, Y., Agassi, M., and Blumberg, D.G. 2003. The spectral reflectance properties of soil structural crusts in the 1.2–2.5 μm spectral region. *Soil Science Society of America Journal* 67: 289–299.

Ben-Dor, E., Levin, T.N., Singer, A., Karnieli, A., Braun, O., and Kidron, G.J. 2006. Quantitative mapping of the soil rubification process on sand dunes using an airborne hyperspectral sensor. *Geoderma* 131: 1–21.

Boardman, J.W. 1993. Automating spectral unmixing of AVIRIS data using convex geometry concepts. In *Proceedings of the Fourth JPL Airborne Geoscience Workshop*, 93-4, Pasadena, CA, pp. 11–14.

Boardman, J.W., Kruse, F.A., and Green, R.O. 1995. Mapping target signatures via partial unmixing of AVIRIS data. In *Proceedings of the Fifth JPL Airborne Earth Science Workshop*, 95-7, Pasadena, CA, pp. 23–26.

Bryant, R.G. 1996. Validated linear mixture modelling of Landsat TM data for mapping evaporite minerals on a playa surface: Methods and applications. *International Journal of Remote Sensing* 17: 315–330.

Casado de Otaola, S. and Montes del Olmo, C. 1995. *Guía de los lagos y humedales de España*. J.M. Reyero. Madrid, Spain.

Castañeda, C. and Herrero, J. 2005. The water regime of the Monegros playa-lakes as established from ground and satellite data. *Journal of Hydrology* 310: 95–110.

Castañeda, C. and Herrero, J. 2008. Assessing the degradation of saline wetlands in an arid agricultural region in Spain. *Catena* 72: 205–213.

Chabrillat, S., Pinet, P.C., Ceuleneer, G., Johnson, P.E., and Mustard, J.F. 2000. Ronda peridotic massif: Methodology for its geological mapping and lithological discrimination from airborne hyperspectral data. *International Journal of Remote Sensing* 21: 2363–2388.

Chabrillat, S., Goetz, A.F.H., Krosley, L., and Olsen, H.W. 2002. Use of hyperspectral images in the identification and mapping of expansive clay soils and the role of spatial resolution. *Remote Sensing of Environment* 82: 431–445.

Cirujano, S. 2000. Humedales de La Mancha—Flora y vegetación. In *Humedales de Ciudad Real*, Ed. V. García Canseco. Editorial Esfagnos, Talavera de la Reina, Spain, pp. 124–131.

Clark, R.N., Swayze, G.A., Gallagher, A., Gorelick, N., and Kruse, F. 1991. Mapping with imaging spectrometer data using the complete band shape least-squares algorithm simultaneously fit to multiple spectral features from multiple materials. In *Proceedings of the Third Airborne Visible/Infrared Imaging Spectrometer (AVIRIS) Workshop*, 91-2 Pasadena, CA, pp. 2–3.

Crowley, J.K. 1991. Visible and near-infrared (0.4–2.5 μm) reflectance spectra of playa evaporite minerals. *Journal of Geophysical Research* 96(B10): 16231–16240.

Crowley, J.K. 1993. Mapping playa evaporite minerals with AVIRIS data: A first report from Death Valley, California. *Remote Sensing of Environment* 44: 337–356.

Crowley, J.K., Hubbard, B.E., and Mars, J.C. 2003. Analysis of potential debris flow source areas on Mount Shasta, California, by using airborne and satellite remote sensing data. *Remote Sensing of Environment* 87: 345–358.

Dehaan, R. and Taylor, G. 2002. Field-derived spectra of salinized soils and vegetation as indicators of irrigation-induced soil salinization. *Remote Sensing of Environment* 80: 406–417.

Dehaan, R. and Taylor, G. 2003. Image-derived endmembers as indicators of salinisation. *International Journal of Remote Sensing* 24: 775–794.

Drake, N.A. 1995. Reflectance spectra of evaporite minerals (400–2500 nm): Applications for remote sensing. *International Journal of Remote Sensing* 16: 2555–2571.

Dutkiewicz, A. 2006. Evaluating hyperspectral imagery for mapping the surface symptoms of dryland salinity. PhD dissertation, The University of Adelaide, Australia.

Farifteh, J. 2007. Imaging spectroscopy of salt-affected soils: Model-based integrated method. PhD dissertation, International Institute for Geo-information Science and Earth Observation (ITC), Enschede, and Utrecht University, the Netherlands.

Farifteh, J. and Farshad, A. 2002. Remote sensing and modeling of topsoil properties, a clue for assessing land degradation. In *Proceedings of the 17th World Congress of Soil Science*, Bangkok, Thailand, vol. 52, pp. 865.1–865.11.

Farifteh, J., Farshad, A., and George, R.J. 2006. Assessing salt-affected soil using remote sensing, solute modelling, and geophysics. *Geoderma* 130: 191–206.

Farifteh, J., Van der Meer, F., Atzberger, C., and Carranza, E.J.M. 2007. Quantitative analysis of salt-affected soil reflectance spectra: A comparison of two adaptive methods (PLSR and ANN). *Remote Sensing of Environment* 100: 59–78.

FAO, ISRIC, and ISSS. 1998. *World Reference Base for Soil Resources*. World Soil Resources Report 84. Food and Agriculture Organization of the United Nations, Rome, Italy.

Green, A.A., Berman, M., Switzer, P., and Craig, M.D. 1988. A transformation for ordering multispectral data in terms of image quality with implications for noise removal. *IEEE Transactions on Geoscience and Remote Sensing* 26: 65–74.

Howari, F.M., Goodell, P.C., and Miyamoto, S. 2002. Spectral properties of salt crusts formed on saline soils. *Journal of Environmental Quality* 31: 1453–1461.

Isidoro, D., Quílez, D., and Aragués, R. 2006. Environmental impact of irrigation in La Violada District (Spain): I. Salt export patterns. *Journal of Environmental Quality* 35: 766–775.

Jensen, J.R. 1996. *Digital Change Detection. Introductory Digital Image Processing: A Remote Sensing Perspective*. Prentice-Hall, New Jersey, NJ.

Kaufmann, H., Segl, K., Chabrillat, S., Hofer, S., Stuffler, T., Mueller, A., Richter, R., Schreier, G., Haydn, R., and Bach, H. 2006. EnMAP, a hyperspectral sensor for environmental mapping and analysis. In *Proceedings of IEEE International Geoscience and Remote Sensing Symposium*, Denver, CO, pp. 1617–1619.

Koch, M. 2000. Geological controls of land degradation as detected by remote sensing: A case study in Los Monegros, north-east Spain. *International Journal of Remote Sensing* 21: 457–473.

Koch, M., Schmid, T., and Gumuzzio, J. 2001. The study of anthropogenic affected wetlands in semi-arid environments applying airborne hyperspectral data. In *Teledetección Medio Ambiente y Cambio Global*, Eds. J. Rosell Urrutia and J. Martínez-Casanovas. Editorial Milenio, Universitat de Lleida, Spain, pp. 297–301.

Koch, M., Schmid, T., Gumuzzio, J., and Mather, P. 2005. Use of image spectroscopy to assess the impact of land use changes in a semi-arid karstic landscape: Los Monegros, Spain. In *Imaging Spectroscopy. New Quality in Environmental Studies*, Eds. B. Zagajewski and M. Sobczak. Wydawnictwa Przemyslowe, WEMA Sp., Warsaw, Poland, pp. 147–155.

Koch, M., Schmid, T., Gumuzzio, J., and Mather, P. 2006. Wetland feature extraction and change detection study of a playa lake environment in NE Spain using hyperspectral and multispectral images. In *Second Recent Advances in Quantitative Remote Sensing*, Ed. J.A. Sobrino. Publicaciones de la Universitat de Valencia, Spain, pp. 284–288.

Kruse, F.A., Lefkoff, A.B., Boardman, J.W., Heidebrecht, K.B., Shapiro, A.T., Barloon, P.J., and Goetz, A.F.H. 1993. The spectral image processing system (SIPS)—Interactive visualization and analysis of imaging spectrometer data. *Remote Sensing of Environment* 44: 145–163.

M.A.P.A. 2007. Visor del SIGPAC, Ministerio de Agricultura, Pesca y Alimentación, España. http://sigpac.mapa.es/fega/visor/ (accessed November 18, 2007).

Maas, E.V. 1990. Crop salt tolerance. In *Agricultural salinity assessment and management*, Ed. K.K. Tanji. American Society of Civil Engineers, New York, NY.

Mather, P.M. 2004. *Computer Processing of Remotely-Sensed Images: An Introduction* (3rd edn.). John Wiley & Sons, Chichester, UK.

Metternicht, G.I. and Zinck, J.A. 2003. Remote sensing of soil salinity: Potentials and constraints. *Remote Sensing of Environment* 85: 1–20.

Mougenot, B. 1993. Effets des sels sur la réflectance et télédétection des sols salés. *Cahiers ORSTOM, Série Pédologie*, vol. XXVIII(1): 45–54.

Okin, G.S., Roberts, D.A., Murray, B., and Okin, W.J. 2001. Practical limits on hyperspectral vegetation discrimination in arid and semiarid environments. *Remote Sensing of Environment* 77: 212–225.

Oliver, G. and Florin, M. 1995. The wetlands of La Mancha, Central Spain: Opportunities and problems concerning restoration. In *Bases ecológicas para la restauración de humedales en la Cuenca Mediterránea*, Eds. G. Oliver, F. Moline and J. Cobos. Junta de Andalucía, Consejería de Medioambiente, Sevilla, Spain, pp. 197–216.

Research Systems Inc. 2006. The environment for visualizing images ENVI, Version 4.3. Research Systems, Inc., Boulder, CO.

Rhoades, J.D. 1990. Overview diagnosis of salinity. Problems and selection of control practices. In *Agricultural Salinity Assessment and Management*, Ed. K.K. Tanji. American Society of Civil Engineers, New York, NY, pp. 19–41.

Samper-Calvete, F.J. and García-Vera, M.A. 1998. Inverse modeling of groundwater flow in the semiarid evaporitic closed basin of Los Monegros, Spain. *Hydrogeology Journal* 6: 33–49.

Sánchez, J.A., Pérez, A., and Martínez-Gíl. 1998. Combined effects of groundwater and aeolian processes in the formation of the northernmost closed saline depressions of Europe: North-east Spain. *Hydrological Processes* 12: 813–820.

Schmid, T.F. 2005. *Integrated remote sensing approach to detect changes in semi-arid wetland areas in central Spain*. Colección Documentos CIEMAT, Madrid, Spain.

Schmid, T., Koch, M., Gumuzzio, J., and Mather, P.M. 2004. A spectral library for a semi-arid wetland and its application to studies of wetland degradation using hyperspectral and multispectral data. *International Journal of Remote Sensing* 25: 2485–2496.

Schmid, T., Koch, M., and Gumuzzio, J. 2005a. Multisensor approach to determine changes of wetland characteristics in semiarid environments (Central Spain). *IEEE Transactions on Geoscience and Remote Sensing* 43: 2516–2525.

Schmid, T., Koch, M., Gumuzzio, J., and Medel, I. 2005b. Field and imaging spectroscopy to determine soil degradation stages in semi-arid terrestrial ecosystems. In *Imaging spectroscopy. New quality in environmental studies*, Ed. B. Zagajewski y M. Sobczak. Wydawnictwa Przemyslowe, WEMA Sp., Warsaw, Poland, pp. 75–183.

Schmid, T., Gumuzzio, J., Koch, M., Mather, P., and Solana, J. 2007. Characterizing and monitoring semi-arid wetlands using multi-angle hyperspectral and multispectral data. In *Proceedings of Envisat Symposium 2007*, ESA SP-636, Montreux, Switzerland, pp. 1–6.

Schoeneberger, P.J., Wysocki, D.A., Benham, E.C., and Broderson, W.D. 2002. *Field Book for Describing and Sampling Soil*, Version 2.0. USDA, Natural Resources Conservation Service, National Soil Survey Center, Lincoln, NE.

Soil Survey Staff. 2006. *Keys to Soil Taxonomy* (10th edn.). USDA, Natural Resources Conservation Service, Washington, DC.

Tanji, K.K. 1990. Nature and extent of agricultural salinity. In *Agricultural Salinity Assessment and Management*, Ed. K.K. Tanji. American Society of Civil Engineers, New York, NY, pp. 3–17.

Van der Meer, F.D. 2000. Imaging spectrometry for geological applications. In: *Encyclopedia of Analytical Chemistry*, Ed. R.A. Meyers. John Wiley & Sons, Chichester, UK.

Whiting, M.L. and Ustin, S.L. 1999. Use of low altitude AVIRIS data for identifying salt affected soil surfaces in Western Fresno County, California. In *Proceedings of the Eighth JPL Airborne Visible Infrared Imaging Spectrometer (AVIRIS) Workshop*, Pasadena, CA, pp. 99–64.

8 Characterization of Salt-Crust Build-Up and Soil Salinization in the United Arab Emirates by Means of Field and Remote Sensing Techniques

Fares M. Howari and Philip C. Goodell

CONTENTS

8.1 INTRODUCTION

Soil salinity occurs naturally in many regions (Tanji, 1990, 1996), while in other areas soil salinization is produced by man. It is estimated that more than half of the irrigated land in arid and semiarid regions of the world is affected to some degree by salinization and that millions of hectares of agricultural land have been abandoned because of salinity build-up (Szabolcs, 1979; Goudie, 1990; Massoud, 1990; Tanji, 1990; Ghassemi et al., 1995). Salinization is the accumulation of soluble salts of sodium, magnesium, and calcium in soils (Sumner, 2000). This happens in poorly drained land under semiarid and arid climates where large quantities of salt have been leached from higher areas in the drainage basins. These leached salts concentrate in slow-flowing groundwater and are brought to the soil surface through high evapotranspiration (Goudie, 1990; Sumner and Naidu, 1998).

Water table depth varies naturally, usually seasonally, and more so in areas altered by man. High water tables result in waterlogging, secondary salinization, and seepage of saline groundwater into irrigation canals. The percolation of irrigation water through salt-affected soils can raise the salt content and the level of the groundwater (Tanji, 1990; Ghassemi et al., 1995). A classic example where salinization has caused severe damage to agricultural land is that of Mesopotamia, where fertile land has been transformed into a salt-caked waste. The fertile soils that once supported "the Gardens of Babylon" do not grant more than an irregular crop of salt-tolerant barley (Meybeck et al., 1989).

In the United Arab Emirates (UAE), accurate and up-to-date information on the spatial extent and variability of soil salinity is lacking, in spite of its relevance for the management of sustainable agriculture and the maintenance of the green landscapes that protect the urban settlements from the frequent sand storms. Such information would also help improve the salinity management in watersheds and ecosystems, especially in the northern part of the UAE.

Remote sensing has advantages over the conventional ground methods used to map and monitor soil salinity. Remote sensing utilizes the electromagnetic energy reflected from targets to obtain information about the Earth's surface with different levels of detail. The visible and infrared portions of the electromagnetic spectrum are especially relevant for the identification of the minerals that originate from salt crusts (Hunt et al., 1971b; Crowley, 1991; Ben-Dor et al., 1999; Clark, 1999; Howari et al., 2002). Several aspects involved in the remote sensing of salt-affected soils are discussed in this chapter. Multispectral data usually provide useful information on surface soil characteristics, while microwave provides information for a shallow depth of bare soils. Other salinity applications of remote sensing data are discussed by Tóth et al. (1991), Metternicht and Zinck (2003), Khan et al. (2005), and Farifteh et al. (2007).

Spectral signatures of the soil surface provide information on salt types and may also provide indication on salt quantity and associated grain size or even crust thickness (Howari, 2003; Ghrefat et al., 2007). However, the effectiveness of remote sensing is more limited at low levels of soil salinity and when the salt crust is not visible or when salt is mixed with other soil constituents (Ben-Dor, 2002; Howari et al., 2002). Thus, an indirect method such as tracing soil salinity and salt crust by means of vegetation cover becomes an option, because plants, especially when they are not salt-tolerant, are good indicators of the conditions that occur below the soil surface. Therefore, salinity and its severity may be inferred from selected plant species.

In the case of mixed salt crusts, numerical models have been used to unmix the mixed pixels and identify the endmembers (Ray and Murray, 1996; Dehaan and Taylor, 2003). Unfortunately, several of these models do not account for the physical factors that influence the spectral signatures, including mineral composition, variation in soil moisture, organic matter content, soil texture, and others. With the development of the hyperspectral sensors, spectral features related to characteristic absorption bands of salt minerals can be mapped with more detail (Cloutis, 1996; Ghrefat et al., 2007). This enables better detection and mapping of soil salinity.

In this chapter, we attempt to show the usefulness of remote sensing for identifying the origin of salts and types of salt-affected soils, as well as the factors that affect the detection of salt-affected soils, taking the UAE as a representative example of arid regions.

8.2 MATERIALS AND METHODS

8.2.1 STUDY AREA

The UAE is located in the southeastern corner of the Arabian Peninsula, between latitude 22°40′ and 26°00′ and longitude 51°00′ and 56°00′, where the interior platform of the Arabian shelf occurs at the subsurface. The area is bordered by the Persian Gulf to the north, Saudi Arabia to the south and west, and Oman and the Gulf of Oman to the east. The seven member states are Abu Dhabi (Abu Zaby), Ajman, Dubai, Al Fujayrah, Ra's al Khaymah, Sharjah, and Umm al Qaywayn. Figure 8.1 shows the location of the UAE and in particular that of Abu Dhabi and Al Ain where the present study has been carried out. Two physiographic units are displayed in white in Figure 8.1. The most extensive unit lies in the coastal zone, in the proximity of wet sabkhas. Elevation is close to zero but increases to the south. The second physiographic unit lies in the eastern UAE, adjacent to the mountainous area of the country, and is referred to as the Al Ain area. In most of the seaward strip, the relief is flat

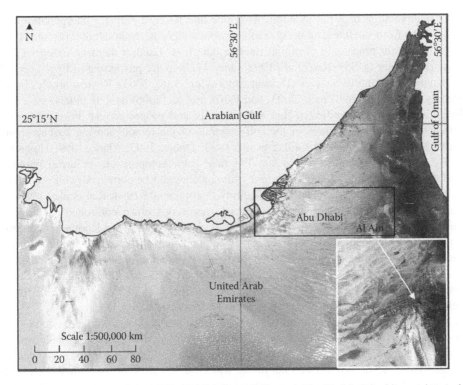

FIGURE 8.1 Location map of the UAE, highlighting Al Ain and Abu Dhabi; the white patches and spots indicate salt-affected land.

with white salt at the surface and the depth to the groundwater is 0.5–1 m, with a gradient of approximately 1:5000 toward the Arabian Gulf (Alsharhan and Kendall, 2002).

Air temperature commonly reaches 45°C–50°C in summer but can be as low as 0°C in winter. Soil surface temperature can be higher than 40°C in the winter months but not much higher than 50°C in the summer months. High temperature, coupled with the aridity of the UAE coast, partially explains the widespread occurrence of carbonates and evaporites in the coastal plain and the formation of inland salt flats (Alsharhan and Kendall, 2002). Infrequent rainfall usually occurs in autumn, winter, and spring; along with strong winds, it can spread salt from the salt flats to nearby natural and reclaimed soils. The average monthly rainfall is higher in the mountains (140–200 mm) and the East Coast (100–140 mm) than in the gravel plains (100–120 mm) and the West Coast (<60 mm). Thus, there is a continuous gradient of increasing precipitation from the southwest toward the northeast.

The combined effect of strong winds and high temperature results in significant evaporation and causes high soil salinity. Evaporation rates are as high as 124 cm year^{-1}. High salinity values recorded in some parts of the southern Gulf suggest that evaporation is greatest in summer, especially in small lagoons. Water lost in evaporation is not compensated by fluvial input and rainfall, thus accelerating salt-crust build-up (Evans et al., 1969). Halite (NaCl) is the major type of salt in the study area and is often leached from the surface sediments. Anhydrite ($CaSO_4$) occurring on the soil surface, above the intertidal zone, is usually eroded and transported via surface runoff or wind storm to the surrounding soils.

8.2.2 Background on Remote Sensing of Salt-Affected Soils

Knowledge about the interaction between energy, soil surface, and the physicochemical characteristics of salts helps improve the study of soil salinity in arid regions. Historically, the application of remote sensing to the detection and mapping of salt-affected soils has gone through the following steps: (1) interpretation of aerial photographs (Paine and Kiser, 2003); (2) interpretation of color composite films from satellite data using visual or optical density methods (Sharma and Bhargava, 1988); (3) computer processing of digital satellite data from Landsat thematic mapper (TM) and multispectral scanner (MSS) (Rao et al., 1998); and (4) computer processing of hyperspectral data and associated unmixing techniques (Dehaan and Taylor, 2002, 2003). Review articles have been provided by Metternicht and Zinck (2003) and Farifteh et al. (2006), among others.

Previous research has indicated that salts in soils and evaporites can be distinguished by examining the absorption features of the soil minerals in the remote sensing spectra (Hunt and Salisbury, 1970; Crowley, 1991; Csillag et al., 1993; Drake, 1995; Clark, 1999; Howari et al., 2002). By visual interpretation of Landsat TM false-color compositions or aerial photographs, image elements such as dull white and bright white patches can be correlated with ground data to determine the degree and type of salinity–alkalinity (Verma et al., 1994; Rao et al., 1998).

Studies on hyperspectroscopy and reflectance spectroscopy under controlled conditions show that different salinity levels are separable. Csillag et al. (1993) analyzed laboratory spectroradio-meter data using a modified stepwise principal component band selection method to separate 13 classes of soil salinity status. They reported the importance of high spectral resolution in recognition accuracy. The salinity status decreased from 91% to 90% to 88% as bandwidth increased by 10, 20, and 40 nm, respectively. The same authors identified 35–42 narrow bands between 500 and 2400 nm, which led to a recognition accuracy of 100%.

At high salinity levels and with salt concentration features at the soil surface, remote detection of salt crusts using airborne or satellite-borne data becomes possible. Howari (2003) used image-processing techniques, such as supervised classification and spectral extraction and matching, to investigate types and occurrence of salts in the Rio Grande valley in the United States–Mexico border area. The sequence of activities used to extract information on salt-affected soils is given in Figure 8.2. The findings show the presence of gypsum and halite as the dominant minerals in the salt crusts in the Rio Grande valley.

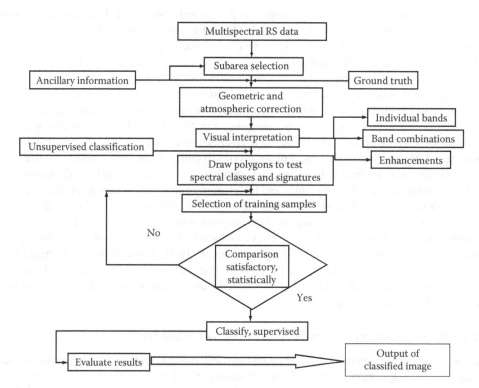

FIGURE 8.2 Sequence of activities in image classification and information extraction, routinely used in remote identification of salt-affected soils.

TABLE 8.1
Applied Image-Processing Techniques

Sensor Data	Acquisition Date	Image-Processing Technique	Application	Accuracy for Salt Identification
Landsat ETM	2000	Band ratioing, spectra extraction, band combinations, and visual interpretation	Image classification and mapping of salt features	60%–75%
IKONOS	2001 and 2003	Visual interpretation and classification	Mapping of salt features	>80%
Ground-based spectroradiometric data (GER 3700)	2003	Spectral matching between generated spectra and published spectral libraries	Salt identification and satellite image calibration	>90%

8.2.3 APPROACH TO REMOTE INVESTIGATION OF SALT-AFFECTED SOILS

A variety of image-processing methods is available to extract information from images (Mulders, 1987; Verma et al., 1994; Metternicht and Zinck, 1997; Rao et al., 1998; Howari, 2003). For their simplicity, supervised and spectral extraction techniques are often used, especially if high salinity or salt crusts are expected to occur. When supervised classification is used, the accuracy can be evaluated at each level of improvement using kappa coefficients. This is a statistical measure of the agreement, beyond chance, between two maps, for instance the classification output map and the ground-truth map. The data presented in this chapter have been subjected to several image-processing techniques for classification enhancement as well as for band ratioing.

Spectral classification and spectral extraction for the presented Landsat enhanced thematic mapper (ETM)+ and IKONOS images were performed using the image-processing softwares ENVI and ERDAS (Table 8.1). The images were calibrated and georeferenced. Details on the aforementioned image-processing procedures can be found in Vincent (1997), Metternicht and Zinck (1997), Howari (2003), and Fernández-Buces et al. (2006). To confirm the results, ground truthing and field checking took place, using conductivity meters for salinity measurement and conversion factors to relate conductivity to salinity. Also, x-ray diffraction (XRD) was used to confirm the identification of the salt types and mineral components in the investigated soils.

8.3 RESULTS AND DISCUSSION

8.3.1 ORIGIN AND LEVELS OF SALT

In the UAE, salt-affected soils can be of primary or secondary origin. Primary salinity is found in (1) soils developed near coastal flats or in arid areas where saline groundwater rises to the soil surface by capillary transport, and (2) ancient marine deposits consisting of evaporites. The secondary source of soil salinity is mostly due to irrigation return flow or inefficient irrigation methods that use saline groundwater.

The primary source of salt lies basically in recent and ancient marine deposits and mainly relates to tidal flats, sabkhas, and terraces in the coastal region along the Arabian Gulf. A sequence of continental and shallow-water marine sedimentary rocks deposited from Tertiary to Quaternary acts as a significant source of salt to the UAE soils, especially in the Abu Dhabi Emirate. The Tertiary unit consists of mudstone layers and evaporites (gypsum, anhydrite, and dolomite) belonging to the lower Fars Formation. An underlying subunit that contains clastic sediments (sandstones and siltstones), intercalated with layers of mudstones and anhydrite, is another potential natural source of salts. These rock formations load the groundwater with salts and, when they are exposed to physical and chemical weathering, the resulting products further increase soil salinity.

These primary factors, coupled with poor agricultural practices and high aridity, cause alarming levels of soil salinity. As a result of salinity stress, millions of trees planted during the last decade are dying off or being damaged. Dust storms and wind erosion increase with the spread of soil salinity. Rising soil salinity is even affecting date palm plantations.

According to the Center for the Study of Built Environment (2007), salinity in soils may be categorized in the following manner:

1. Extremely saline sites (over 16 dS m^{-1} [deciSiemens per meter] or 9600 ppm [parts per million]): The salinity level may rise up to that of seawater (30,000 ppm). Some of the groundwater in the north of the UAE has reached salinity levels of 25 dS m^{-1}, or half of that of seawater. Plants adapted to such high salinity levels are halophytes like mangrove plants, desert saltbushes (e.g., Haloxylon), or trees including various types of Tamarix and Acacia.
2. Very saline sites (8–16 dS m^{-1} or 4800–9600 ppm): This is equivalent to soil salinity levels found in sabkhas and in depleted inland aquifer areas. Plants such as the date palm and casuarina trees tolerate such salinity levels in well-drained soils and under sustainable water supply, i.e., where there is no increase in soil salinity due to the application of saline irrigation water.
3. Moderately saline sites (4–8 dS m^{-1} or 2400–4800 ppm): This is the salinity level of most groundwater in the UAE and increasingly also in terrains that are regularly irrigated. Many perennial plants, shrubs and trees can sustain growth in moderately saline soils provided that adapted irrigation methods and soil amendments be used to keep salinity levels in check.
4. Slightly saline sites (2–4 dS m^{-1} or 1200–2400 ppm): These sites are mostly found inland, in desert or semidesert situations where there is no regular irrigation with saline water. Although the salinity level is low, the vegetation cover is sparse because soils are very sandy and poor in organic matter and nutrients.

8.3.2 Temporal Variations of Land-Cover and Landscape Features

Arid land with different surface conditions and weathered materials may show a great variability in spectral reflectance. Heterogeneity and similarity between surface condition of soils, salt-affected soils and land-cover, provide the insight necessary for classification and interpretation of remotely sensed data. The spectral reflectance characteristics of the plants in arid areas may be also a good salinity indicator. Natural vegetation in the UAE is scarce because of high soil salinity and lack of water. With increasing salinity, chlorophyll concentration in plants decreases significantly and, in consequence, the color of the vegetation cover changes. The land-cover and landscape elements encountered in the vicinity of the agricultural areas include saline water bodies, wet and dry sabkhas, salt crusts, scattered vegetation, sand dunes, and carbonate-rich soils. Figure 8.3 shows the temporal variations of these classes in Abu Dhabi between 1993 and 2000. The extent of white salt-crusted patches has increased by about 15% in this period of time. The salinity in these areas is mainly derived from natural sources.

There is a general zoning of landscape elements from the coast to inland, starting with wet coastal sabkhas, followed by dry inland sabkhas, then carbonate-rich soils, and finally sand dunes. During the period 1993–2000, all these elements have expanded and moved inland. As observed in Figure 8.3, saline and hypersaline groundwater emerges at the soil surface and evaporates in several areas along the Abu Dhabi coast; salts are left behind on the soil surface. Groundwater is hypersaline because it is partly polluted by the intrusion of seawater. Over time, salts accumulate in the

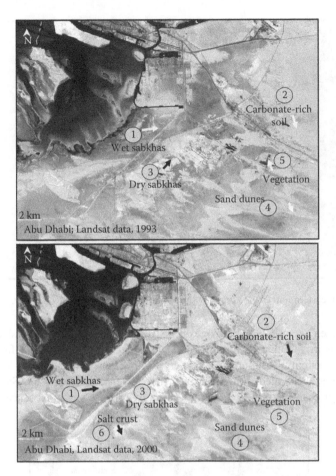

FIGURE 8.3 Temporal variations of land-cover and landscape elements in the coastal area of Abu Dhabi.

discharge areas and eventually the salt concentration becomes so high that plant growth is restricted. The lack of vegetation or the presence of scattered vegetation on salt-affected soils made it possible to remotely detect soil crusts in several locations. In Abu Dhabi and other parts of the UAE, salt crusts are often mixed with calcium carbonate crust or with carbonate-rich fractions on sandy land. In some other locations, deep soils occur with gravel at depth of more than 50 cm and high quantities of calcium carbonate (e.g., in the Dhaid, Ghuraif, and Al Maddam areas). The texture varies between sandy loam and loamy sand. Most of the soils in the Abu Dhabi Emirate are suitable for agriculture, but their potential use is constrained by high evaporation, that causes high soil salinity, and by the scarcity of irrigation water.

8.3.3 Remote Detection of Secondary Salinity

Figure 8.4 shows a secondary source of soil salinity that is due to irrigation return flow or inefficient irrigation methods using saline groundwater. In this location, salts leached from the nearby agricultural farms have accumulated at very shallow depths and are brought to the soil surface through high evapotranspiration. Other causes of irrigation-related salinity include overirrigation of farmland; poor drainage; irrigating unsuitable soils; water ponding for long periods; and seepage from irrigation channels, drains, and water storage sites. This increases the leakage of surface water to the groundwater system and causes the water table to rise, together with the salt that may have accumulated in the subsoil layers, as shown in Figure 8.4.

FIGURE 8.4   Southeastern part of Al Ain showing irrigation lands affected by salinity.

Figure 8.5 presents two IKONOS images near Al Ain city. The top image shows, to the south, an area of current irrigation and farming (dark tones) and, to the north, a farming area being abandoned due to salinity build-up and salt crust formation (white tones). The bottom image shows a completely abandoned farm with even more intense salt build-up. Salinity can also be detected from color changes of the vegetation caused by osmotic stress. Absence of vegetation is a

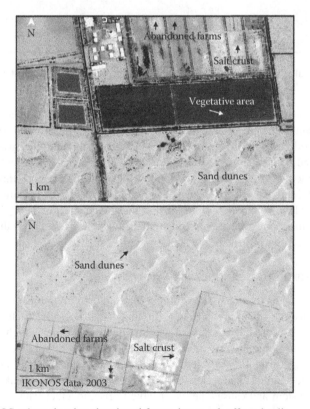

FIGURE 8.5   IKONOS subset showing abandoned farms due to salt-affected soils.

reflection of extreme stress. The dark tones of greenness in the vegetation-covered areas on the images were analyzed as an indication for soil quality. As observed, the green area is decreasing in size due to salt-crust build-up. Without field checking it would have been difficult to confirm that the white patches in Figure 8.5 are actually salt crusts. Misclassification can occur between carbonate-rich and evaporite-rich areas or, in general, between different ground-targets that exhibit similarly high reflectance values. To overcome this problem, it is recommended to incorporate multisensor/multispectral data, contextual landscape information and various data layers for better spatial discrimination of surface features in salt-affected areas (Metternich and Zinck, 1997). At dry landscape sites, calcareous soils are often mistaken for saline soils by the remote sensors, but the combination of field and spectral data can help improve their respective identification. For instance, saline soils are usually found on mid- to lower-slope positions, and reflectance spectroscopy can differentiate between crust types as highlighted by several studies (Howari, 2003, 2006; Fernández-Buces et al., 2006).

8.3.4 HYPERSPECTROSCOPY AND BAND RATIOS

Hyperspectral studies, airborne or ground-based, can offer a solution to potential misclassifications but with some challenges. For example, the salt crusts in Abu Dhabi are of different types with high concentrations of Na^+, Mg^{2+}, Ca^{2+}, and SO_4^{2-}. Figure 8.6 shows that such crusts are separable using their reflectance spectra. Additional factors, frequently interlinked, might make the identification of salt crusts somewhat cumbersome; they include soil layering, salt cover, grain size, moisture content, atmospheric effect as well as washout or rainfall impact. Detecting and mapping salinity from remote sensing data is also challenging because most salt minerals (e.g., halite) are spectrally featureless and because the signals received from salt-affected soils are relatively weaker than those of other soil components, especially those exhibiting responses in the electronic and vibration range of the electromagnetic spectrum (Farifteh et al., 2007).

The visible and near-infrared reflectance spectra of the salt crust samples collected from Abu Dhabi were confirmed by XRD patterns, as discussed in Howari (2006). The spectra of some samples are dominated by the gypsum spectral bands and the spectra of other samples contain features characteristic of calcite and gypsum. Calcium carbonate spectra have plateau-shaped spectral profiles from 500 to 1300 nm. From 1400 to 2000 nm, calcite develops two successive absorption features as shown in Figure 8.6. Gypsum spectra are identical to those published in the

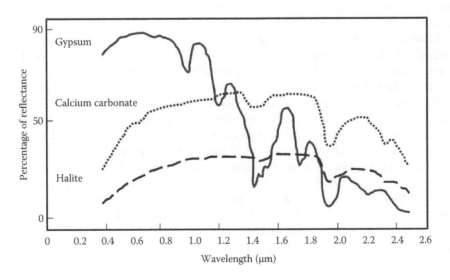

FIGURE 8.6 Spectral reflectance pattern of salt types occurring in the investigated area.

literature (Drake, 1995; Clark, 1999). In the spectra of gypsum crusts, the overtones or combination tones from fundamental vibrations of the water molecules produced a series of bands affecting the infrared spectrum between 1000 and 2300 nm (e.g., 1464, 1550, 1750, 1978, and 2300 nm) (Figure 8.6). Halite itself does not induce absorption bands in the visible, near-infrared and thermal infrared regions (Hunt et al., 1971a,b; Mougenot et al., 1993). Water bands (1400, 1900, and 2250 nm) were observed in the spectra of halite crusts due to moisture and fluid inclusions (Figure 8.6), and these bands could be considered as absorption features of halite. Similar results were obtained by Mulders (1987), Hunt et al. (1971), and Mougenot et al. (1993). These spectral features could eventually be used to resolve confusion issues among salt types and/or other soil surface characteristics.

Metternicht (1996) addressed the issue of spectral confusion between saline crusts and other surface targets in a study carried out in Cochabamba, Bolivia. The performance of the visible range in separating unaffected areas from saline and saline–alkaline is less satisfactory than that of the infrared range. According to Metternicht and Zinck (2003):

> The thermal range improved the separability of the alkaline areas because alkalinity is controlled by the amount of free carbonates, which have a strong absorption feature in TM band 6. Overall, the microwave region (L-band) has low amounts of spectral confusion and provides the best separability between saline and alkaline areas, non-affected and saline–alkaline areas, as well as among saline and saline–alkaline areas. Thus, the different regions of the electromagnetic spectrum have complementary capabilities for spectral separability of the salinity–alkalinity classes. This calls for data fusion to increase the efficiency of the remote sensing data.

Spectral unmixing techniques (Ray and Murray, 1996; Dehaan and Taylor, 2002, 2003) have been suggested to deal with the spectral confusion issue. Another option is the variables isolation approach by mobilizing contextual knowledge and applying accepted straightforward, simple remote identification techniques to determine the occurrence of components for the purpose of isolating the ones of interest. For example, several band ratios are widely implemented in investigation of surface minerals in soils, geological formations or other targets (Davis et al., 1993; Peña and Abdelsalam, 2006; McKee et al., 2007). Using such ratios, it can be assumed that a mineral detected in a crust is mixed with other crust components to a degree that it suppresses the spectral features of these other components. Band ratioing is used because some band ratios are sensitive to certain minerals and not to others. For instance, grayscale images produced using the ratios of bands 6/4 and 5/7 revealed the most information on the mineralogy of high albedo minerals in sandy soils and sand dunes (Pease et al., 1999).

On Landsat 5 and 7 images of Abu Dhabi and Al Ain, there are several areas that appear to have salt-affected soils. The application of the band ratioing technique revealed the presence of carbonate and quartz. This observation decreased the possibility to find halite and gypsum as the main components in the scene. Yet, separating quartz from carbonate raises a new challenge. The most useful image for distinguishing carbonate minerals from others such as quartz was generated using a 5/7 ratio. According to Pease et al. (1999), this ratio provides higher values for pixels corresponding to mixed carbonate and quartz. Intermediate values of image gray-tones are generated for high carbonate sands and low values (black) for quartz-rich sands. In Figure 8.7, the relative concentrations of quartz and carbonate in a sandy material are shown in two different landscapes and sites. The first is close to Al Ain and the second is close to Abu Dhabi. The grayscale changes from black to light gray with decreasing quartz and increasing carbonate contents. The darkest areas represent the highest quartz content and the lightest areas represent the highest carbonate content, while mixed compositions are represented by medium gray-tones. Other studies used ratios directly to map salt-affected soils. For example, Nield et al. (2007) mapped gypsiferous areas using the normalized difference ratio of bands 5 and 7 with a threshold >0.11. This approach correctly predicted 87% of

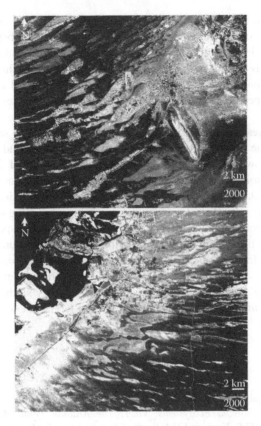

FIGURE 8.7 Ratio of near-infrared to mid-infrared (bands 5/7) of Landsat ETM (2000). The figure shows relative concentrations of quartz and carbonate in a sandy material. The grayscale changes from black to light gray with decreasing quartz and increasing carbonate contents. The darkest areas represent the highest quartz content and the lightest areas represent the highest carbonate content. Mixed compositions are represented by medium gray-tones.

the field-observed gypsic soil areas. The above authors concluded that normalized difference ratio models can be developed for digital soil mapping in areas where land surface features produce differences in Landsat spectral band reflectance.

8.4 CONCLUSIONS

Because of the arid conditions prevailing in the UAE, mapping of salt-affected soils using remote sensing data can be fast, timely, and relatively cheap. Yet, several interlinked factors influence the identification of salt-affected soils and often cause spectral confusions between salt-related features and other soil surface characteristics. This chapter concludes that salt-affected soils in the UAE are mostly associated with high reflectivity and uneven growth or absence of plant cover. On the grayscale images, the salt-affected soils often show a patchy pattern and smooth texture, with no or scattered vegetation. The land elements encountered in the studied images include saline water bodies, wet sabkhas, salt crusts, dry sabkhas, scattered vegetation, sand dunes, and carbonate-rich soils. The temporal variations of these classes in Abu Dhabi indicate that the extent of white patches, which are expected to be salt-crust areas, has increased about 15% during the period 1993–2000.

Halite (NaCl) is the major type of salt in the study area, often leached from the surface sediments in the coastal areas or concentrated on the surface from shallow groundwater. The net weathering products from halite and other evaporite minerals end up via surface runoff or windstorm in the surrounding soils. Misclassification mainly occurred between carbonate-rich and evaporite-rich areas or, in general, between different ground-targets that exhibit similarly high albedo values. The study reports ways and techniques to overcome this problem, including the variables isolation approach by applying contextual knowledge, band ratioing, incorporation of multisensor and multispectral data, advanced hyperspectral techniques, and spectral unmixing techniques.

Due to the advantages that remote sensing offers, its application to salt-affected soils received special attention from the remote sensing community, soil scientists, and agricultural experts. The present investigation shows that integration of remote sensing and field data can be used successfully in mapping saline soils. Supervised classification of Landsat data and the analysis of temporal variations help separate classes of salt-affected soils under arid conditions. However, the possibility of satisfactorily detecting salt-affected soils by means of remote sensing is enhanced by the use of multisensor high-resolution data. Future research should focus on hyperspectral data and techniques to deal with the mixture problem and with identifying additional measurable and hidden variables affecting the absorption features of soil spectra. Also, research should incorporate more indicators and techniques such as salinity indices, band ratioing, vegetation index, crop performance, leaf angle orientation, and chlorosis degree, in the study of areas only slightly affected by salinity or exposed to incipient salinization, which should be the new target in assessing and monitoring soil degradation.

REFERENCES

Alsharhan, A.S. and Kendall, C.G. 2002. Holocene carbonate evaporites of Abu Dhabi, and their Jurassic ancient analogs. In: *Sabkha Ecosystems, Vol.1: The Arabian Peninsula and Adjacent Countries*, M.A. Kahn, B. Boer, G.S. Kust, and H.J. Barth (Eds.), Springer-Verlag, Berlin, pp. 187–202.

Ben-Dor, E. 2002. Quantitative remote sensing of soil properties. *Advances in Agronomy* 75: 173–243.

Ben-Dor, E., Irons, J.R., and Epema, G. 1999. Soil reflectance. In: *Remote Sensing for Earth Sciences: Manual of Remote Sensing*, Vol. 3, 3rd ed., A.N. Rencz (Ed.), John Wiley & Sons, New York, pp. 111–188.

Center for the Study of Built Environment (2007). Accessed February 2008. Available at http://csbe.org/.

Clark, R.N. 1999. Spectroscopy of rocks and minerals and principles of spectroscopy. In: *Remote Sensing for Earth Sciences: Manual of Remote Sensing*, Vol. 3, 3rd ed., A.N. Rencz (Ed.), John Wiley & Sons, New York, pp. 3–52.

Cloutis, E.A. 1996. Hyperspectral geological remote sensing: Evaluation of analytical techniques. *International Journal of Remote Sensing* 17: 2215–2242.

Crowley, J.K. 1991. Visible and near-infrared (0.4–2.5 micron) reflectance spectra of playa evaporate minerals. *Journal of Geophysical Research* 96: 16231–16240.

Csillag, F., Pásztor, L., and Biehl, L.L. 1993. Spectral selection for characterization of salinity status of soils. *Remote Sensing of Environment* 43: 231–242.

Davis, P.A., Breed, C.S., McCauley, J.F., and Schaber, G.G. 1993. Surficial geology of the Safsaf region, south-central Egypt, derived from remote-sensing and field data. *Remote Sensing of Environment* 46: 183–203.

Dehaan, R.L. and Taylor, G.R. 2002. Field-derived spectra of salinized soils and vegetation as indicators of irrigation-induced soil salinization. *Remote Sensing of Environment* 80: 406–417.

Dehaan, R. and Taylor, G.R. 2003. Image-derived spectral endmembers as indicators of salinization. *International Journal of Remote Sensing* 24: 775–794.

Drake, N.A. 1995. Reflectance spectra of evaporite minerals (400–2500 nm): Applications of remote sensing. *International Journal of Remote Sensing* 16: 2555–2571.

Evans, G., Schmidt, V., Bush, P., and Nelson, H. 1969. Stratigraphy and geologic history of the sabkha, Abu Dhabi, Persian Gulf. *Sedimentology* 12: 145–159.

Farifteh, J., Farshad, A., and George, R.J. 2006. Assessing salt-affected soils using remote sensing, solute modelling, and geophysics. *Geoderma* 130: 191–206.

Farifteh, J., van der Meer, F., Atzberger, C., and Carranza, E.J.M. 2007. Quantitative analysis of salt-affected soil reflectance spectra: A comparison of two adaptive methods (PLSR and ANN). *Remote Sensing of Environment* 110: 59–78.

Fernández-Buces, N., Siebe, C., Cram, S., and Palacio, J.L. 2006. Mapping soil salinity using a combined spectral response index for bare soil and vegetation: A case study in the former lake Texcoco, Mexico. *Journal of Arid Environments* 65: 644–667.

Ghassemi, F., Jakeman, A.J., and Nix, H.A. 1995. *Salinisation of Land and Water Resources: Human Causes, Extent, Management and Case Studies*, The Australian National University, Canberra, Australia and CAB International, Wallingford, Oxon, UK.

Ghrefat, H.A., Goodell, P.C., Hubbard, B.E., Langford, R.B., and Aldouri, R.E. 2007. Modeling grain size variations of aeolian gypsum deposits at White Sands, New Mexico, using AVIRIS imagery. *Geomorphology* 88: 57–68.

Goudie, A.S. 1990. Soil salinity—causes and controls. In: *Techniques for Desert Reclamation*, A. Goudie (Ed.), John Wiley & Sons, UK, pp. 110–111.

Howari, F.M. 2003. The use of remote sensing data to extract information from agricultural land with emphasis on soil salinity. *Australian Journal of Soil Research* 41: 1243–1253.

Howari, F.M. 2006. Spectral analyses of sabkha sediments with implications for remote sensing on Mars. *International Journal of Astrobiology* 5: 47–56.

Howari, F.M., Goodell, P.C., and Miyamoto, S. 2002. Spectral properties of salt crusts formed on saline soils. *Journal of Environmental Quality* 31: 1453–1461.

Hunt, G.R. and Salisbury, J.W. 1970. Visible and near-infrared spectra of minerals and rocks. I. Silicate minerals. *Modern Geology* 1: 283–300.

Hunt, G.R., Salisbury, J.W., and Lenhoff, C.J. 1971a. Visible and near-infrared spectra of minerals and rocks. IV. Sulphides and sulphates. *Modern Geology* 3: 1–4.

Hunt, G.R., Salisbury, J.W., and Lenhoff, C.J. 1971b. Visible and near-infrared spectra of minerals and rocks. III. Oxides and hydroxides. *Modern Geology* 2: 195–205.

Khan, N.K., Rastoskuev, V.V., Sato, Y., and Shiozawa, S. 2005. Assessment of hydrosaline land degradation by using a simple approach of remote sensing indicators. In: *Proceedings Interregional Conference on Environment and Water No. 6 (2003)*, Vol. 77(1–3), Albacete, Spain, pp. 96–109.

Massoud, I.I. 1990. *Salinity and Alkalinity As Soil Degradation Hazards*. FAO, Rome.

McKee D., Cunningham, A., and Dudek, A. 2007. Optical water type discrimination and tuning remote sensing band-ratio algorithms: Application to retrieval of chlorophyll and Kd (490) in the Irish and Celtic Seas Estuarine. *Coastal and Shelf Science* 73: 827–834.

Metternicht, G.I. 1996. Detecting and monitoring land degradation features and processes in the Cochabamba Valleys, Bolivia: A synergistic approach. ITC dissertation 36, International Institute for Aerospace Survey and Earth Sciences, Enschede, the Netherlands.

Metternicht, G. and Zinck, J.A. 1997. Spatial discrimination of salt- and sodium-affected soil surfaces. *International Journal of Remote Sensing* 18: 2571–2586.

Metternicht, G.I. and Zinck, J.A. 2003. Remote sensing of soil salinity: Potentials and constraints. *Remote Sensing of Environment* 85: 1–20.

Meybeck, M., Chapman, D.V., and Helmer, R. 1989. *Global Environment Monitoring System: Global Freshwater Quality; A First Assessment*. Basil Blackwell, Oxford, UK.

Mougenot, B., Epema, G.F., and Pouget, M. 1993. Remote sensing of salt-affected soils. *Remote Sensing Reviews* 7: 241–259.

Mulders, M.A. 1987. *Remote Sensing in Soil Science*. Developments in Soil Science, Elsevier, Amsterdam, the Netherlands.

Nield, S.J., Boettinger, J.L., and Ramsey, R.D. 2007. Digitally mapping gypsic and natric soil areas using Landsat ETM data. *Soil Science Society of America Journal* 71: 245–252.

Paine, D.P. and Kiser, J.D. 2003. *Aerial Photography and Image Interpretation*, 2nd ed., Wiley's Higher Education.

Pease, P.P., Bierly, G.D., Tchakerian, V.P., and Tindale, N.W. 1999. Mineralogical characterization and transport pathways of dune sand using Landsat TM data, Wahiba Sand Sea, Sultanate of Oman. *Geomorphology* 29: 235–249.

Peña, S.A. and Abdelsalam, M.G. 2006. Orbital remote sensing for geological mapping in southern Tunisia: Implication for oil and gas exploration. *Journal of African Earth Sciences* 44: 203–219.

Rao, B.R.M., Dwivedi, R.S., Sreenivas, K., Khan, Q.I., Ramana, K.V., and Thammappa, S.S. 1998. An inventory of salt-affected soils and waterlogged areas in the Nagarjunsagar Canal Command Area of Southern India, using space-borne multispectral data. *Land Degradation and Development* 9: 357–367.

Ray, T.W. and Murray, B.C. 1996. Nonlinear spectral mixing in desert vegetation. *Remote Sensing of Environment* 55: 59–64.

Sharma, R.C. and Bhargava, G.P. 1988. Landsat imagery for mapping saline soils and wet lands in North-west India. *International Journal of Remote Sensing* 9: 39–44.

Sumner, M.E. 2000. *Handbook of Soil Science*, CRC Press, Boca Raton, FL.

Sumner, M.E. and Naidu, R. 1998. *Sodic Soils: Distribution, Properties, Management, and Environmental Consequences*, Oxford University Press, New York.

Szabolcs, I. 1979. Introduction. In: *Review of Research on Salt-Affected Soils*, UNESCO, Paris.

Tanji, K.K. 1990. The nature and extent of agricultural salinity problems. In: *ASCE Manuals and Reports on Engineering Practice*, 71, American Society of Civil Engineering, New York, pp. 1–17.

Tanji, K.K. 1996. *Agricultural Salinity Assessment and Management*, American Society of Civil Engineering, New York.

Tóth, T., Csillag, F., Biehl, L.L., and Michéli, E. 1991. Characterization of semivegetated salt-affected soils by means of field remote sensing. *Remote Sensing of Environment* 37: 167–180.

Verma, K.S., Saxena, R.K., Barthwal, A.K., and Deshmukh, S.N. 1994. Remote sensing technique for mapping salt-affected soils. *International Journal of Remote Sensing* 15: 1901–1914.

Vincent, R.K. 1997. *Fundamentals of Geological and Environmental Remote Sensing*. Prentice-Hall, Upper Saddle River, NJ. Available at http://csbe.org/water_conserving_landscapes/publications/saline_soils/index.htm (Center for the Study of Built Environment, 2007).

9 Assessment of Salt-Affected Soils Using Multisensor Radar Data: A Case Study from Northeastern Patagonia (Argentina)

Héctor F. del Valle, Paula D. Blanco, Walter Sione, César M. Rostagno, and Néstor O. Elissalde

CONTENTS

9.1 INTRODUCTION

Latin America has 12 million ha of irrigated land. About 1.7 million ha are in Argentina, of which 600,000 ha are salt-affected (Siebert et al., 2006). In the lower Chubut valley (LCV), northeastern Patagonia, primary and secondary salinization affects 20,717 ha, representing 55% of the valley (Laya, 1981) (Table 9.1). Irrigation in the LCV began in 1865 when a group of Welsh colonists first settled in the valley. This early settlement triggered the colonization of the whole of Patagonia in the following decades. In fact, without the pioneering irrigation work in the LCV, the history of the southern edge of South America would have been a very different one.

TABLE 9.1

Salinity and Sodicity Classes in the LCV

Salinity and Sodicity Classes	Soil Texture	Vegetation Indicators	Surface (ha)	Percent
Nonaffected ECe: 0–4 dS m^{-1} SAR: 0–6	Sandy loam	Tilled field; no evidence of salt effect	10,448	25.3
Slightly affected ECe: 4–8 dS m^{-1} SAR: 6–10	Silt loam Clay	Tilled field; slight evidence of salt effect	8,158	19.7
Moderately affected	Silt loam	Tilled field; subsoil saline or sodic or both; halophytic grass species (*D. spicata*, *C. aphilla*) usually present	2,276	5.5
ECe: <4; 8–16 dS m^{-1} SAR: 10–15	Clay loam Clay			
Strongly affected	Clay	Halophytic shrub and grass species common in bare areas (*S. divaricata*, *A. lampa*, *S. ambigua*, *D. spicata*, *C. aphilla*)	11,334	27.4
ECe: <4; 16–32 dS m^{-1} SAR: 15–30	Silt loam			
Very strongly affected	Silt loam	Only halophytic shrubs (*S. divaricata*, *A. lampa*, *S. ambigua*); bare salt-encrusted surface	9,107	22.1
ECe: <4; >32 dS m^{-1} SAR: >30	Sandy clay loam Clay			
			41,323	100.0

The Chubut river starts in the Andes, more than 500 km westward of the study area, and then flows southeast and finally joins the Atlantic Ocean. Along its way from the mountain to the sea, the river contributes to feeding the groundwater. The LCV has a total surface area of 41,323 ha and is approximately 90 km long and 4–10 km wide. It is the southernmost irrigated valley of Argentina, where forage, vegetables, fruits, and flowers are cultivated under intensive land management. The LCV irrigation system is over 100 years old and has become largely inefficient because of seepage from channels and water loss from farmland. Furrow and flood irrigation are the most commonly used irrigation methods. Water cost is relatively low, basically because there is more water available than can be used in the irrigation schemes. Better irrigation practices can minimize the salinity and sodicity problems but require detecting high water tables at an early stage. The research described in this chapter was carried out in the irrigated portion of the LCV (Figure 9.1).

Salinity and sodicity adversely affect the biological soil functions in the LCV and cause degradation of soil and water resources. Several factors contribute to salt concentration in the soil, including insufficient drainage network, poor maintenance of collectors and main drains, high water table, soil compaction, natural cycles of seawater intrusion through the lower Chubut river, manure and fertilizer application, and soil parent material. Scabland, saltland, white alkali, black alkali, and sour ground are local terms commonly used to describe salt-affected areas (Reichart, 1961).

In 1981, a conventional soil survey was carried out in the LCV using aerial photographs (scale 1:10,000) and intensive fieldwork (Laya, 1981). Natural and human-induced salinity and sodicity were recognized as being the main soil degradation forms. Dryland salinity results from changes in

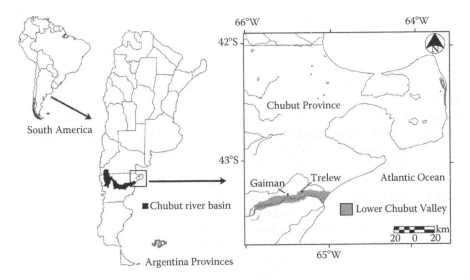

FIGURE 9.1 Geographic location of the LCV.

the water balance of soilscapes following the removal of native vegetation and the introduction of agricultural practices. The main causes of irrigation salinity include inefficient water use and poor drainage. This diagnosis with conventional techniques was time-consuming and laborious; it is much easier now using remote sensing. A major challenge of remote sensing, as a potential alternative technique, is to detect different levels of soil salinity and sodicity (Taylor et al., 1994; Bannari et al., 2007).

The use of radar images for the detection of salt-affected soils is a promising domain of remote sensing (Taylor et al., 1996; Metternicht, 1998; Aly et al., 2004). The basic principle rests on the relationship between the amount of salt in the soil, the soil moisture content, and the dielectric properties of this mixture. Advances in the field of L-band microwave sensing have stimulated the development of new techniques with predictive mapping capabilities (Sreenivas et al., 1995). New satellite-borne sensors using the L-band are expected to play a major role in enhancing the capability of orbital sensors to assess salt-affected soils. Examples are the current PALSAR (Phased Array Type L-band Synthetic Aperture Radar) on board the ALOS (Advanced Land Observing Satellite) and the forthcoming SAOCOM (Satellites for Observation and Communications) to be developed by Argentina in partnership with Belgium and Italy (CONAE, 2007).

Given the context of salinity management and risk assessment in the LCV, the general objectives of this study were (1) to evaluate the usefulness of radar-derived parameters for detecting and mapping salt-affected soils within the irrigated portion of the valley, and (2) to classify main landuses/landcovers using an object-oriented image classification approach. The specific objectives were to

1. Analyze the contribution of the C- and L-bands for accurate mapping of salt-affected areas, in relation to the SAR system parameters (mainly wavelength, number of looks and polarization mode) and the environmental setting of the study area.
2. Determine an optimal combination of wavelength and polarization for generating color composite images that can be used in visual interpretation of salt-degraded land.
3. Test and implement object-oriented image analysis algorithms for the recognition and classification of salt-affected soils from radar-derived textural measures.
4. Investigate the scattering properties of different salt-affected soil classes.

9.2   MATERIALS AND METHODS

9.2.1   Location and Environmental Characteristics of the Test Area

The study area covering approximately 11,700 ha (Figure 9.2) is located in the central part of the LCV, between 43°16′S–43°21′S and 65°18′W–65°31′W, in the Chubut province (Patagonia, Argentina). The climate is arid, with high evaporation enhanced by windy conditions (Coronato, 1994). Annual rainfall rarely exceeds 200 mm, with a long-term average of 233 mm (1961–2006). Rains occur mostly during the cool season (April–October), with May and October being the rainiest months. Extreme temperatures range from 40°C to −12°C, with an annual average of 13°C. The relative air humidity varies from 30% in summer to 70% in winter. The frost-free period is 100–140 days, while frost can occur from March to November. Average annual evaporation is 722 mm, more than three times the annual rainfall. Wind speed is highest in spring (September–November) and summer (December–February), causing frequent dust storms.

The LCV is part of the Chubut Basin depression, filled with Quaternary and Tertiary (Miocene) marine sediments. Surface-water inflow is practically the only source of water to the valley and supplies almost all the water used for irrigation, given that groundwater resources are often brackish. In addition, surface water supplies most of the recharge to the aquifers of the valley by infiltration of excess irrigation water and canal leakage. Recharge by seepage from the streams that cross the permeable alluvial fans and from infiltration of rainwater is less significant. There are unconfined and confined aquifers. Evapotranspiration, seepage from drainage canals and pumping wells are the main factors of discharge from the unconfined aquifers. Discharge from the confined aquifers is due mainly to wells, some springs, and groundwater rise through clayey sediments into the unconfined aquifers.

The vegetation belongs to the southern sector of the Monte Desert (desert shrub) biome (León et al., 1998). In the study area, the most widespread community is a shrub steppe of *Suaeda divaricata*, *Atriplex lampa*, and *Salicornia ambigua*. Other frequent species are the shrubs *Chuquiraga avellanedae*, *Prosopis alpataco*, and *Larrea divaricata*, and grasses such as *Distichlis spicata* and *Cassia aphilla*. Introduced trees such as *Populus*, *Salix*, *Eucalyptus*, and *Tamarix* are used as windbreaks. Cultivated vegetation includes fruit-tree plantations, annual crops, and mixed perennial-annual pastures.

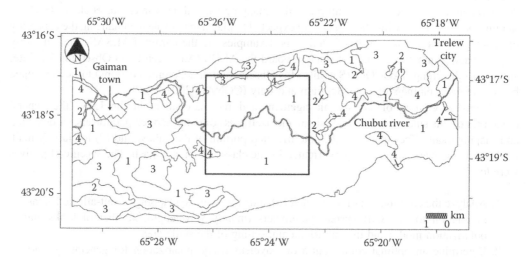

FIGURE 9.2   Saline and sodic soils in the study area (Laya, 1981). (1) Nonsaline to slightly saline–sodic soils; (2) Moderately saline and slightly sodic soils; (3) Moderately to strongly saline and strongly sodic soils; and (4) Strongly to very strongly saline and very strongly sodic soils. Box area indicates the location of the test site where a landuse/landcover classification using an object-oriented approach was undertaken.

Soils vary widely in physical and chemical properties, especially the salt-affected ones (Laya, 1981). In general, soils have clayey texture, low to moderate infiltration rate, and slow hydraulic conductivity particularly in the subsoil. The amount of salts varies with depth. Some soils have high salt content near the surface entailing the formation of whitish salt crusts and glazed polyhedral cracks. Locally, the terrain surface consists of loose, powdery, fine-crumby, or granular soil material with salt crystals. Some soils have nonsaline topsoils, without salt crystals, but have subsurface horizons at less than 1 m depth with electrical conductivity (ECe) higher than 4 dS m^{-1}. In some cases, the soils are saline and sodic from the surface downwards, while in other cases salinity and sodicity start only in the subsoil. The main soil taxa (Laya, 1981) are Fluventic Aquicambids, Typic Haplosalids, Typic Aquisalids, Typic Natrargids, Vertic and Typic Torrifluvents, Vertic and Typic Torriorthents, and Typic Salitorrerts (USDA, 2003).

9.2.2 DATABASE AND DIGITAL IMAGE PROCESSING

A detailed flowchart of the methodology is shown in Figure 9.3, and the main steps are described hereafter.

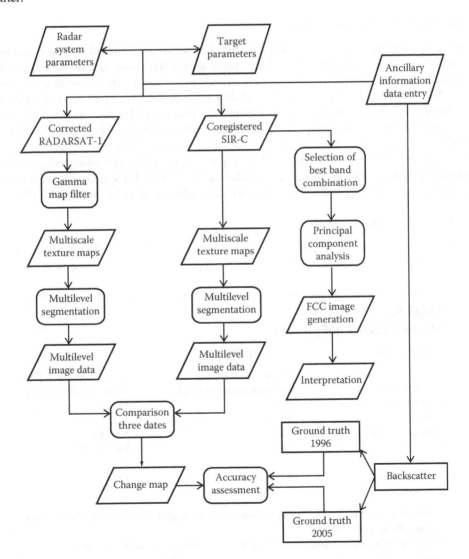

FIGURE 9.3 Flowchart of the methodology.

9.2.2.1 SAR Image Selection

The irrigation system of the study area operates from September to May, according to crop requirements, but is more intensively used during the summer (December–February) when evaporation is highest and soil salinity and sodicity increase largely. The irrigation water requirement decreases in autumn (March–May). Waterlogging is severe at the beginning of the crop-growing season (October), but usually tapers off and reabsorbs at the time of crop harvest. However, in many low-lying areas, waterlogging continues throughout the harvest season.

Considering all these aspects and the need to minimize the effect that soil moisture can pose on the backscattering behavior of salt-related surface features, we considered that January (midsummer) and April (mid-autumn), at the end of the crop-growing season, were the best months for the selection of radar imagery aimed at the detection of secondary salinization. Irrigation events are better detected in October (spring).

Table 9.2 shows the main characteristics of the SAR imagery analyzed in this study. At the time of the SAR image acquisition, no significant rainfall had been recorded during the previous month, and relative air humidity was typically low.

9.2.2.2 SAR Image Preprocessing

To ensure accurate data overlay at pixel level, the metallic roofs of the farmland buildings were used as reference points for geometric adjustment. They appear as very bright objects in the SAR image due to the high returns caused by the corner reflector effect, and can be identified by their rectangular or square shapes. Additionally, geometric differences and distortions in data processing were recognized and calculated. The images were rectified and transformed to a Transverse Mercator cartographic projection, applying a cubic convolution interpolation technique, while retaining pixel sizes of 8 and 25 m for RADARSAT-1 and SIR-C, respectively. The root-mean-square (RMS) transformation errors were about 4 and 10 m (RADARSAT-1 and SIR-C, respectively). Finally, all backscattering coefficients (σ^0) were expressed in decibels (dB). The characteristic RADARSAT-1 speckle was reduced using a Gamma-MAP filter, with a 3×3 window size and one iteration, as done by Metternicht (1993).

9.2.2.3 Digital Elevation Model

Digital elevation models (DEM) provide strong prior evidence of parts of the terrain that are likely to be or become saline (Caccetta, 1999). A DEM developed at the remote sensing data center (DFD)

TABLE 9.2

SAR Instrument Configuration

Characteristics	RADARSAT-1	SIR-C
Date	January 18, 2005	April 13/October 5, 1994
Orbit direction	Descending (west looking)	Ascending (east looking)
Incidence angle (degree)	44.5	49.5/55.8
SAR band	C	C, L
Wavelength (cm)	5.6	5.8, 23.9
Polarization[a]	HH	HH, HV, VV, VH
Number of looks	1	4
Pixel spacing (m)	6.25	12.5
Resolution (m)	8	25[b]

[a] Backscatter coefficients corresponding to these regimes are $\sigma^0_{HH}(\theta)$, $\sigma^0_{VV}(\theta)$, and $\sigma^0_{HV}(\theta)$. It is worth noting that $\sigma^0_{HV}(\theta) = \sigma^0_{VH}(\theta)$.

[b] The multilook data were processed to 8 m pixel size to match the RADARSAT-1 data.

TABLE 9.3

Correlation Coefficients between the SIR-C Bands

	C-HH	C-HV	C-VV	L-HH	L-HV	L-VV
C-HH	1	0.97 (0.87)	0.93 (0.99)	0.89 (0.87)	0.87 (0.84)	0.86 (0.88)
C-HV		1	0.86 (0.91)	0.87 (0.84)	0.86 (0.85)	0.84 (0.85)
C-VV			1	0.87 (0.88)	0.84 (0.86)	0.87 (0.88)
L-HH				1	0.98 (0.89)	0.95 (0.99)
L-HV					1	0.89 (0.91)
L-VV						1

Note: Values without brackets are for the April SIR-C image and values in brackets are for the October SIR-C image.

of the German Aerospace Center (DLR) was used in this study (del Valle et al., 2002). The DEM was generated using ascending and descending tandem images from the ERS-1/2 SAR satellites, with different interferometric baselines. Tandem data were selected based on usual criteria related to the coherence and baseline characteristics (Coulson, 1995). The horizontal DEM accuracy is 25 m, sampled at 1 m along the elevation axis.

9.2.2.4 Selection of Best SIR-C Band Combination

The SIR-C images of April and October 1994, including two different wavelengths (C- and L-bands) and three polarization modes (HH, VV, and VH), were highly correlated (Table 9.3). This suggests that the intrinsic dimensionality of the data is small, certainly smaller than the number of physical variables that potentially affect the signal (Weeks et al., 1997). Thus, given the high correlation (>0.9) between the SIR-C bands for the two acquisition dates, a principal component analysis (PCA) was applied to remove redundant information (Drury, 1993). The analysis revealed that the first two components explained 72.6% and 14.2% of the April SIR-C image variability. For the October SIR-C image, the first two principal components accounted for 79.1% and 10.3% of the variability. For the two dates analyzed, the C-HV, L-HH, and L-HV bands of the SIR-C images showed the highest contribution to the first two principal components (Table 9.4). Thus, they were selected as the raw bands to be included in the production of false-color composites (FCC) showing optimal combination of wavelength and polarization modes.

9.2.2.5 Extraction of Derived Textural Measures

The extraction of textural measures in this research used only the C-HH bands from RADARSAT-1 and SIR-C. The HH polarization, modified by vegetation density and radar frequency, tends in general to interact to a lesser extent with crop canopy and penetrates more effectively to the underlying soil (Taylor et al., 1996). The selection of textural features was based on their capability to discriminate eight primary landuse/landcover classes, namely flood irrigation, waterlogged areas,

TABLE 9.4

Eigenvectors of the PCA of the SIR-C Images

	C-HH	C-HV	C-VV	L-HH	L-HV	L-VV
PC1	0.31 (0.40)	−0.54 (0.52)	−0.36 (0.03)	0.29 (0.06)	−0.63 (−0.71)	0.06 (0.23)
PC2	0.26 (0.34)	−0.34 (0.52)	0.34 (−0.34)	0.82 (0.42)	0.15 (0.55)	0.02 (−0.15)

Note: Values without brackets are for the April SIR-C image and values in brackets are for the October SIR-C image.

infrastructure features (canal network, roads), cropland 1 (without surface symptoms of salt-affected soils), croplands 2 and 3 (with surface symptoms of salt-affected soils), shrub steppe with bare ground and bare ground with shrubs and grass meadow. As the fruit-tree and windbreak target has a complex structure in the study area, the incoming waves have multiple interactions within this feature. Large differences in the classification of irrigated fruit trees occur. The spacing between fruit trees and others crops affects the total spectral behavior for both radar systems. Mean Euclidean distance, variance and skewness were the image parameters (Erdas Inc., 2006) considered most useful for texture analysis. The texture-analyzed images were subsequently used as input to the classification routine described hereafter.

9.2.2.6　Object-Oriented Approach

Several segmentation-based schemes have been used in hierarchical classifications of polarimetric data for land applications (Dobson et al., 1996; Benz and Pottier, 2001; Lombardo and Oliver, 2002). We used an object-oriented multiscale image analysis method embedded in the software eCognition (Definiens, 2003). Spectral, shape and statistical characteristics, as well as relationships between linked levels of the image objects, can be used in the rule base to combine objects into meaningful classes (Benz et al., 2004). The segmentation implemented in this research used as input only the textural measures obtained in Section 9.2.2.5. The object-oriented approach considers three parameters for image segmentation, namely (a) scale, (b) color (gray levels)/shape ratio, and (c) form/spatial properties, as related to a smoothness/compactness ratio.

In this research, scale parameters of 20 and 200 were selected because this segmentation fitted best the information class extraction in the SIR-C and RADARSAT-1 data, respectively. Color and shape parameters can be weighted from 0 to 1. Here, after several iterations, the segmentation weights for color and shape were established to a ratio of 0.8:0.2 for the relative importance of reflectance versus shape, and 0.1:0.9 for compactness versus smoothness.

Accuracy was assessed by means of an error matrix based on stratified and randomly selected sites across the study area (Congalton and Green, 1999). Ground truth was obtained from field campaigns carried out in 1996 (del Valle et al., 1997) and 2005. Sampling sites were located using a GPS with 4 m accuracy. Several sampling points were revisited and photographed to document the soil surface characteristics. Salt-affected soils observed during the field campaign in 2005 were related to the surface characteristics expected to determine the backscatter ($\sigma°$). Each sampled site was described in terms of landuse, vegetation and soil morphology (USDA, 2003). Soil samples were collected within a perimeter of 25 m from each central observation point. The field and laboratory data set consisted of 200 composite topsoil samples of 5 cm depth. Laboratory determinations included particle size distribution, pH, electrical conductivity of the saturation extract, and sodium adsorption ratio.

A set of 188 specific sampling sites with a variety of salt-affected soil features was selected in the SAR imagery to allow direct comparison of the radar reflectivity. Polygons of 100–140 pixels were delineated within the images for each sampling site, and their backscattering characteristics were extracted.

9.2.2.7　Analysis of the SAR Backscatter

Variables related to the radar design and the environmental setting were analyzed to better understand the behavior of the radar signal backscattering. Four factors appeared to be significant when determining the variations in the backscattering coefficient for the environmental setting: soil type (soil texture), tillage conditions or soil surface aspect, soil moisture or surface water, and salt-affected soils. The backscattering intensity changes with these parameters and produces a range of image brightness variations, expressed as changes in the pixel gray levels of the images analyzed.

From visual interpretation of the processed images, it appears that surface roughness depends on the tillage conditions and the presence of direct indicators of soil salinity at the soil surface. The qualitative interpretation applied in this research did not require detailed roughness measurements, but knowledge about the landscape and land management variables that most likely influence changes in image brightness. Figure 9.4 shows some examples of surface roughness, while Table 9.5 describes the main saline and sodic classes in terms of textural composition and surface roughness, following an example presented in Metternicht (1998) for the detection of salt-affected areas using JERS-1 L-band imagery.

9.3 RESULTS AND DISCUSSION

9.3.1 MULTIFREQUENCY AND MULTIPOLARIZATION APPROACH

Farm paddocks and flood irrigation/waterlogged areas could be clearly delineated on the SIR-C composite images (Figures 9.5 and 9.6). The SIR-C images have lower spatial resolution than the RADARSAT-1 image, but they have also reduced speckle effect because of the larger number of image looks (e.g., four looks for SIR-C vs. one look of RADARSAT-1 imagery). Usually, multilook images exhibit lower image speckle (Xiao et al., 2003). However, for an improvement in radiometric resolution using multiple looks, there is an associated degradation in spatial resolution (Jordan et al., 1995).

Figure 9.5 shows a FCC of the SIR-C (C-HV, L-HH and L-HV) acquired on April 13, 1994, almost at the end of the irrigation period (postharvest). Tilled fields with different levels of soil moisture and vegetation types can be clearly distinguished. According to field observations, this is mostly due to changes in soil surface morphology rather than changes in the vegetation cover. The sensitivity to variations in soil moisture and vegetation cover appears to be higher for the L-HV (blue) band than for the C-HV (red) band. Wet low-lying areas and ditches tend to be bright (white and light blue tones), providing additional information in terms of surface roughness and dielectric characteristics (Mohamed et al., 2003; Aly et al., 2004).

The L-band reveals more detail of the braided drainage and irrigation network, whereas only a few of the larger ditches are apparent in the C-band. Most of the paddocks present low brightness values when the HV polarization mode is used with either the C-band or the L-band. According to Taylor et al. (1996) and Shao et al. (2003), this is indicative of the reduced volume scattering that occurs at these wavelength and polarization mode.

Areas with high soil water content and ditches exhibit strong backscattering because of the high dielectric constant of water in the microwave range (Figure 9.6). Areas recently irrigated can be detected because of the high backscatter exhibited in the imagery. This relates to an external moist crop canopy, suggesting that SAR can be used for real-time irrigation monitoring. Soil and canopy moisture also revealed the difference in water content between irrigated and nonirrigated fields. The presence of water-saturated areas leads to increased double-bounce scattering that enhances the strength of the ground-vegetation interaction term, as found by Dobson et al. (1995). This feature has been exploited for evaluating the water content of the upper soil layer by microwave remote sensing (Ulaby and Siqueira, 1995). The backscatter from nonsaturated sites was low. Conversely, paddocks with inundated crops showed high brightness values, whereas sites with the water table at the surface presented intermediate backscattering.

In the C-HV band (assigned to the red channel in Figures 9.5 and 9.6), farm paddocks show a wide range of bright signatures. The green and blue colors of the L-band (HH and HV) reflect a variety of distinctive signatures, which depend on target characteristics such as surface roughness, scattering mechanisms and dielectric properties, as reported by Taylor et al. (1996).

Irrigated land, as well as canals, drains and roads, show high brightness values in the L-HH assigned to the green channel of the FCC. In contrast, salt-affected areas have a distinctive red and blue mottle due to relatively high brightness values in the C-HV and L-HV bands assigned to the red and blue channels of the FCC, respectively. High backscattering values shown in the L-band are thought to be due to increased soil moisture content close to the surface and to increased vegetation

FIGURE 9.4 Examples of surface roughness in the study area. (a) Scattering from an agricultural ground target; soil/canopy (windbreaks) interactions (bright targets); no evidence of salt effect; (b) forage crops behaving as smooth surfaces in the SAR imagery; nonsaline to slightly saline soils; (c) tall halophytic shrub steppe with bare soil (dryland salinity); rough surfaces; very strongly saline–sodic soils; (d) irrigation canal bordered by trees (*Salix, Populus*) and crops; both direct canopy and canopy–water interactions are observed in the images; (e) cracked soil surface with salt efflorescence; strongly saline and strongly sodic soils; (f) tillage row direction (crop planting); variable soil roughness; slightly saline soils; (g) flood irrigation; saline soils; and (h) puffy soil surfaces and loosely structured crusts; variable soil roughness.

TABLE 9.5

Main Characteristics of the Salt-Affected Soil Classes in the Study Area

Salt-Affected Soil Classes	N	Mean ± Standard Deviation			Dominant Soil Texture	DEM m a.s.l.				Characteristics
		pH	ECe dS m^{-1}	SAR		Mean	Standard Deviation	Maximum	Minimum	
Nonaffected	41	7.1±0.5	1.2±0.9	3.9±1.7	Sandy loam	24	6	39	14	Fields tilled to 25–30 cm depth; variable soil surface roughness
Saline	57	7.5±0.3	5.4±1.2	7.9±1.5	Silt loam	21	3	27	16	Fields tilled to 25–30 cm depth, slight soil surface roughness; glassy salt crusts rich in sodium (chloride and sulfate)
Saline–sodic	52	8.4±0.4	12.7±3.1	13.6±2.7	Clay loam	21	3	27	17	Variable soil surface roughness (smooth and rough); puffy soil surfaces and loosely structured crusts
Sodic	38	9.4±0.5	3.1±0.3	30.8±7.4	Sandy clay loam	21	2	26	17	Rough soil surfaces; puffy soil surfaces (chloride and sulfate)
	188									

Note: According to the Rayleigh criterion, a surface is considered smooth if the RMS height of the microrelief is less than one-eighth of the radar wavelength divided by the cosine of the incidence angle: $h < \lambda/8 \cos \theta$; where h is the RMS height, λ is the wavelength, and θ is the incidence angle. Because this criterion does not consider an intermediate category of surfaces between definitely smooth and definitely rough, it was modified by Peake and Oliver (1971) to include factors that define the upper and lower values of RMS surface smoothness or roughness. The modified Rayleigh criterion considers a surface to be smooth where: $h < \lambda/25 \cos \theta$, and rough where: $h > \lambda/4.4 \cos \theta$.

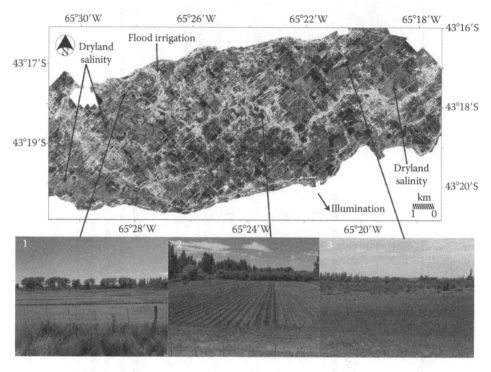

FIGURE 9.5 Multipolarization FCC of the SIR-C image acquired on April 13, 1994, at C-HV (red), L-HH (green), and L-HV (blue). Orbit direction: ascending (east looking) pass. (1) Forage crops; (2) vegetable crops; and (3) wetter, low-lying areas which tend to coincide with abandoned river channels.

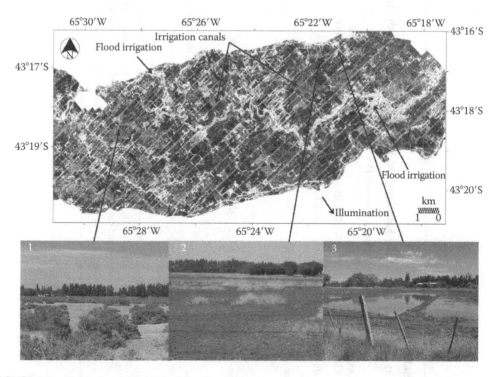

FIGURE 9.6 Multipolarization FCC of the SIR-C image acquired on October 5, 1994, at C-HV (red), L-HH (green), and L-HV (blue). Orbit direction: ascending (east looking) pass. (1) Dryland salinity; (2) saline–sodic soil; and (3) flood irrigation.

cover along the irrigation ditches. Both features entail increased volume scattering and a higher cross-polarized return (Paine, 2003).

9.3.2 RADAR DATA CLASSIFICATION

Good discrimination of landuse/landcover classes affected by salinity (e.g., croplands 2 and 3) is shown in Figure 9.7, corresponding to the object-oriented classification conducted over a portion of

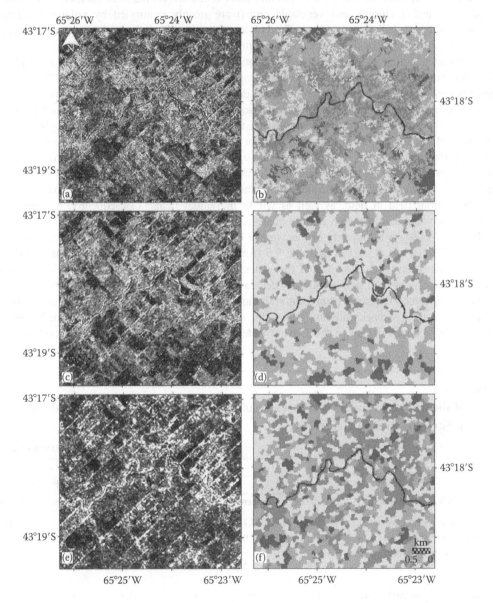

FIGURE 9.7 Subsets of images. (a) From RADARSAT-1 (January 2005), (c) and (e) from SIR-C (April and October 1994, respectively), and their corresponding object-oriented classifications (b), (d), and (f). Cyan: flood irrigation/waterlogged areas. Yellow: croplands 1 and 2 (forage, vegetables and fruit trees); nonsaline to slightly saline–sodic soils. Orange: cropland 2 (forage and vegetable crops); moderately saline and slightly sodic soils. Red: cropland 3 (forage crops)/halophytic shrubs with bare ground; strongly saline and very strongly sodic soils.

the study area. Both radar sensors (RADARSAT-1 and SIR-C) are sensitive to soil moisture and surface roughness, but with different identification capabilities.

Large backscatter values in the radar images result in white pixels (flood irrigation/waterlogged areas), whilst areas of low return are black (cropland interspersed with remnant native vegetation, mainly shrubs of *S. divaricata*). Different halophytic species are indicators of varying levels of soil salinity (Laya, 1981). Rows of trees (windbreaks) and infrastructure features have also high roughness and strong backscatter. Large differences in the classification of irrigated fruit trees also occur. The spacing between fruit trees and other crops affects the total backscatter behavior. Croplands 1 and 2 (forage and vegetable crops) are mainly controlled by volume scattering providing uniform medium gray tones. The volume scattering depends on the size of the canopy structural elements relative to the radar wavelength (Ferrazzoli and Guerriero, 1995).

Backscatter is high in the SAR images of January and October (Figure 9.7a and e) in places where full canopy cover dominates the ground target and moisture levels are kept high through the application of irrigation water. After crop harvest in April (Figure 9.7c), backscatter values drop due to the removal of vegetation and the lowering of soil moisture. The backscattering characteristics of soils appear to be a function of particle size distribution (clay content), surface roughness, soil moisture and salt content, as mentioned by Taylor et al. (1996). The extent of the classes resulting from the different input maps is reported in Table 9.6. In general, the maps resulting from the object-oriented classification using textural data show high overall accuracy and kappa statistic. Comparative observations between 1994 and 2005 denote that salt-affected soil targets vary spatially with differences from field to field, as well as within individual fields. Consequently, mapping and monitoring dryland and irrigation salinity presents an enormous challenge.

Table 9.7 presents a summary of the average backscatter (σ^0) of the image characteristics per salt-affected soil class described in Table 9.5 (for the whole study area). Salt content in the soil increases from class 1 to class 4, and the general backscatter trend is to decrease in the same direction. Thus, higher soil salinity/sodicity decreases soil backscattering. This observation is useful as it helps segregate saline soils from sodic soils, and that is most important from a reclamation point of view. The average backscattering values for all salt-affected soil classes were higher in the L-band (low frequency) than in the C-band (intermediate frequency) of SIR-C, when the same

TABLE 9.6

Extent of the Landuse/Landcover Classes Resulting from the Thematic Classification of a Test Site (Box Area in Figure 9.2)

	SIR-C				RADARSAT-1	
	1994				2005	
	April		October		January	
Landuse/Landcover Classes	Hectares	Percent	Hectares	Percent	Hectares	Percent
Flood irrigation/waterlogged areas	111	4.7	683	29.2	482	20.6
Croplands 1 and 2; nonsaline to slightly saline–sodic soils	1287	55.0	834	35.6	418	17.9
Cropland 2; moderately saline and slightly sodic soils	800	34.2	686	29.3	1303	55.6
Cropland 3/halophytic shrubs/bare ground; strongly saline and very strongly sodic soils	142	6.1	137	5.9	137	5.9
	2340	100.0	2340	100.0	2340	100.0
Overall accuracy (%)	80.2		79.4		81.4	
Kappa statistics (KHAT, %)	79.1		78.3		79.8	

TABLE 9.7

Average Backscattering (σ⁰) and Standard Deviation for the Salt-Affected Soil Classes Described in Table 9.5

Salt-Affected Soil Classes	N	SIR-C April 1994						SIR-C October 1994						RADARSAT-1 January 2005
		C-HH	C-HV	C-VV	L-HH	L-HV	L-VV	C-HH	C-HV	C-VV	L-HH	L-HV	L-VV	C-HH
Nonaffected	41	−17.7	−20.3	−17.0	−16.4	−19.0	−16.9	−15.9	−19.6	−16.6	−15.4	−18.7	−15.9	−15.6
		6.0	6.6	5.8	6.2	6.0	6.1	5.8	8.1	5.2	9.1	5.4	8.8	5.3
Saline	57	−20.0	−24.9	−18.8	−19.9	−22.5	−17.4	−19.5	−25.2	−20.1	−18.1	−19.8	−18.6	−16.2
		4.8	4.3	5.4	5.1	4.3	5.4	5.7	4.3	5.5	4.9	4.2	4.1	5.4
Saline–sodic	52	−25.0	−25.0	−19.8	−22.5	−24.4	−18.7	−20.2	−26.0	−20.8	−20.1	−21.6	−20.5	−17.4
		4.7	5.3	6.1	4.9	4.8	6.4	5.2	3.5	4.8	4.3	3.8	3.7	5.2
Sodic	38	−24.5	−26.4	−20.9	−24.2	−24.3	−20.4	−22.5	−26.3	−22.9	−21.5	−23.8	−22.0	−17.7
		5.4	4.0	5.4	3.6	5.0	4.5	4.5	4.2	3.9	6.2	3.7	5.8	5.1
	188													

polarization modes were compared (Taylor et al., 1996). The low-frequency band has a high penetration depth, thus allowing to detect both salinity and soil moisture (Shao et al., 2003). Soil moisture changes strongly influence C-band data, especially for high values and in the case of sparse vegetation (Notarnicola et al., 2006).

The relative backscatter importance per frequency, polarization and soil class is as follows:

1. SIR-C of April 1994, C- and L-bands
   (a) VV > HH > HV (saline, saline–sodic and sodic soils)
   (b) HH > VV > HV (nonaffected soils)
2. SIR-C of October 1994, C- and L-bands
   (a) HH > VV > HV (all soil classes)
3. RADARSAT-1 of January 2005, C-HH band
   (a) RADARSAT-1 > SIR-C October > SIR-C April (all soil classes)

9.3.3 RADAR SYSTEM AND TARGET PARAMETERS

The radar response in terms of backscattered intensity and the interaction with salt-affected soils are dependent upon a number of system and target parameters. All parameters interact and influence each other. In this sense, both radar systems (RADARSAT-1 and SIR-C) used in this study showed different identification capabilities.

In spite of having a coarser spatial resolution, the SIR-C image quality varies markedly due to differences in pulse bandwidth, providing better results and more interpretable imagery than RADARSAT-1. The C-band and L-band from SIR-C showed to be appropriate for detecting salt-affected soils directly or indirectly. However, the L-band performed better than the C-band, especially in the information concerning surface roughness and soil moisture content or waterlogged areas. Image brightness in the L-band increased with increasing moisture content. The resulting scattering was affected by both soil surface and subsurface properties.

The ascending (east looking) and descending (west looking) orbit detects different features and different facets of the roughness elements from the ground targets (Blumberg and Freilikher, 2001). The ascending orbit of SIR-C was more useful than the descending orbit of RADARSAT-1 to analyze surface roughness and infrastructure features.

Surface roughness is one of the major factors influencing the scattering properties of the targets in the study area. On the RADARSAT-1 and SIR-C images acquired at a larger incidence angle of 44.5° and 49.5°/55.8°, respectively, the dark tones of the halophytic shrub species with bare ground contrasted strongly with the brighter returns from the surrounding irrigated croplands.

The dielectric constant is an indicator of the moisture level of a target. Soil–water content is an important parameter for agriculture in the LCV. The radar images utilized in the study area with different number of looks, geometry and polarization, were essential to interpret the soil water content and soil surface roughness.

9.4 CONCLUSION AND RESEARCH PERSPECTIVE

Dryland and irrigation salinization in the LCV is far from being a uniform process. Longer time series of observations are required to monitor salinity as a hazard to crop production. Remote sensing and salinity studies are still scarce in Argentina, but radar information is increasingly used for improved assessment of local salinity and sodicity processes.

The use of SAR imagery for mapping salt-affected soils in the LCV has great potential. Results from this research show that the two radar systems analyzed have different identification capabilities. The approach of combining two different sensors is a sound one, given that orbital sensors able to acquire multifrequency and multipolarization (like- and cross-polarization) data are not yet available. The case study illustrates some considerations on the tradeoffs and priorities in acquisition parameters for future missions such as SAOCOM, as well as the potential for interpretation of existing data.

In order of priority, our proposal for future research includes

1. Evaluation of the response to soil moisture and salinity shown in the SARAT images (Airborne Synthetic Aperture Radar project) from Argentina's National Commission on Space Activities (CONAE) on a temporal and incidence-angle basis.
2. Evaluation of the relationships between dielectric constant, soil salt content and backscattering coefficient, using theoretical and semiempirical backscattering models.
3. Characterization of the surface roughness in different seasons, taking into account soil processes such as wind erosion, surface swelling and desiccation, among others, that operate over relatively short timescales.
4. Assessing the influence of cropland and natural vegetation on the backscatter response both qualitatively, by comparing image texture with field observations, and quantitatively with field measurements of vegetation/soil parameters.

ACKNOWLEDGMENTS

We are very grateful to G. Metternicht and A. Zinck for helping to improve this manuscript. Argentina's National Commission on Space Activities (CONAE) supplied the RADARSAT-1 ASAR image, within the framework of the Airborne Synthetic Aperture Radar (SARAT) project. The Argentina-Germany Technical Cooperation provided the SIR-C images within a GTZ-Universität München-CENPAT (CONICET)-UNPSJB project. The authors acknowledge the technical assistance given by F. Coronato and A. San Martín.

REFERENCES

Aly, Z., F. Bonn, and R. Magagi. 2004. Modelling the backscattering coefficient of salt-affected soils: Application to Wadi El-Natrun bottom, Egypt. In: *European Association of Remote Sensing Laboratories (EARSel), eProceedings* 3(3): 372–381.

Bannari, A., A.M. Guedon, A. El-Harti, F.Z. Cherkaqui, A. El-Ghmari, and A. Saquaque. 2007. Slight and moderate saline and sodic soils characterization in irrigated agricultural land using multispectral remote sensing. In: *Proceedings of the Photogrammetry, Remote Sensing and Spatial Information Sciences*, Davos, Switzerland, pp. 445–450.

Benz, U. and E. Pottier. 2001. Object-based analysis of polarimetric SAR data in alpha-entropy-anisotropy decomposition using fuzzy classification by eCognition. In: *IEEE International Geoscience and Remote Sensing Symposium (IGARSS)* 3: 1427–1429.

Benz, U.C., P. Hoffmann, G. Willhauck, I. Lingenfelder, and M. Heynen. 2004. Multi-resolution, object-oriented fuzzy analysis of remote sensing data for GIS-ready information. *Journal of Photogrammetry and Remote Sensing* 58: 239–258.

Blumberg, D.G. and V. Freilikher. 2001. Soil water-content surface roughness retrieval using ERS-2 SAR data in the Negev Desert, Israel. *Journal of Arid Environments* 4: 449–464.

Caccetta, P.A. 1999. Some methods for deriving variables from digital elevation models for the purpose of analysis, partitioning of terrain and providing decision support for what-if scenarios. Australian Commonwealth Scientific and Research Organization (CSIRO), Mathematical and Information Sciences (MIS), Technical Report number CMIS 99/164, Perth, Australia.

CONAE. 2007. SAOCOM 1-A satellite. http://www.conae.gov.ar/eng/satelites/saocom.html.

Congalton, R.G. and K. Green. 1999. *Assessing the Accuracy of Remotely Sensed Data: Principles and Practice*. Lewis Publishers, New York.

Coronato, F. 1994. Clima del nordeste del Chubut. In: *Guía de campo Península Valdés y Centro Noreste del Chubut. CADINQUA. Séptima Reunión de Campo*, Eds. A. Súnico, P. Bouza, C. Cano, H.F. del Valle, L. Videla, and A. Monti, 13–20, CENPAT-CONICET, Puerto Madryn, Chubut, Argentina.

Coulson, S.N. 1995. SAR interferometry. In: Multi-disciplinary online documentation. URL: http://earth.esa.int/docpot.

Definiens, 2003. eCognition, user guide, version 4.0. URL: http://www.definiens-imaging.com/.

del Valle, H.F., N.O. Elissalde, and D.A. Gagliardini. 1997. ERS-1/SAR and SPOT data for irrigated land evaluations in a sector of the Chubut inferior valley (Central Patagonia, Argentina). In: *International Seminar on the Use and Applications of ERS in Latin America*, ESA (European Space Agency), ESA SP-405, pp. 89–92.

del Valle, H.F., A. Buck, and H. Mehl. 2002. Digital elevation models as tools in soil research in northeastern Patagonia. In: *Proceedings XVIII Argentine Soil Science Meeting*, CDrom. Puerto Madryn, Chubut, Argentina.

Dobson, M., F. Ulaby, and L. Pierce. 1995. Land-cover classification and estimation of terrain attributes using synthetic aperture radar. *Remote Sensing of Environment* 51: 199–214.

Dobson, M.C., L.E. Pierce, and F.T. Ulaby. 1996. Knowledge-based land-cover classification using ERS-1/JERS-1 SAR composites. *IEEE Transactions on Geoscience and Remote Sensing* 34(1): 83–99.

Drury, S.A. 1993. *Image Interpretation in Geology*, 2nd ed. Chapman & Hall, London.

Erdas Inc. 2006. Erdas Imagine, version 9.1. Leica Geosystems. URL: http://ww.erdas.com/.

Ferrazzoli, P. and L. Guerriero. 1995. Radar sensitivity to tree geometry and woody volume: A model analysis. *IEEE Transactions on Geoscience and Remote Sensing* 33: 360–371.

Jordan, R.L., B.L. Huneycutt, and M. Werner. 1995. The SIR-C/X-SAR synthetic aperture radar system. *IEEE International Geoscience and Remote Sensing Symposium (IGARSS)* 33(4): 829–839.

Laya, H. 1981. Formulación de un plan integral de manejo hídrico para el Valle Inferior del río Chubut. Levantamiento semidetallado de suelos. Consejo Federal de Inversiones (CFI), provincia de Chubut, Convenio VIRCH, volumenes 1 y 2. Buenos Aires, Argentina.

León, R.J.C., D. Bran, M. Collantes, J.M. Paruelo, and A. Soriano. 1998. Grandes unidades de vegetación de la Patagonia extrandina. *Ecología Austral* 8(2): 275–308.

Lombardo, P. and C.J. Oliver. 2002. Optimal classification of polarimetric SAR images using segmentation. In *IEEE National Radar Conference*, Long Beach, CA, pp. 8–13.

Metternicht, G. 1993. A comparative study on the performance of spatial filters for speckle reduction in ERS-1 data. In: *Remote Sensing: From Research to Operational Applications in the New Europe*, Ed. R. Vaughan. Springer-Verlag, Dundee, Scotland, pp. 275–284.

Metternicht, G. 1998. Fuzzy classification of JERS-1 data: An evaluation of its performance for soil salinity mapping. *Ecological Modelling* 111: 61–74.

Mohamed, Z.A.R., F. Bonn, L.P. Giugni, and A. Mahmood. 2003. Potentiality of RADARSAT-1 images in the detection of salt-affected soils in the arid zone: Wadi El-Natrun Bottom, Egypt. In: *IEEE International Geoscience and Remote Sensing Symposium (IGARSS)* IV: 2777–2779.

Notarnicola, C., M. Angiulli, and F. Posa. 2006. Use of radar and optical remotely sensed data for soil moisture retrieval over vegetated areas. *IEEE Transactions on Geoscience and Remote Sensing* 44: 925–935.

Paine, J.G. 2003. Mapping near-surface salinization using long-wavelength AIRSAR. Report prepared for National Aeronautics and Space Administration under NASA Grant No. NAG5-7582, SENH98-0113.

Peake, W.H. and T.L. Oliver. 1971. The response of terrestrial surfaces at microwave frequencies. Electroscience Laboratory, 2440–7, Technical Report AFALTR-70301. Ohio State University, Columbus, OH.

Reichart, M.L. 1961. Estudio y reconocimiento de los suelos del Valle Inferior del río Chubut, desde el punto de vista agrológico. Anexo 1, Segunda Parte, Programa de Desarrollo del VIRCH, Rawson, Chubut, Argentina.

Shao, Y., Q. Hu, H. Guo, Y. Lu, Q. Dong, and C. Han. 2003. Effect of dielectric properties of moist salinized soils on backscattering coefficients extracted from RADARSAT image. *IEEE Transactions on Geoscience and Remote Sensing* 41: 1879–1888.

Siebert, S., J. Hoogeveen, and K. Frenken. 2006. Irrigation in Africa, Europe and Latin America—update of the digital global map of irrigation areas to version 4. Frankfurt Hydrology Paper 05, Institute of Physical Geography, University of Frankfurt, Germany, and FAO, Rome, Italy.

Sreenivas, K., L. Venkataratnam, and P.V.N. Rao. 1995. Dielectric properties of salt-affected soils. *International Journal of Remote Sensing* 16(4): 641–649.

Taylor, G.R., B.A. Bennett, A.H. Mah, and R. Hewson. 1994. Spectral properties of salinised land and implications for interpretation of 24 channel imaging spectrometry. In: *Proceedings of the First International Remote Sensing Conference and Exhibition*, Vol. 3, Strasbourg, France, pp. 504–513.

Taylor, G.R., A.H. Mah, F.A. Kruse, K.S. Kierein-Young, R.D. Hewson, and A.B. Bennett. 1996. Character-ization of saline soils using airborne radar imagery. *Remote Sensing of Environment* 57: 127–142.

Ulaby, F.T. and P. Siqueira. 1995. *Polarimetric SAR Soil Moisture Inversion Algorithms*. The University of Michigan, Radiation Laboratory, Technical Memorandum 12–95.

USDA. 2003. *Keys to Soil Taxonomy*, 9th ed. United States Department of Agriculture (USDA), Natural Resources Conservation Service (NRCS), Washington, DC.

Weeks, R., M. Smith, K. Pak, and A. Gillespie. 1997. Inversions of SIR-C and AIRSAR data for the roughness of geological surfaces. *Remote Sensing of Environments* 59: 383–496.

Xiao, J., J. Li, and A. Moody. 2003. A detail-preserving and flexible adaptive filter for speckle suppression in SAR imagery. *International Journal of Remote Sensing* 24(12): 2451–2465.

10 Application of Landsat and ERS Imagery to the Study of Saline Wetlands in Semiarid Agricultural Areas of Northeast Spain

Carmen Castañeda and Juan Herrero

CONTENTS

10.1 INTRODUCTION

Wetlands are among the most valuable ecosystems on earth and are considered priority sites in environmental policy due to their biodiversity and hydrological function. However, during the last century, many wetlands have been lost or degraded because of human pressure (Dahl and Johnson, 1991; Davis and Froend, 1999; Finlayson and Rea, 1999; Tiner, 2002). Most of them are located on private lands, and agricultural activities have been responsible for about 80% of wetland loss (OECD, 1996), a percentage likely to increase in the future (Millennium Ecosystem Assessment, 2005).

A regional overview of wetland degradation in Europe indicates that these areas are under great pressure from urban and industrial development (Stevenson and Frazier, 1999). Since the Ramsar Convention in 1971 (Ramsar Convention Secretariat, 2006), various initiatives have been undertaken to increase conservation awareness about wetland habitats. The MedWet long-term collaborative program has emphasized the many valuable functions of wetlands, especially as habitat for birds (Costa et al., 1996), and stressed the importance of inventories to monitor changes and explain their causes and consequences (Tomàs-Vives, 1996).

Many of the world's most spectacular wetlands are in the drylands of Australia, India, the United States, South America, South Africa, and Arabia. Wetlands are significant hydrological features in drylands. They are strongly linked to groundwater fluctuations and range from temporary to perennial. Wetlands in dry environments vary greatly in size, from small pans of less than 0.1 km^2 to large intermittent lakes of thousands of kilometer square, and in salinity, from freshwater to hypersaline. A broad spectrum of habitats is tied to their cyclic dry–wet periods. Most wetlands are geological relics and thus good indicators of paleoenvironments. They are exceptional study areas for geomorphology, hydrology, sedimentology, and pedology. Our understanding of wetlands in dry environments is relatively limited, as they are poorly mapped (Finlayson et al., 1999) and most of them have not been systematically surveyed or assessed for cultural heritage significance.

Inland saline wetlands in arid environments, including closed lakes, dry lakes, salt pans, inland sabkhas, and playa-lakes, are scarce in Europe and are not well represented in inventories and conservation plans. In Spain, playa-lakes and other closed saline depressions are spread over the main Tertiary basins (Peña and Marfil, 1986; Gutiérrez-Elorza et al., 2002; Desir et al., 2003; Díaz de Arcaya et al., 2005; Rodríguez et al., 2006) and are linked to the substrate lithology and groundwater dynamics under continental climatic conditions.

In the center of the Ebro Basin, Spain, many saline wetlands are isolated environments surrounded by irrigated or dry-farming areas. In this chapter, we describe remote sensing-based studies carried out in two different saline wetlands, namely, the Monegros wetlands and the Gallocanta Lake. Being the largest European playa-lake, the Gallocanta Lake was designated a Ramsar site as a wetland of international importance. Though subject to different environmental conditions, these two wetland areas have suffered from the intensification of agricultural activities during the last decades.

For 10 years, the European Union blocked the installation of irrigation systems in Monegros, with the consequence that the irrigation scheme initially planned for the whole area was approved in 2004 for only half of the area. Irrigation with water from the Pyrenees will modify the hydrological balance, and both fresh and polluted water will be discharged into these wetlands and alter their flooding periods. Such an effect has been observed in other Spanish saline wetlands that experience inflow of freshwater and permanent flooding. In these cases, the increase in birds can be a false positive effect, offset by the extinction of extremophiles and other valuable organisms.

The management of the Gallocanta Lake area must address two main issues. The first issue concerns the changes in the hydrological regime. Groundwater extraction for irrigation had to be restricted to avoid the desiccation of the playa-lake. From a maximum of 2.3 m recorded in 1974, water depth has substantially decreased in recent decades. The second issue is establishing effective area delimitation. The recovery of the natural vegetation in fringe areas is possible only if restrictions are enforced on agricultural uses. Playa-lake delimitation is not an easy task because of the difficulty in drawing limits between land and water, given the fluctuations of the water level proper to wetlands in semiarid environment. The playa-lake limits recently established by the government are not based on scientific criteria, but rather on administrative and political arrangements. The protected site is restricted to the actual saline playa-bottom because the historically flooded conterminous areas have not been included.

Inland saline wetlands around the world have been studied using the approaches of sedimentology, geochemistry, hydrology, and limnology. Their mapping, however, has been overlooked, and in any case their study has been marginal and often conducted for scientific purposes only. Standard field procedures for wetland delineation aimed at environmental monitoring and conservation purposes, like those established by the U.S. Army Corps of Engineers (Environmental Laboratory, 1987) or those mentioned in the MedWet initiative for wetland delineation (Costa et al., 1996), are difficult to apply in playa-lake areas. Such procedures may require specific protocols (Lichvar et al., 2004, 2006), as well as inferential study and professional judgment (Brosttoff et al., 2001), and specific diagnostic tools and indicators.

Environmental concern is changing the societal value of inland saline wetlands. Wetlands in semiarid regions have been surveyed and classified in the framework of national wetland inventories, including isolated playas and salt flats in the United States (Tiner et al., 2002), endorheic systems in South Africa (Dini et al., 1998), and playas in Australia (Duguid et al., 2002). Wetland inventories implement remote sensing and geographic information system (GIS) techniques to characterize land uses and habitats for conservation planning (Tiner, 2004). When soil and vegetation maps at adequate scales are not available, remote sensing is an alternative means to generate data for wetland monitoring because of its synoptic view and revisiting capability. The application of remote sensing to inland wetlands in arid environment has been restricted to playa-lakes or large inland sabkhas (Epema, 1992; Harris, 1994; Bryant, 1999). The techniques implemented in these studies are not well suited for the mapping of small saline wetlands due to the spatial resolution of the sensors or the digital image-processing techniques used. The study of the saline wetlands of NE Spain requires an approach adapted to their median size (7 ha in Monegros), the irregular and rapid changes in their appearance, and the lack of field data.

The objective of this chapter is to show the application of remote sensing to characterize inland saline wetlands in NE Spain. Two saline wetland sites with different characteristics were studied, the Monegros saladas and the Gallocanta Lake area (Figure 10.1). We developed a remote sensing

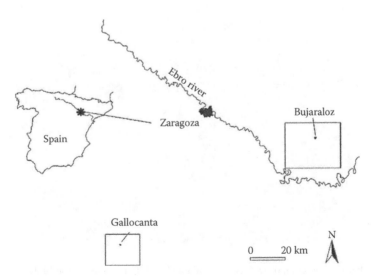

FIGURE 10.1 Location of the two study areas: Monegros saladas (Bujaraloz area) and Gallocanta Lake (Gallocanta area).

approach adapted to varying scales, ground knowledge, and environmental conditions. Remote sensing was integrated with ground knowledge to map saline areas, characterize their facies, analyze their hydrological regime, and assess their conservation status, taking into account the impact of the new irrigation schemes.

10.2 MONEGROS PLAYA-LAKES

The Monegros playa-lake area lacks a map showing the magnitude, frequency, and extent of flooding in natural saline wetland depressions of high ecological value. Such information is basic for environmental protection and will allow predicting the extent and evolution of saline areas when the irrigation scheme starts being fully functional. Remote sensing was used here for this purpose.

10.2.1 STUDY AREA SETTING

The area is an arheic plain, including 36,000 ha of farmland, with scattered saline wetlands locally named "saladas." Figure 10.2 is a false-color composite (5,4,3) of a Landsat Enhanced Thematic Mapper (ETM+) image from July 24, 2004 that shows the saladas scattered over the same area represented in Figure 10.1. They are located between 320 and 400 m a.s.l. (meters above sea level) on a gypsiferous Miocene plateau more than 100 m above the Ebro River. The area is one of Europe's most arid zones (Herrero and Snyder, 1997), with a mean annual rainfall of 355 mm and a mean annual ET_0 of 1255 mm. The mean monthly temperature is 14.4°C with 45°C and −15°C being the absolute maximum and minimum, respectively, registered in the area since 1985.

The saladas, which at present occupy less than 3% of the area, range from less than 2 ha to more than 200 ha. They occur in smooth, flat karstic depressions and flat-bottom valleys. Some of them are playa-lakes and others are closed saline depressions with scientific and environmental value as natural habitats of endemic microorganisms (Casamayor et al., 2005), plants (Domínguez et al., 2006), and animals (Melic and Blasco, 1999). The area is protected by the European Birds and Habitats

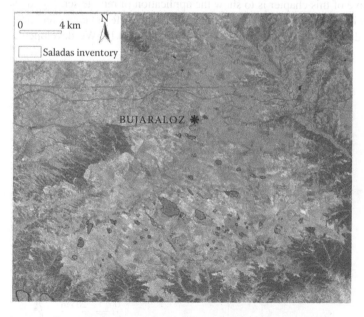

FIGURE 10.2 False-color composite (5,4,3) of a Landsat ETM+ image from July 24, 2004, showing the Monegros saladas spread over the dry-farming area south of Bujaraloz, located on the border of the irrigated area.

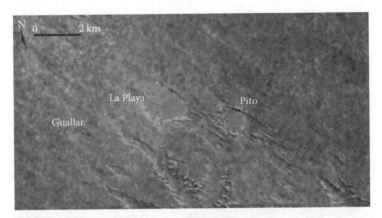

FIGURE 10.3 Elevation model obtained from the difference between the SRTM and a local DEM, showing the NW-SE orientation of the north and south escarpments of the Monegros playa-lakes.

Directives, but only a part has been proposed for inclusion in the Mediterranean region of the Natura 2000 network, the cornerstone of the EU nature protection policy that is currently being elaborated.

Most of the largest saladas are bordered by a sharp NW-SE escarpment less than 20 m high. The difference between the the Shuttle Radar Topography Mission (SRTM) elevation model of the study area and a local elevation model enables identification of the smallest elevation features in the area (Figure 10.3). The higher and the longer the escarpment, the better preserved the saladas, since height and spatial continuity increase confinement and hamper agricultural invasion. Many escarpments have been smoothed to facilitate plowing, which contributes to sedimentation in the saladas. The smaller saline depressions are rounded and have gentle margins and wet bottoms with halophytes indicating water discharge, but are rarely flooded.

Usually, saline wetlands strongly contrast with the rest of the landscape. Topography is flat and the dark topsoil remains permanently wet even during dry periods, though slightly hardened and sometimes covered by salt efflorescence (Figure 10.4). The core area of the playa-lakes exposed to flooding is usually barren, but the rims are colonized by salt-tolerant plants. Intermittent salty water occurrence is critical to the conservation of the salada habitat. In general, water hinders agricultural use but, as soon as surface water evaporates, some of the smaller saladas are plowed to increase the total acreage eligible for agricultural subsidies.

Soils outside the saline depressions are shallow, calcareous, or gypseous with low organic matter contents. Soils are saline toward the borders of the wetlands with salinity often exceeding the thresholds for crop production. The wetland bottoms are covered by fine sediments, rich in calcium carbonate, gypsum, and more soluble salts, and more than 2 m thick in the center of the depressions. The electrical conductivity in the saturated paste extract of the topsoil layer (0–60 cm) is often more than 70 dS m^{-1}. The occurrence and extent of shallow brine during a few weeks per year depend on the scarce rainfall and groundwater dynamics. Spread over the wetland bottom by the prevailing NW wind and hardly ever reaching more than 50 cm in depth, the brine determines the distribution of the vegetation. Evaporation and subsequent salt concentration increase water salinity up to 105 dS m^{-1}.

Comprehensive mapping of the wetlands is lacking, but remote sensing can provide a basic synoptic view of the area and contribute to the detection of the wetlands.

10.2.2 MAPPING FLOODED SALINE AREAS

10.2.2.1 Remote Sensing Data Set

Karstification and dissolution of gypsiferous rock formations have produced the wetlands and saline depressions in Monegros. In the absence of soil maps in Monegros, the parallel studies carried out

FIGURE 10.4 Field photographs of the largest playa-lakes in Monegros: (a) La Playa, (b) Piñol, (c) Valdefrancín, (d) Pito, (e) Pueyo, and (f) Salineta.

by Martínez et al. (2003) in Mexico and by Martínez and Castañeda (2004) in Monegros, enabled us to evaluate three Landsat ETM+ images from May, June, and August for discrimination of gypsiferous formations in Monegros.

The capability of SAR images to detect surface features related to changes in wetness and roughness (Taylor et al., 1996) has also been explored by means of three Precision Image (PRI), Synthetic Aperture Radar (SAR), and European Remote Sensing Satellite (ERS-2) images from 2002 covering early spring (March 2), late spring (June 15), and summer (August 24), which is the period that experiences the greatest changes. ERS images were used alone, as well as integrated with the Landsat ETM+ images.

SAR images are advantageous for delineation of wetlands in vegetated areas (Ozesmi and Bauer, 2002). C- and L-band SAR images enable discriminating wetlands from their surroundings because of (a) the very low backscatter produced by open water surfaces and (b) the very high backscatter created by the volumetric scattering of flooded vegetation (Rosenqvist et al., 2007). Taft et al. (2004) used three C-band SAR images and simultaneous ground reference data related to wetness and vegetation cover to classify the radar backscatter of different habitats in small wetlands located within agricultural land. The lack of ground truth simultaneous to the ERS-2 passes was overcome due to previous knowledge on soil surface changes, hydrological regime, and agricultural practices, together with the information extracted from a series of Landsat images described in the next sections.

10.2.2.2 Image-Processing Techniques Applied

A maximum likelihood supervised classification of the Landsat ETM+ images was applied by varying the number of classes and dates involved in the classification. Training and verification areas were selected from lithological units, irrigated areas, and wetlands based on the visual interpretation of false-color composites (1,4,7 and 2,4,7) of the Landsat ETM+ scenes.

The multitemporal SAR composition enhances changes in backscattering, and the thematic classification of these images using contextual methods enables the differentiation of soil cover types at landscape scale (Chust et al., 2004). The ratio of exponentially weighted average (ROEWA) operator was applied for image segmentation to take into account the speckle of radar images. After selecting the most suitable edge-strength map, according to the size of the wetlands and the natural and farming features in the area, an image of closed skeleton contours was produced by means of the watershed algorithm (Fjørtoft et al., 1997). In order to preserve the contours and linear structures, the adaptive filter linear vectorial minimum mean square error filter based on segmentation (LVMMSES) (Fjørtoft et al., 1997) was applied for statistical calculation using the regions defined by image segmentation.

10.2.2.3   Discussion of Results

Identification of gypsiferous land was improved when including the Landsat ETM+ thermal band in the multitemporal classification and when grouping the geological formations by main rock components. Gypsum-rich areas could be discriminated only when the gypsiferous horizon was exposed at the surface. The occurrence of stony areas related to interbedded limestone layers, brought to the surface by deep plowing, led occasionally to overestimate the extent of gypsum-rich areas (Figure 10.5).

The uncertainty included in the backscattering information and the speckle contribution made it difficult to interpret the radar backscattering levels at each date. The C-band backscattering value of the salada bottom was very variable, as it was dependent on the surface roughness. Under dry surface conditions, the vegetated bottom areas showed backscattering values similar to those of the surrounding agricultural areas. Although the flat bare bottoms of the largest wetlands were

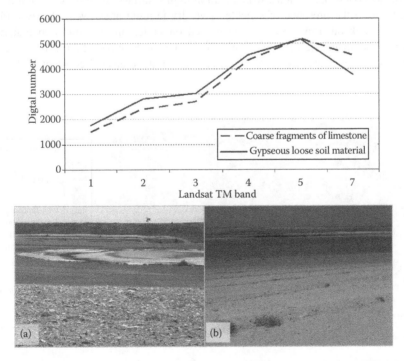

FIGURE 10.5   Similarity between the spectral signatures of different soil surface features. The abundance of limestone fragments on the surface (a) can create spectral confusion with gypseous material (b).

FIGURE 10.6 Smooth playa-lake bottoms can show increased backscatter in the ERS C-band when the water surface is rippled by strong winds. The salada bottoms have gray levels similar to those of the surrounding fields.

discriminated on each date, C-band alone failed to differentiate between standing water and extremely smooth surfaces of salt pans because of similar specular response. Moreover, the backscatter of water bodies could vary from very low to very high, if windy conditions created water ripples (Figure 10.6). This uncertainty hampers the comparison of radar response data with the thematic classes established with optical data (Castañeda et al., 2005a).

In the dry-farming areas, annual plowing depends on the Common Agricultural Policy subsidies. This condition results in frequent and unpredictable changes in the extent of the cropped areas and the fields left aside, with subsequent changes in surface roughness (e.g., size of clods produced by plowing). The SAR imagery separated the largest wet and flat saladas of playa-lake type from the heterogeneous regions, which exhibit higher surface roughness, though their borders were blurred due to the presence of speckle. Two dry-farming areas were clearly differentiated, namely, (1) the rock outcrops and soils developed on lutites and gypsum-rich materials in the northern part of the image and (2) a limestone area in the southern part (Figure 10.7). The different backscattering of these areas may be due to differences in the dielectric constant of the terrain.

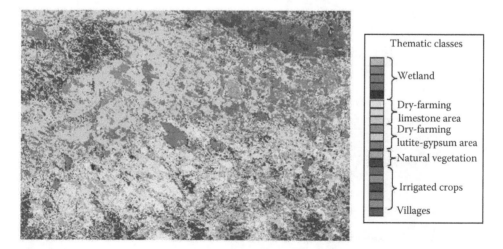

FIGURE 10.7 Fusion of multitemporal ERS and Landsat classifications from 2002, showing the main land covers discriminated in the area (with subclasses).

The fusion of ERS data with a set of three Landsat ETM+ images acquired in the same season improved the classification accuracy of each data set separately, with an 88% kappa coefficient of agreement (κ) and 89.2% of pixels correctly classified. The inclusion of radar improved the detection of landscape structures related to linear features and changes in soil surface roughness, and highlighted the relief. Furthermore, field edges and their radiometric differences were enhanced. Discrimination of the natural vegetation increased except for the xerophytes in flat areas, which appeared undifferentiated from nonirrigated crops. Wetland bottoms were enhanced with clearer determination of the borders, though the discrimination of the wetland facies was not significantly improved. For environmental monitoring purposes, field observations simultaneous to image acquisition are needed to identify the surface aspects that cause the variability of the radar classes.

10.2.3 Delimitation and Inventory of Saline Wetlands

10.2.3.1 Remote Sensing Data Set and Processing Techniques

An integrated approach including remote sensing, ancillary data (e.g., topography), and field visits was used for the inventory of the Monegros wetlands (Castañeda and Herrero, 2008a). The saladas boundaries were delineated using a Landsat image acquired on April 2, 1997 (Castañeda et al., 2005b) with the assistance of topographic maps at scales of 1:25,000 and 1:50,000. Year 1997 was selected because it was the wettest in the last two decades with an annual rainfall of 535 mm against an average of 350 mm for the period 1974–2006.

Noncartographic information describing the location of the saladas was also collected, most of it being unpublished but invaluable documents that contribute to gaining general understanding of the area and developing field survey itineraries. Close to 100 wetlands were reported with 76% named and the rest referred to with the local generic term "clota." Our remotely sensed inventory was contrasted in a GIS with the geographical coordinates and the toponymy extracted from the ancillary data sources. This resulted in a geospatial database of saladas labeled Inventory 2003, where our 53 surveyed wetlands were given precise names, boundaries, and coordinates.

10.2.3.2 Discussion of Results

Intensive agricultural use and irrigation works made it difficult to identify the spatial extent of some saladas in the field. However, the largest and the wettest saladas could be clearly recognized in the image, as well as those preserving their escarpment and native vegetation. A total of 39 saladas, ranging from 1.8 to 200 ha, was initially identified in the remotely sensed imagery. Subsequent field surveys aided by orthophotographs at scale 1:10,000 enlarged that inventory to 53 saladas, with a total wetland extent of 860 ha (Figure 10.8). Only 31 saladas, corresponding to those with more permanent water and clear boundaries, are shown in the official topographic maps. These maps are produced using photogrammetric techniques and aerial photographs from a single date, and thus miss some saladas that can only be detected from multitemporal data sets. The saladas mapped have an average area of 16 ha, with 77% having less than 20 ha. Saladas with more than 20 ha belong to the playa-lake type.

10.2.4 Mapping Facies Variability Inside Wetlands

10.2.4.1 Remote Sensing Data Set and Processing Techniques

The land cover types of the saline wetlands were studied using a series of 52 Landsat Thematic Mapper (TM) and ETM+ images spanning 20 years from 1984 to 2004. Ten images belong to winter–early spring, 14 to spring, and 33 to summer seasons. After orthorectification and resampling to 25 m pixel size, images were classified using the ISODATA clustering method (Swain, 1973). The classes were grouped according to their spectral and visual characteristics to produce a map for

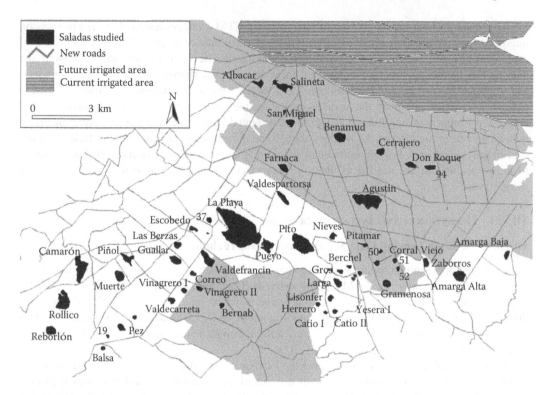

FIGURE 10.8 The 53 inventoried saladas with names and limits, as extracted from unpublished studies and field visits in 2003. The limits of the current and future irrigated areas are shown.

each date that identifies the extent of the water surface and associated covers. Field knowledge was crucial to assign thematic meaning to the spectral classes obtained from image classification.

Standard definitions of the playa-lake land covers applicable to this study were not available. Accordingly, we used the English term facies (Castañeda et al., 2005a), with a meaning similar to the French term "état de surface" defined by Escadafal (1992), to characterize the surface of the arid soils using field observations and remote sensing.

A set of five facies associated with flooding and drying events were systematically identified in the field and in the images, describing the location in the landscape, pattern, surface area, appearance, and spectral signature. They are listed by order of decreasing moisture as follows: water, watery ground, wet ground, vegetated ground, and dry, bare ground. The list could be enlarged on the basis of more detailed field studies under varying climatic conditions and with the interpretation of images from other sensors. The definition of facies overcomes the lack of a conceptual framework to assess the ecological status of the saladas, and the adopted criteria make distinguishing these facies easy, both in the field and on Landsat images. Together with the spectral behavior (Figure 10.9), additional criteria contributed to the accurate definition of the facies, namely, their entity (i.e., persistence in the images and occurrence in most saladas), their significance, and their separability (i.e., the ease of field and remote recognition).

10.2.4.2 Discussion of Results

The extent of each facies, which in practice can only be estimated from remote sensing, and their evolution from 1984 to the present (Figure 10.10) are the key to appraising the conservation status of these unique habitats and detecting environmental alterations. In the entire saladas area (about 900 ha), the mean extent of the flooding surface (i.e., water plus watery ground) was only 13%,

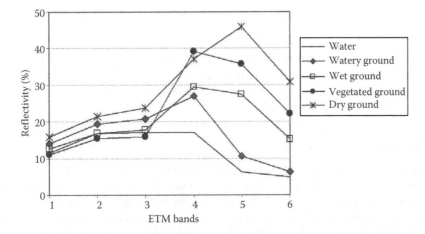

FIGURE 10.9 Spectral signatures of the facies established in the saline wetlands of Monegros, obtained by unsupervised classification of a Landsat image from April 27, 2003.

reaching a maximum of 53% in January 1987 and being totally absent in May 1991. Some bias can be assumed given the lack of satellite images during cloudy periods. The average extent of dry bare ground was 11%, whereas vegetated ground had a mean extent of 26%. Wet ground, the facies with the highest average surface, had a mean of 50%, reaching a maximum of 71% in August 1985 and a minimum of 6% in March 2000. Temporal variations in the surface area occupied by each facies were noticeable, but they were small for the 20-year span of this study. Based on our experience, most of the changes in the surface area of water and watery ground are episodic, and the duration of

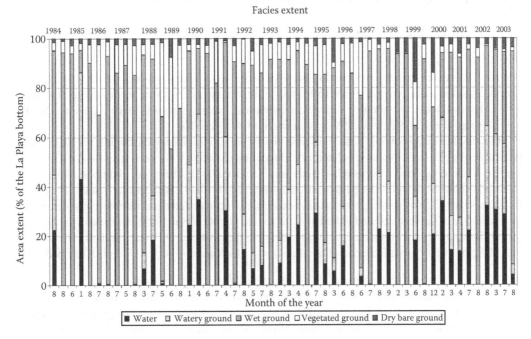

FIGURE 10.10 Area extent of the land facies from 1984 to 2004 at La Playa site (month numbers on the X-axis).

these episodes can only be established by organized monitoring with more frequent observations, either in the field or by remote sensing.

The total evaporative surface, represented by the extent of the facies contributing to the water discharge in the saladas, showed no significant change during the period studied. Increasing the number of images up to 52 did not modify the temporal trend already reported in Castañeda et al. (2005a) with 26 images, but confirmed that the vegetated grounds increased over the study period, probably due to agricultural intensification.

10.2.5 HYDROLOGICAL REGIME

The flood regime of the Monegros playa-lakes is related to the climatic and hydrogeological features of the basin. The area functions as a single, closed basin where the solubility, permeability, and complex disposition of the geological materials make the hydrological connections complex. The saladas play an important role in the regional hydrology, having close connection with two main aquifers (Castañeda and García-Vera, 2008). They are discharge areas with shallow water table in the low-lying parts of the playa-lakes and other saline depressions, allowing water evaporation from the capillary fringe.

10.2.5.1 Remote Sensing Data Set and Processing Techniques

Remote sensing has yielded excellent synoptic information and quantitative data on water occurrence and on the hydrologic behavior of the main playa-lakes. The extent of flooded areas was extracted from a series of 52 Landsat images acquired from 1984 to 2004. The total flooded area in the saline wetlands, computed as water and watery ground, was added up for each image date. The sum of the extent of these two facies was regarded as relevant from the environmental and hydrological points of view. A yearly pattern of maximum and minimum water extent and its relationship with weather, and the similarity and dissimilarity of the playa-lakes flooding behavior were determined applying a hierarchical cluster analysis to the main playa-lakes. The persistence of water during the two decades studied was assessed by means of statistical analysis on a pixel basis.

10.2.5.2 Discussion of Results

10.2.5.2.1 Multitemporal Variability of Flooded Areas
The net water extent varied considerably between years, even when comparing years for the same season. Flooded areas covered from 15 to 400 ha in winter and from 60 to more than 340 ha in spring. Areas flooded in summer varied somewhat less, ranging from 9 to 19 ha, except for the rainiest years (1993, 1997, and 1999) with a range of 115–240 ha.

The satellite data showing water presence or absence cycles allowed grouping the main playa-lakes according to their dynamic behavior or inundation patterns, with a result similar to that of the analysis based on ground records (Castañeda and Herrero, 2005). The larger eight playa-lakes were clustered according to their flooded area for each date (Figure 10.11), with similar values of over 85% for seven saladas. This grouping suggests the existence of groundwater connections controlled by the hydrogeology of the area, as established by Samper and García-Vera (1998). Salineta in the northern part of the study area is connected to two aquifers—a fact that causes lower similarity (69%). This playa-lake has water more frequently than the others, which are dominated by annual dry periods.

10.2.5.2.2 Relationship between Water Surface, Rainfall, and Evaporation
The relationships between the flooded area detected by remote sensing and the accumulated rainfall were analyzed for periods of 15, 30, 90, 180, and 365 days previous to the image acquisition date. There is a significant ($P < 0.05$) relationship for the 180-day accumulated rainfall. In the best cases, only 30% of the flooded area variation can be explained by the rainfall variation from the previous 180 days. Salineta is an exception, having no significant relationship with any rain period analyzed.

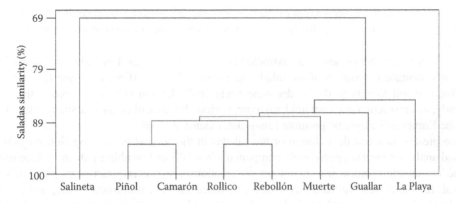

FIGURE 10.11 Similarities between playa-lakes based on water ponding areas as detected from 52 Landsat images.

Evaporation appears as the key factor influencing the removal of water from the lakes. The linear regression between flooded area and accumulated ET_0 for 15, 30, 90, 180, and 365 days previous to image acquisition yields r^2 values that decrease as the period of accumulated ET_0 increases. The best relationship occurs for ET_0 (15), with r^2 values between 0.4 and 0.5. For the most favorable cases, 50% of the water area variations can be explained by ET_0 variations. Similar results were obtained for the relationship between flooded areas and the difference $(P - ET_0)$. Salineta has significant relationships only for the periods of 90 and 180 days.

10.2.5.2.3 Persistence of Flooded Areas from 1984 to 2004
Annual flooded areas were produced by computing the water and watery ground facies extents from each Landsat image. In some of the saladas, flooded areas were present in 48% of the images studied. In order to map the typical location of flooded areas, we applied a frequency analysis to the time series of 52 facies maps on a pixel basis. The resulting frequency map represented the spatial distribution of the water persistence in the time series studied. Five classes were then established, from very low to very high water persistence (Table 10.1). The maximum water persistence corresponds to Salineta, occupying almost all its low-lying area. La Playa and Camarón are the saladas having the next highest water persistence.

10.2.6 Wetland Degradation

The saladas, long considered as wasteland or barren terrain from an agricultural productivity point of view, have been submitted to farming intensification that is constrained only by spots of flooding and extreme salinity. Nearby traditional dry-farming is currently changing to irrigation, and that increases wetland degradation at a time when little or no information is available to environmental

TABLE 10.1

Five Classes of Water Persistence in the Saladas of Monegros as Established from 52 Landsat Images between 1984 and 2004

Water Persistence	Very Low	Low	Medium	High	Very High
Percentage of dates with water	4	8–16	20–36	40–64	68–100
Water surface extent (ha)	110.5	143	82.6	30.2	7.0
Number of saladas with water	25	16	11	3	1

managers. Our systematic assessment of the saladas status has allowed the quantification of features otherwise difficult to measure in the field, and to build a baseline for agroenvironmental monitoring purposes.

The historical water presence, as extracted from satellite images, has been used to establish a series of systematic descriptors of the saladas to assess their status (Castañeda and Herrero, 2006). Flooding and soil salinity in the border areas traditionally limited cultivation around the saladas. Currently, this restriction does not hold anymore, perhaps because of better machinery and subsidies from the European Union that promote low-input, extensive crops.

The preservation and degradation of the saladas in the context of the irrigation projects were assessed analyzing selected perceptible components in a GIS and combining them in three indexes: (1) the conservation index was based on observations about escarpment, cropping, and stone removal; (2) the current vulnerability index was based on size, water occurrence, and proximity to roads; (3) the future vulnerability index was created adding the proximity to the planned irrigation area. The resulting values of the indexes were scaled from very poor to very good conservation and from very low to very high vulnerability (Castañeda and Herrero, 2008b). The relationship between the vulnerability of the saladas and the water occurrence obtained from remote sensing yields negative Pearson correlation coefficients, highlighting the impact of increasing human influence. The highest vulnerability values occur in saladas with less than 10 ha, which are also those with less water occurrence. The conservation or the vulnerability is not reflected in any single salada feature, such as size or flood frequency, because of the sharp rise in agricultural intensification in recent years. This assessment allows the establishment of priorities for surveillance, conservation, and restoration. In general, the loss of saladas could be mitigated through specific policy measures.

10.3 GALLOCANTA STUDY AREA

In the Gallocanta study area, remote sensing was used to identify the land cover types and their variability in recent decades and thus determine the functional area of this playa-lake. The study has provided thematic maps of the saline area and its surroundings, and a quantification of the flooding extent. So far, only the water depth has been surveyed by the Ebro Water Authority for groundwater balance purposes. Remotely sensed information can help in the recognition of the functional areas of the Gallocanta Lake, a basic step for area delimitation, hydrological survey, vegetation conservation strategies, and management of the peripheral agricultural areas.

10.3.1 STUDY AREA SETTING

Gallocanta is the largest playa-lake in Europe, located 115 km south of Zaragoza in an intermountain tectonic depression at 1000 m a.s.l. (Figure 10.1). The current extent of the lake-bed is about 14 km^2, but ancient lake sediments were found more than 10 km away from the present shoreline (Gracia, 1995). Annual rainfall ranges from 320 to 650 mm, with a mean of 488 mm. The maximum monthly temperature occurs in July with a mean of 21.1°C, and the minimum in January with a mean of 2.9°C. The absolute maximum and minimum temperatures are 39°C and −21°C, respectively. A regular northwestern wind frequently reaches more than 80 km h^{-1}. Average evaporation measured in class A tank is 1257 mm year^{-1}.

The sources of water in the playa-lake are both fresh surface runoff and saline groundwater recharge. A maximum water depth of 2.3 m was recorded in 1974 (CHE, 2002). The lake can dry annually during periods of strong evaporation. The water salinity ranges from 15 to 150 g L^{-1} (Comín et al., 1991).

Figure 10.12 shows a false-color composite (4,3,2) of a Landsat ETM+ image from July 30, 2000, corresponding to the area framed in Figure 10.1. The extent of the playa-lake can be easily recognized in the field and by remote sensing, although spatial delimitation is difficult because of seasonal and interannual water variations from irregular rainfalls under semiarid conditions. Rainfall

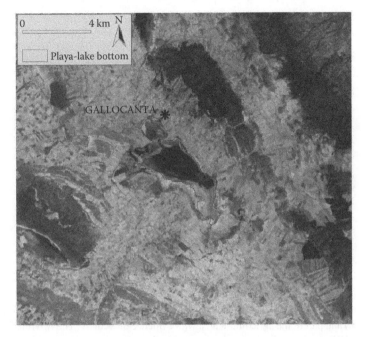

FIGURE 10.12 False-color composite (4,3,2) of a Landsat ETM+ image from July 30, 2000, showing the Gallocanta Lake area and village, surrounded by farming fields irrigated with groundwater.

variation and the flat topography of the bottom of the basin favor pronounced and repeated migrations of the lake margins. The seasonal and interannual water level fluctuations depend on meteorological factors and on long-term climatic oscillations like ENSO (Comín et al., 1991). These fluctuations affect the distribution and status of the plant communities on the lakeshores considered of special interest for conservation under European regulations (European Union, 1997).

The determination of the lake boundaries is a key issue for management purposes. Recently, the Government of Aragon has placed under legal protection an area delimited on an agreement with farmers. The delimitation, based on the current lake appearance, fails to account for water fluctuations related with strings of wet years. Multitemporal remote sensing can be useful for establishing the boundaries between land and water in this fluctuating wetland.

10.3.2 Mapping of Facies and Their Variability

10.3.2.1 Remote Sensing Data and Processing

The facies concept as defined above for Monegros was applied in the Gallocanta Lake area, a similar environment with alternating flooding and drying episodes. We analyzed a series of 27 Landsat images of the playa-lake acquired in different seasons between 1984 and 2000, reflecting thus different moisture conditions in the saladas.

A preliminary outer border matching the 1010 m contour line extracted from a digital elevation model was drawn in order to map the flooding area, the surrounding wet bare surfaces, the fringes of natural vegetation, and the contiguous cultivated areas (Figure 10.13). The enclosed area of 60.45 km^2 is 81.3% larger than the bare lake bottom (11.3 km^2). Areas under agricultural use with surface evidence of wetness and natural vegetation were included.

Visual analysis of false-color composites and field survey conducted in different seasons in 2005 allowed recognizing wet areas around the water body, temporary wetlands adjacent to the main lake, freshwater springs in the vegetated margins, and halophytic vegetation distributed in

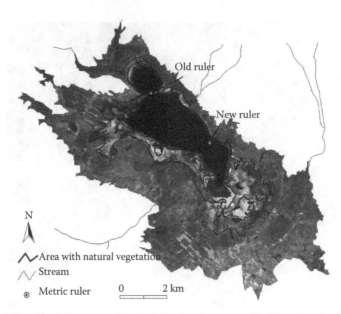

FIGURE 10.13   Gallocanta Lake area enclosed within the 1010 m contour line, showing borders with natural vegetation (in green) and part of the agricultural area with evidence of wetness. The two rulers used by the Ebro Water Authority (CHE) for measuring water depth are also shown.

fringes according to salt tolerance. Images were radiometrically corrected and orthorectified using a digital elevation model of 25 m pixel size and 100 ground-control points. Images were resampled to 25 m. Multispectral and panchromatic bands were fused with ETM+ images using the "à trous" algorithm (González-Audícana et al., 2003). Unsupervised classification using the ISODATA algorithm was applied (Díaz de Arcaya et al., 2005), and the resulting spectral classes were grouped and contrasted with visual analysis of hue-intensity-saturation (HIS), tasselled cap transformation, and a normalized difference vegetation index (NDVI). The experience acquired with the Monegros playa-lakes was crucial for assigning thematic meaning to the spectral classes obtained by digital processing.

10.3.2.2   Discussion of Results

A total of 15 facies was established for the playa-lake and its surroundings (Table 10.2). Nine of them, classified as palustrine and lacustrine, occur in the central area and correspond to the water surface and its fringes. Their extent varies between 16% and 25% of the studied area, determining a fluctuating dry–wet area of 500 ha with high ecological interest. The transitional facies in the rims were classified as mixed classes of wetland and cultivated areas. Among these, the wet ground, detected in 14% of the studied dates, was the most fluctuant with a maximum extent of 33% of the studied area.

The 27 resulting thematic maps corresponded to a wide range of moisture conditions. The flooded area, computed as water and watery ground, was detected in 92% of the image dates analyzed, with a maximum extent of 17% of the total area studied and 84% of the lacustrine central area.

In the thematic map from March 16, 1994 (Figure 10.14), the water body is deep enough to show a uniform dark color over the entire bottom, in contrast to the Monegros playa-lakes where the reflectivity of water is frequently mixed with that of the soil due to shallow water depth. Wetland functional areas outside the playa bed are represented by different facies depending on the image dates.

TABLE 10.2

Surface Area of the Land Facies Established in the Gallocanta Lake from 1984 to 2000 by Means of Landsat Imagery Classification

| | Lacustrine and Palustrine Facies (ha) | | | | | | | | Mixed Facies Wetland/Current Cultivated Area (ha) | | | | | |
| | Water | | Wet Ground | | | Dry Ground | | | Wet Ground | | Dry Ground | | Vegetation | |
Dates	Water + Lacustrine Vegetation	Watery Ground	I	II	III	Without Salt Efflorescence	With Salt Efflorescence	Palustrine Vegetation	Bare	With Sparse and Low-Vigor Vegetation	Bare	With Sparse and Low-Vigor Vegetation	Medium Vigor	High Vigor
July 10, 1984	600.5	140.1	90.3	231.8	0.0	325.6	212.7	0.0	0.0	0.0	3225.7	975.4	242.9	0.0
August 30, 1985	0.0	244.5	172.4	354.4	0.0	385.4	149.4	0.0	459.9	115.4	4042.6	0.0	0.0	120.9
August 17, 1986	0.0	192.4	226.8	285.1	0.0	538.3	151.4	0.0	0.0	115.1	4184.1	205.2	0.0	146.6
August 20, 1987	671.9	132.6	145.2	122.6	0.0	268.8	54.6	0.0	0.0	0.0	3550.3	561.7	411.9	125.4
June 22, 1989	1164.5	31.4	251.8	0.0	0.0	0.0	0.0	57.9	0.0	0.0	43.9	462.7	991.4	3041.4
May 27, 1991	1026.8	80.6	227.3	46.9	0.0	0.0	0.0	50.5	0.0	0.0	0.0	657.3	476.6	3196.5
July 14, 1991	1037.4	0.0	156.1	0.0	38.6	0.0	0.0	0.0	1341.2	1167.1	700.8	2028.6	251.6	636.0
February 7, 1992	1139.1	98.9	103.6	0.0	212.3	0.0	0.0	0.0	0.0	709.7	1849.8	590.5	0.0	0.0
February 2, 1992	968.1	76.2	242.4	0.0	132.1	23.2	0.0	84.3	0.0	0.0	2376.3	1061.0	768.1	313.3
February 25, 1993	964.1	101.6	161.6	0.0	111.4	0.0	0.0	0.0	506.6	329.6	3062.0	808.1	0.0	0.0
March 16, 1994	652.1	185.3	151.7	140.2	0.0	318.8	0.0	0.0	0.0	228.5	312.7	1719.1	1274.7	1061.9
April 4, 1995	0.0	365.6	315.3	135.1	0.0	397.9	190.3	0.0	0.0	280.2	448.4	1959.3	893.3	1059.6
February 27, 1995	0.0	114.4	218.7	659.1	0.0	349.5	0.0	68.0	0.0	0.0	2718.6	1267.3	520.6	128.9
July 11, 1996	0.0	424.2	146.8	369.4	0.0	419.8	67.1	0.0	0.0	0.0	2754.3	1340.0	430.3	93.2
March 27, 1998	568.2	177.4	211.6	0.0	239.0	0.0	0.0	176.6	0.0	0.0	166.4	1372.1	1502.6	1631.0
July 17, 1998	469.4	176.9	190.4	0.0	0.0	309.5	0.0	0.0	0.0	718.4	3433.4	502.8	0.0	244.2
October 21, 1998	172.6	326.1	413.9	0.0	115.3	177.6	0.0	0.0	896.3	0.0	3299.4	387.9	88.1	167.9
March 30, 1999	197.9	331.5	483.5	0.0	158.0	161.1	0.0	0.0	0.0	642.8	961.2	2244.2	514.4	280.3
August 21, 1999	205.8	82.3	619.7	0.0	374.4	0.0	0.0	0.0	0.0	0.0	1076.0	2871.3	444.3	326.6
August 29, 1999	0.0	375.6	260.2	341.6	0.0	371.5	54.3	0.0	0.0	0.0	4214.3	0.0	318.1	109.4
November 9, 1999	0.0	11.7	564.0	379.8	0.0	213.6	0.0	140.8	0.0	931.8	2008.2	888.3	590.9	43.6
January 12, 2000	0.0	364.9	541.3	197.7	285.3	0.0	0.0	55.0	1279.2	1018.3	1902.1	598.9	0.0	0.0
March 8, 2000	0.0	0.0	313.9	218.8	0.0	255.9	411.9	0.0	421.9	0.0	4220.7	223.1	0.0	0.0
July 30, 2000	0.0	431.4	190.8		0.0	147.3	301.8	122.2	0.0	0.0	4071.8	497.9	0.0	62.9
October 18, 2000	0.0	0.0	883.1	403.4	0.0	36.4	0.0	0.0	626.4	23.0	4006.4	0.0	66.1	0.0

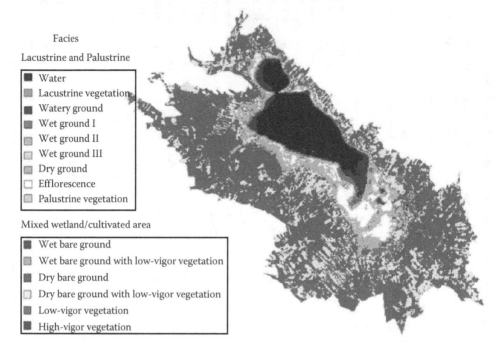

FIGURE 10.14 Land facies distribution in the Gallocanta Lake area on July 10, 1984.

The temporarily flooded fringes and the nonflooded wet areas are very much involved in the wetland dynamics. The surface extent of the main central facies is shown in Figure 10.15. Palustrine vegetation is a fragile fringe of gradual transition that separates the inner from the outer facies, including plant communities protected by the European Habitats Directive. It was detected in 32% of the studied dates. No relation ($r^2 < 0.12$) was found with the rainfall accumulated in the previous year or with the extent of the fluctuations in the flooding area. This fringe was only detected in the absence of salt efflorescence, and its relationship with the wet ground, in terms of surface area, is low and negative. These observations illustrate the difficulty for remote detection of scarce vegetation in playa-lake environments, where the soil surface can be wet or covered by salt efflorescence.

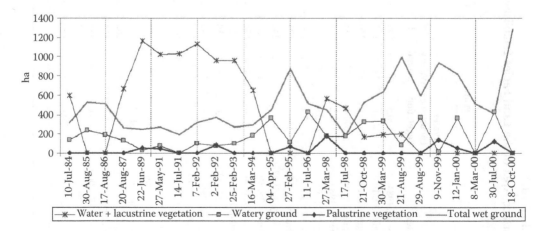

FIGURE 10.15 Extent variations of the inner land facies in the Gallocanta Lake area from 1984 to 2000.

The facies distribution reveals the wetland function and boundaries during recent decades and helps to delineate protected areas and farmland. The water body fluctuations together with the decrease in the water volume, due to groundwater extraction for agricultural use, are critical for migratory bird protection. Because of the ecological significance of the Gallocanta Lake, a delimitation program based on detailed remote sensing and ground data would be of major interest.

10.3.3 Hydrological Regime

10.3.3.1 Remote Sensing Data and Processing

Cycles of water presence and absence were studied from the aforementioned 27 thematic maps. Image acquisition dates were set to be free from rain in the preceding days to allow studying the water level fluctuations of the playa-lake in different seasons. The water surface was measured for each date by computing the facies water and watery ground (Table 10.2). A maximum water depth of 1.13 m was determined as a function of the remotely sensed surface, using the available rating curve (CHE, 2002).

A series of existing water-depth data, measured either by ruler or by limnograph, was used to check the accuracy of the remote sensing-derived water depths. The time spans between ground measures and image acquisition dates were less than 3 days, except for 2 cases spanning less than 15 days and 5 cases without close ground measurements. Rainfall occurring between the image acquisition date and the water-depth measure has been used to interpret the water presence and the status of the soil surface in every image.

10.3.3.2 Discussion of Results

Water depth variations are in closer agreement with the variations of the facies water surface area than with those of the facies water plus watery ground. The latter overestimated the flooding area, especially when the playa-lake was almost dry (Figure 10.16). Part of the incoherence could be explained by the fact that the old ruler (Figure 10.13) was not at the deepest point of the lake, but 14.5 cm above it. As a consequence, the deepest part of the lake could be flooded up to 4 km^2 without being registered by the old ruler and that difference could even be larger during windy

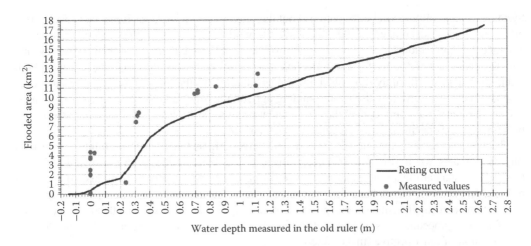

FIGURE 10.16 Zero reading on the old ruler corresponds to 14.5 cm water depth on the new ruler.

periods. In 2000, a new ruler was installed to correct the depths measured with the old ruler. The facies watery ground was not measurable with the ruler but was noticeable on the satellite images. It represents a transition between the continuous water body and the disconnected ponds with very shallow water. In the period studied, the Gallocanta Lake experienced a maximum water extent in February 1992, corresponding to a water depth of 1.136 m determined using the available rating curve (CHE, 2002).

10.4 DISCUSSION

Remote sensing has been applied in the Monegros and Gallocanta saline wetlands with the purpose of characterizing land cover types and, especially, determining the extent of flooded areas. In both cases, we have integrated field knowledge to the spatial and spectral data provided by satellite images. The multitemporal study based on a series of Landsat images was more productive in the Gallocanta Lake area, because the latter is much larger than any of the playa-lakes of Monegros. However, the lack of soil and vegetation maps and in-situ data at the dates of image acquisition had similar effect at both study sites, irrespective of their size. Spectral information from playa-lakes is very rich, and its full interpretation would require detailed and timely acquisition of ground information in test sites. Land facies were established from field observation that required the degradation of the spectral information, i.e., simplification by forming groups of spectral signatures. Satellite images enabled an integrated analysis of the playa-lakes inside their agricultural surroundings, which is difficult to attain from occasional and noncomprehensive field observations. Moreover, the multitemporal approach provided the historical perspective to understand land cover relationships and interpret the playa-lake status and evolution for environmental monitoring.

Visual interpretation of radar images contributes interesting features related to the texture and the structure of the area, even in the case of small playa-lakes. However, the potential of radar images for land cover discrimination using classification techniques requires the availability of computer-friendly programs to overcome operational difficulties related to the radar satellites and to the specific software needed. A multitemporal approach for land cover discrimination was applied in the arid environment of Monegros, taking advantage of the scarcity of natural vegetation. However, the small size and surface variation of the playa-lakes make the interpretation of standard and systematic land covers in multitemporal studies difficult.

10.5 CONCLUSIONS

Systematic and multitemporal information about the selected saline wetlands is available for the first time. The method developed in the Monegros site worked also well in the Gallocanta site. We envisage its application to similar saline environments in arid countries. Both field survey and remote sensing provided useful data. However, in practice, it was the synergistic use of both techniques that significantly improved the understanding of the playa-lakes' appearance over time and their relationship with the surrounding agricultural areas and the groundwater dynamics. A combination of radar and optical data holds great promise, as higher resolution radar sensors are being developed specifically for the study of land surfaces.

The remote sensing techniques applied here have demonstrated their usefulness at ground surveying scales when soil and vegetation maps are lacking, reducing the cost and effort of fieldwork in standard mapping procedures. Limitations depend on the objective pursued and must be taken into account to avoid useless or unsound results. The integrated and iterative use of ground observations and satellite images has provided new insights into the saline wetlands studied here. We feel this procedure has potential for future observations and research tasks and for the interpretation of retrospective image series.

ACKNOWLEDGMENTS

The Spanish Ministry of Education and Science supported this work with the research project AGL2006-01283, and with the grants EX2006-0347 to C. Castañeda and PR2007-0453 to J. Herrero.

REFERENCES

Brosttoff, W., Lichvar, R., and Sprecher, S. 2001. Delineating playas in the arid southwest. A literature review. Technical Report ERDC TR-01-4, US Army Corps of Engineers. Army Engineer Research and Development Center, Hanover, NH.

Bryant, R.G. 1999. Application of AVHRR to monitoring a climatically sensitive playa. Case study: Chott el Djerid, Southern Tunisia. *Earth Surface Processes and Landforms* 24: 283–302.

Casamayor, E., Castañeda, C., Pena Baixeras, A., Vich Homs, M.A., and Herrero, J. 2005. Monegros: Riqueza escondida bajo la sal del desierto. *Investigación y Ciencia* 349: 38–39.

Castañeda, C. and García-Vera, M.A. 2008. Water balance in the playa-lakes of an arid environment, Monegros, NE Spain. *Hydrogeology Journal* 16: 87–102.

Castañeda, C. and Herrero, J. 2005. The water regime of the Monegros playa-lakes established from ground and satellite data. *Journal of Hydrology* 310: 95–110.

Castañeda, C. and Herrero, J. 2006. Using Landsat imagery for reconstructing wetlands inventory and functional status in Monegros, NE Spain. In *Proceedings of the GlobWetland Symposium* (CDROM), ESA/ESRIN, Frascati, Italy.

Castañeda, C. and Herrero, J. 2008a. Assessing the degradation of saline wetlands in an arid agricultural region in Spain. *CATENA* 72: 205–213.

Castañeda, C. and Herrero, J. 2008b. Measuring the condition of inland saline habitats under agricultural intensification. *Pedosphere* 18: 11–23.

Castañeda, C., Herrero, J., and Casterad, M.A. 2005a. Facies identification within the playa-lakes of the Monegros Desert, Spain, with field and satellite data. *CATENA* 63: 39–63.

Castañeda, C., Herrero, J., and Casterad, M.A. 2005b. Landsat monitoring of playa-lakes in the Spanish Monegros Desert. *Journal of Arid Environments* 63: 497–516.

CHE (Ebro Water Authority). 2002. Establecimiento de las normas de explotación de la Unidad Hidrogeológica "Gallocanta" y la delimitación de los perímetros de explotación de la laguna. Technical Report, Zaragoza, Spain.

Chust, G., Ducrot, D., and Pretus, J.L. 2004. Land cover discrimination potential of radar multitemporal series and optical multispectral images in a Mediterranean cultural landscape. *International Journal of Remote Sensing* 25(17): 3513–3528.

Comín, F., Julià, R., and Comín, P. 1991. Fluctuations, the key aspect for the ecological interpretation of saline lake ecosystems. *Oecologia Aquatica* 10: 127–135.

Costa, L.T., Farinha, J.C., Hecker, N., and Tomàs-Vives, P. 1996. *Mediterranean Wetland Inventory: A Reference Manual*, Vol. 1. MedWet/Instituto da Conservaçao da Naturaleza/Wetlands International Publication, Lisbon, Potugal.

Dahl, T.E. and Johnson, C.E. 1991. *Status and Trends of Wetlands in the Conterminous United States, Mid-1970s to Mid-1980s*. US Department of the Interior, Fish and Wildlife Service, Washington, DC.

Davis, J.A. and Froend, R. 1999. Loss and degradation of wetlands in southwestern Australia: Underlying causes, consequences and solutions. *Wetlands Ecology and Management* 7: 13–23.

Desir, G., Gutiérrez-Elorza, M., and Gutiérrez-Santolalla, F. 2003. Origen y evolución de playas en una zona semiárida con arenas eólicas (región de Coca, Cuenca del Duero, España). *Boletín Geológico y Minero* 114(4): 395–407.

Díaz de Arcaya, N., Castañeda, C., Herrero, J., and Losada, J.A. 2005. Cartografía de coberturas asociadas a las fluctuaciones de la Laguna de Gallocanta. *Revista de la Asociación Española de Teledetección* 24: 61–65.

Dini, J., Cowan, G., and Goodman, P. 1998. South African National Wetland Inventory. National Wetland Inventory. http://www.environment.gov.za/soer/nsoer/resource/wetland/index.htm (accessed November 25, 2007).

Domínguez, M., Conesa, J.A., Pedrol, J., and Castañeda, C. 2006. Una base de datos georreferenciados de la vegetación asociada a las saladas de Monegros. In *Proceedings XII Congreso Nacional de Tecnologías de la Información Geográfica*, Granada, Spain.

Duguid, A., Barnetson, J., Clifford, B., Pavey, C., Albrecht, D., Risler, J., and McNellie, M. 2002. Wetlands in the arid Northern Territory. A report to Environment Australia on the inventory and significance of wetlands in the arid NT, Parks and Wildlife Commission of the Northern Territory, Alice Springs, Australia.

Epema, G.F. 1992. Spectral reflectance in the Tunisian desert. PhD dissertation, Wageningen Agricultural University, the Netherlands.

Environmental Laboratory. 1987. Corps of engineers wetlands delineation manual. Technical Report Y-87-1, US Army Engineer Waterways Experiment Station, Vicksburg, MS.

Escadafal, R. 1992. Télédétection de la surface des sols arides. Concept et applications. In: *L'aridité. Une contrainte au développement*, Eds. E. Le Floc'h, M. Grouzis, A. Cornet, and J.C. Bille. Editions Orstom, Paris, France.

European Union. 1997. Council Directive 97/62/EC of 27 October 1997 adapting to technical and scientific progress Directive 92/43/EEC on the conservation of natural habitats and of wild fauna and flora. *Official Journal of the European Communities* 305: 42–65, Luxembourg.

Finlayson, C.M. and Rea, N. 1999. Reasons for the loss and degradation of Australian wetlands. *Wetlands Ecology and Management* 7: 1–11.

Finlayson, C.M., Davidson, N.C., and Stevenson, N. 1999. Report from workshop 4: Wetland inventory, assessment and monitoring—practical techniques and identification of major issues. In *Wetlands—A Source of life*. Wetlands International/IUCN/WWF/Ministry of Environment and Nature Protection of Senegal, Dakar, Senegal.

Fjørtoft, R., Séry, F., Ducrot, D., Lopès, A., Lemaréchal, C., Fortier, C., Marthon, P., and Cubero-Castan, E. 1997. Segmentation, filtering and classification of SAR images. In *Proceedings 8th Latin American Symposium on Remote Sensing (SELPER'97)*, Mérida, Venezuela.

González-Audícana, M., García, R., and Herrero, J. 2003. Fusion of multispectral and panchromatic images using new methods based on wavelet transforms—evaluation of crop classification accuracy. In *Geoinformation for European-Wide Integration: Economic and Environmental Applications*, Ed. T. Benes, Millpress, Rotterdam, the Netherlands.

Gracia, F.J. 1995. Shoreline forms and deposits in Gallocanta Lake (NE Spain). *Geomorphology* 11: 323–335.

Gutiérrez-Elorza, M., Desir, G., and Gutiérrez-Santolalla, F. 2002. Yardangs in the semiarid central sector of the Ebro Depression (NE Spain). *Geomorphology* 44: 155–170.

Harris, A.R. 1994. Time series remote sensing of a climatically sensitive lake. *Remote Sensing of Environment* 50: 83–94.

Herrero, J. and Snyder, R.L. 1997. Aridity and irrigation in Aragón, Spain. *Journal of Arid Environments* 35: 535–547.

Lichvar, R., Gustina, G., and Bolus, R. 2004. Ponding duration, ponding frequency, and field indicators: A case study on three California playas. *Wetlands* 24(2): 406–413.

Lichvar, R., Brostoff, W., and Sprecher, S. 2006. Surficial features associated with ponded water on playas of the arid southwestern United States: Indicators for delineating regulated areas under the Clean Water Act. *Wetlands* 26(2): 385–399.

Martínez-Montoya, J.F., Herrero, J., and Casterad, M.A. 2003. Discriminación de la ocupación del suelo en un área yesosa de México mediante imágenes Landsat ETM+. In: *Teledetección y desarrollo regional*, Eds. U.R. Pérez and C.P. Martínez. X Congreso Nacional de Teledetección, Cáceres, Spain.

Martínez-Montoya, J.F. and Castañeda, C. 2004. Utilidad de las imágenes Landsat ETM+ en la discriminación de litología yesosa en Monegros Sur. *Geotemas* 6(2): 349–351.

Melic, A. and Blasco, J. Eds. 1999. Manifiesto científico por los Monegros. *Boletín de la Sociedad Entomológica Aragonesa* 24.

Millennium Ecosystem Assessment. 2005. Ecosystems and human well-being: Wetlands and water synthesis. A Report of the Millennium Ecosystem Assessment, World Resources Institute, Washington, DC.

OECD. 1996. *Guidelines for Aid Agencies for Improved Conservation and Sustainable Use of Tropical and Subtropical Wetlands*. Organisation for Economic Co-operation and Development, Paris, France.

Ozesmi, S.L. and Bauer, M.E. 2002. Satellite remote sensing of wetlands. *Wetlands Ecology and Management* 10: 382–402.

Peña, J.A. and Marfil, R. 1986. La sedimentación salina actual en las salinas de La Mancha: una síntesis. *Cuadernos de Geología Ibérica* 10: 235–270.

Ramsar Convention Secretariat. 2006. *The Ramsar Convention Manual: A Guide to the Convention on Wetlands (Ramsar, Iran, 1971)*, 4th ed. Gland, Switzerland.

Rodríguez, M., Benavente, J., Cruz-San Julián, J.J., and Moral, F. 2006. Estimation of ground-water exchange with semi-arid playa lakes (Antequera region, southern Spain). *Journal of Arid Environments* 66: 272–289.

Rosenqvist, A., Finlayson, C.M., Lowry, J., and Taylor, D. 2007. The potential of long-wavelength satellite-borne radar to support implementation of the Ramsar Wetlands Convention. *Aquatic Conservation: Marine and Freshwater Ecosystems* 17: 229–244.

Samper-Calvete, F.J. and García-Vera, M.A. 1998. Inverse modelling of groundwater flow in the semiarid evaporitic closed basin of Los Monegros, Spain. *Hydrogeology Journal* 6: 33–49.

Stevenson, N.J. and Frazier, S. 1999. Status of national wetland inventories in Africa. In: *Global Review of Wetland Resources and Priorities for Wetland Inventory*, Eds. C.M. Finlayson and A.G. Spiers. Supervising Scientist Report 144, Canberra, Australia.

Swain, P.H. 1973. Pattern recognition: A basis for remote sensing data analysis (LARS information note 111572). Laboratory for Applications of Remote Sensing, Purdue University. West Lafayette, Indiana.

Taft, O.W., Haig, S.M., and Kiilsgaard, C. 2004. Use of radar remote sensing (RADARSAT) to map winter wetland habitat for shorebirds in an agricultural landscape. *Environmental Management* 33: 750–763.

Taylor, G.R., Mah, A.H., Kruse, F.A., Kierein-Young, K.S., Hewson, R.D., and Bennett, B.A. 1996. Characterization of saline soils using airborne radar imagery. *Remote Sensing of Environment* 57: 127–142.

Tiner, R.W. 2002. Watershed-based wetland planning and evaluation. A collection of papers from the Wetland Millennium Event (August 6–12, 2000; Quebec, Canada). Association of State Wetland Managers, Berne, NY.

Tiner, R.W. 2004. Remotely-sensed indicators for monitoring the general condition of "natural habitat" in watersheds: An application for Delaware's Nanticoke River watershed. *Ecological Indicators* 4: 227–243.

Tiner, R.W., Bergquist, H.C., DeAlessio, G.P., and Starr, M.J. 2002. *Geographically Isolated Wetlands: A Preliminary Assessment of Their Characteristics and Status in Selected Areas of the United States.* US Department of the Interior, Fish and Wildlife Service, Northeast Region, Hadley, MA.

Tomàs-Vives, P. (Ed.). 1996. *Monitoring Mediterranean Wetlands: A Methodological Guide.* MedWet publication, Wetlands International, Slimbridge, UK, and ICN, Lisbon, Portugal.

11 Mapping Soil Salinity Using Ground-Based Electromagnetic Induction Technique*

Florence Cassel S., Dave Goorahoo, David Zoldoske, and Diganta Adhikari

CONTENTS

---

* Mention of trademark or proprietory products in this chapter does not constitute a guarantee or warranty of the products by the authors and does not imply its approval to the exclusion of other products that may also be suitable.

11.1　INTRODUCTION

Soil salinity is a critical problem in many arid and irrigated agricultural areas of the world because of saline parent material, intensive irrigation, shallow water table, and inadequate drainage that prevents the leaching of soluble salts. Excessive salinity negatively affects crop productivity and eventually results in land degradation. Therefore, it is important to evaluate the extent and progression of soil salinity in order to select agricultural, irrigation, and drainage management practices that will help alleviate salinization. Soil salinity has traditionally been assessed from soil sampling and laboratory analyses that are labor-intensive, time-consuming, and expensive. The electromagnetic (EM) induction technique has become a very useful and cost-effective tool to monitor and diagnose soil salinity over large areas, because it allows for rapid, above-ground measurements with noninvasive sampling. Additionally, when coupled with a GPS and data logging capabilities, a mobilized EM system can provide automated and georeferenced measurements of soil salinity over vast areas.

This chapter covers the principle of the EM technology and its advantages and limitations for salinity assessment and mapping. It also describes commonly used EM sensors; explains the acquisition, calibration, and mapping of the EM data; and provides examples of studies conducted using the EM technology.

11.2　SOIL SALINITY ASSESSMENT

11.2.1　Importance of Salinity Assessment

Soil salinity is a severe agricultural and environmental problem in many arid and semiarid regions around the globe (Hillel, 2000). Elevated salinity levels adversely affect crop production, soil quality, and water quality, and eventually result in soil erosion and land degradation (Rhoades and Loveday, 1990; Corwin and Lesch, 2003). Although salinity mostly affects irrigated lands, ecosystems such as wetlands and forests are also being degraded by escalating salinity levels.

Evaluating the worldwide extent of salt-affected areas is difficult. It is generally estimated that approximately 955 million ha of land are subjected to primary salinity and that 77 million ha are affected by secondary salinization (Metternicht and Zinck, 2003). Salinity affects about 20% of the irrigated lands, which represents approximately 50 million ha (Corwin and Lesch, 2003). According to Postel (1999), salinization increases worldwide at a rate of up to 2 million ha a year. For instance, in California, salinity affects large areas because of the inherently clayey and saline nature of the soils, intensive irrigation with saline water that results in rising water tables, high evapotranspiration, and inadequate drainage of subsurface saline waters. About 2 million ha of irrigated farmlands in the state are estimated to be saline soils or affected by saline irrigation water (Letey, 2000). In the western part of the San Joaquin Valley in Central California, the net daily salt inflow into the region during the irrigation season is approximately 1.46 million tons (San Joaquin Valley Drainage Implementation Program, 1998).

With the extent of salinization that causes productivity loss and severe environmental problems, monitoring and managing salinity have become essential at the field, farm, local, and regional scales in order to sustain agricultural lands and natural ecosystems. Particularly, assessing the spatial and temporal changes in soil salinity is needed to understand the causes of salinization and recommend appropriate water management and soil reclamation practices. In irrigated agricultural areas, such assessment is necessary to reduce secondary salinization and sustain land quality.

11.2.2 TRADITIONAL ASSESSMENT METHODOLOGIES

Traditionally, soil salinity has been assessed by collecting in situ soil samples and analyzing those samples in the laboratory to determine their solute concentrations or electrical conductivity (EC). Laboratories use the water extract of the soil samples to measure EC in $mS\ m^{-1}$ or $dS\ m^{-1}$. The water extracts are obtained using different ratios of soil to water weights (i.e., 1:1, 1:2, or 1:5 ratios) or saturated pastes (EC_e). Each ratio gives a different EC value although the soil salinity is the same, and therefore interpretation of the laboratory data needs to be conducted cautiously. Calculation of soil EC on saturated paste-extracts is widely accepted, because the guidelines describing crop response and tolerance to salinity are based on soil EC_e (Rhoades et al., 1999). Hanson et al. (1999) provide such guidelines for a variety of fiber, grain, grass, forage, vegetable, fruit, and woody crops, as well as for ornamental shrubs, trees, and ground covers. However, determination of soil EC using soil to water ratios is often preferred when normalization of the data is needed for analysis.

Although the traditional surveys based on soil sampling and laboratory determinations have conventionally been used to monitor and quantify soil salinity, they involve destructive sampling and only provide measurements at particular locations, rendering salinity assessment over large areas very costly and time-consuming. Additionally, traditional surveys do not adequately account for the spatial variability of soil salinity because of the small number of measurements collected (Corwin and Lesch, 2003). These limitations reduce the accuracy of maps derived from such data. Therefore, the need for extensive but cost-effective appraisal becomes evident if soil salinity is to be evaluated precisely at the field scale or over large geographical areas.

11.2.3 FIELD SURVEY INSTRUMENTS

The ability to map and monitor soil salinity quickly has been greatly enhanced with the development of field survey instruments. Three types of portable instruments are commercially available for salinity appraisal: (1) remote EM induction sensors; (2) electrode sensors, including surface array or insertion probes; and (3) time-domain reflectometry (TDR) sensors. Both remote EM induction and electrode sensors provide depth-weighted apparent electrical conductivity (EC_a) measurements that can be converted to soil salinity (Rhoades et al., 1976, 1989; Corwin and Rhoades, 1990, 2005). The main distinction between the two instrument types is that EM sensors are based on the conductivity principle with the use of induced current flows, whereas electrode sensors are constructed on the resistivity principle where electric currents are directly injected into the ground. The TDR technology is based on the time required for a voltage pulse to travel along a probe and back, which is dependent on the dielectric constant of the measured media. This technique was first developed for measuring volumetric water content in soils (Topp and Davis, 1981; Topp et al., 1982), but it was further applied for calculating EC_a (Dalton, 1982; Dalton et al., 1984; Wraith et al., 2005). Although the TDR technology has been demonstrated to provide EC_a measurements comparable to those obtained with the other two methods (Mallants et al., 1996; Wraith, 2002), the stationary nature of its instrument has prevented mapping of large areas.

Among the three technologies, the EM sensors are the most suited for field-scale agricultural applications as well as for environmental and hydrogeological studies (Corwin and Rhoades, 1982; Cook et al., 1989; Hoekstra et al., 1992; Sudduth et al., 2003; Paine et al., 2004). The importance attributed to these sensors is explained by their reliability, robust construction, compact design, noncontacting nature, and relative ease of use, as well as the accuracy and close relation of EC_a measurements to salinity. Such characteristics enable a large volume of data to be acquired quickly over extensive areas. The EM sensors are particularly suitable for compacted soils where electrode sensors may not make good contact with the ground or may disturb vegetation or crops. Numerous researchers have demonstrated the usefulness of EM surveys as a rapid and economical technique to provide 3D quantification of soil salinity levels (Hendrickx et al., 1992; Lesch et al., 1992; McNeill, 1992; Rhoades et al., 1999). Since the EM remote sensing technique is most extensively used for salinity mapping, it will be the focus of this chapter.

11.3 GROUND-BASED ELECTROMAGNETIC INDUCTION TECHNIQUE

11.3.1 ELECTROMAGNETIC (EM) INDUCTION PRINCIPLE

EM induction is based on the principle that electric currents can be applied to the soil through induction and that the magnitude of the induced current loops is directly proportional to the depth-weighted EC_a of the soil (McNeill, 1980; Rhoades and Corwin, 1981). EM induction devices are composed of a transmitter coil and a receiver coil installed at each end of a nonconductive bar. An alternating, time-varying electrical current in the transmitter coil creates a primary magnetic field in the ground (Figure 11.1). This field induces circular eddy current flow loops, which in turn generate a secondary magnetic field. A receiver coil, placed at a fixed distance from the transmitter, measures both the primary (H_p) and secondary (H_s) fields. The EM devices quantify two components of the induced magnetic field: the first is the quadrature phase that gives the soil EC_a by measuring the ratio of the two magnetic fields and the second is the in-phase that detects buried metallic objects.

In general, the quadrature (q) component of the ratio of H_s to H_p is a complicated function of the intercoil spacing, the operating frequency of the alternating current, and the ground conductivity. However, under conditions of operation at low induction number ($N \ll 1$), homogeneous half-space, and instrument held at the soil surface, the ratio becomes a much simpler function:

$$\left(\frac{H_s}{H_p}\right)_q \cong \frac{i\omega\mu_o\sigma_a s^2}{4} \qquad (11.1)$$

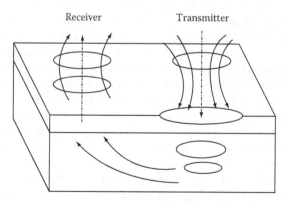

FIGURE 11.1 Schematic representation of the EM induction principle. (After Ehsani, R. and Sullivan, M., Soil electrical conductivity (EC) sensors. Extension Fact Sheet AEX-565–02, Ohio State University, Columbus, OH, 2002.)

where

$\omega = 2\pi f$ (rad s^{-1}) with $f =$ operating frequency (Hz)

μ_o is the permeability of free space ($4\pi \times 10^{-7}$ H m^{-1})

σ_a is the ground conductivity (mS m^{-1})

s is the intercoil spacing (m)

$i = \sqrt{-1}$

Since the ratio between the two magnetic fields is linearly proportional to the ground conductivity, the soil EC$_a$ can be calculated as follows (McNeill, 1980):

$$\sigma_a = \frac{4}{\omega \mu_o s^2} \left(\frac{H_s}{H_p} \right)_q \tag{11.2}$$

The induction number N is identified as the ratio of the intercoil spacing s to the electrical skin depth δ. Thus, small induction numbers are obtained when the coils are closely spaced compared to the skin depth. The electrical skin depth is a parameter used to quantify the approximate depth of exploration of the induced current and is defined as

$$\delta = \sqrt{\frac{2}{\mu_o \omega \sigma_a}} \tag{11.3}$$

Therefore, as the operating frequency of the EM instrument is increased, the exploration depth of the measurements decreases. In a homogeneous half-space, the skin depth is defined as the depth at which the primary magnetic field has been attenuated to $1/e$ (i.e., 37%) of its original strength (McNeill, 1980).

In addition to the transmitter-to-receiver coil separation and the operating frequency, the effective exploration depth also depends on the coil orientation relative to the soil surface. Coils can be oriented either vertically or horizontally, inducing coplanar or coaxial current loops, respectively (Figure 11.2). Measurements made in the horizontal dipole configuration are particularly sensitive to the shallow subsurface conductivity, whereas vertical measurements have a greater sensitivity to deeper conductive layers. However, it is important to note that the ratio described in Equation 11.1 is independent of the dipole configuration.

The sensitivity of the EM response with depth, mathematically translated into a nonlinear function, represents an important characteristic of the EM technology. Figure 11.3 illustrates this depth-weighted nonlinearity by showing the relative contribution to H_s from a homogeneous conductive layer dZ located at a normalized depth Z (where Z is the actual soil depth z divided by the intercoil spacing s). The figure reveals that, for vertical dipoles, the soil material found at a depth of approximately 0.4 s provides the maximum contribution to H_s and that the material near the surface makes a very small contribution. The equations $\Phi(Z)$ governing these relative EM responses are defined for the vertical and horizontal dipoles as (McNeill, 1980)

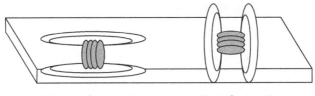

Horizontal orientation   Vertical orientation

FIGURE 11.2   Coil orientation of EM devices.

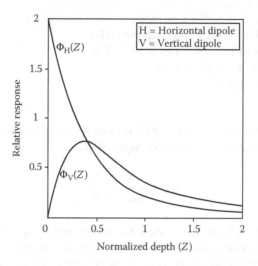

FIGURE 11.3 Comparison of the horizontal $\Phi_H(Z)$ and vertical $\Phi_V(Z)$ relative instrument responses versus normalized depth Z. Normalized depth Z is defined as actual soil depth z (m) divided by intercoil spacing of instrument s (m). (From McNeill, J.D., Electromagnetic terrain conductivity measurements at low induction numbers. Technical Note TN-6, Geonics Ltd., Mississauga, Ontario, Canada, 1980.)

$$\Phi_V(Z) = \frac{4Z}{(4Z^2 + 1)^{3/2}} \qquad (11.4)$$

$$\Phi_H(Z) = 2 - \frac{4Z}{(4Z^2 + 1)^{1/2}} \qquad (11.5)$$

If we consider the contribution from the different layers below a depth Z, the cumulative EM response can be related to the relative contributions of each layer with the following equation:

$$R(Z) = \int_0^\infty \Phi(Z)dZ = 1 \qquad (11.6)$$

For both dipole configurations, the cumulative response functions R(Z) are expressed as

$$R_V(Z) = \frac{1}{(4Z^2 + 1)^{1/2}} \qquad (11.7)$$

$$R_H(Z) = (4Z^2 + 1)^{1/2} - 2Z \qquad (11.8)$$

Thus, the EC_a measured by the EM instruments can be computed mathematically as follows (McNeill, 1980; Cook and Walter, 1992):

$$EC_a = R(Z)EC(Z) = \int_0^\infty \Phi(Z)EC(Z)dZ \qquad (11.9)$$

Since the EC_a readings represent the average conductivity of the different layers within the depth of investigation, the measurements are referred to as depth-weighted or bulk EC_a.

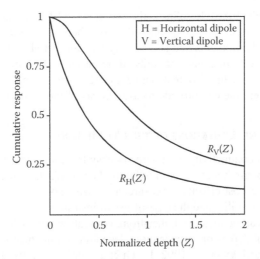

FIGURE 11.4 Cumulative horizontal $R_H(Z)$ and vertical $R_V(Z)$ instrument responses versus normalized depth Z. Normalized depth Z is defined as actual soil depth z (m) divided by intercoil spacing of instrument s (m). (From McNeill, J.D., Electromagnetic terrain conductivity measurements at low induction numbers. Technical note TN-6, Geonics Ltd., Mississauga, Ontario, Canada, 1980.)

Figure 11.4 depicts the cumulative response for both vertical and horizontal dipole configurations. The function represents the relative contribution to H_s from all layers below a normalized depth Z. For example, in the vertical configuration, all layers below a depth of two intercoil spacings contribute 25% to H_s. Additional details on the principle of the EM technology and assumptions presented above for the calculation of EC_a are described in detail in McNeill (1980). Predictions of EC_a in layered soils and at different instrument heights above the surface are also discussed in Borchers et al. (1997) and Hendrickx et al. (2002).

11.3.2 SOIL FACTORS AFFECTING EM MEASUREMENTS

The bulk EC_a, as measured by EM devices, is affected by the conductors found in the ground as well as the physical and chemical properties of the soil matrix. The ground conductors include dissolved electrolytes in the soil water, conductive minerals found in the rocks and soil clays, and buried metallic objects (Mc Neill, 1980). The presence of all these conductors increases the ground conductivity. However, in the absence of metallic objects, the ground conductivity is primarily electrolytic since most soil and rock minerals are poor electrical conductors. The electrolytes, either inherent to the soil or introduced through anthropogenic activities, are found in the form of cations adsorbed to the clay minerals (i.e., Ca, Mg, Na, H, K) and by-products of chemical weathering (i.e., K and HCO_2 released during the dissolution of minerals), as well as anions and cations (i.e., NO_3, NH_4, etc.) released by agrochemicals or present in applied waters. The conductivity of all these electrolytes is proportional to the total number of ions in solution, their charge, and velocity. In addition to electrolytes, several soil physical properties, including porosity (shape, size, and number of pores and interconnecting passages), moisture content (extent to which the pores are filled with water), and pore-water temperature also greatly affect the ground conductivity.

Therefore, in principle, the EM readings increase with a rise in ion concentration, cation exchange capacity (CEC), volumetric water and clay content, and soil water temperature (Rhoades et al., 1976, 1999; McNeill, 1980). In saline soils, variations in EM measurements are primarily due to changes in ionic concentrations in the soil water, and therefore good correlations are generally obtained between EM measurements and soil salinity (De Jong et al., 1979; Cameron et al., 1981; Williams and Baker, 1982). In nonsaline soils, differences in EM measurements are mainly a function of soil texture, moisture content, and CEC (Rhoades et al., 1976; Kachanoski et al., 1988). Thus, the EM technique can provide measurements of those soil properties.

11.3.3 ADVANTAGES AND LIMITATIONS OF THE EM METHOD

Measurements of ground conductivity are relatively easy to acquire and record using EM devices. The EM method offers many benefits, the most important being the remote nature of the devices, which provides rapid and cost-effective measurements with no ground contact. Other important advantages include the portability and the ease-of-use of instruments, the precision with which small changes in conductivity can be measured, the strong correlation with salinity, and the possibility of collecting data at different depths of investigation, depending on the instrument operating frequency and coil orientation (Hendrickx et al., 1992; Lesch et al., 1995a; Johnston et al., 1997).

The main disadvantage of the EM technique relates to the relative nature of the EC_a measurements. Since the EM response is dependent on several soil properties, including salinity, water content, and clay content, conversion of the EC_a measurements is necessary to obtain absolute soil salinity estimates (EC_e). The conversion involves calibration of the EM data through soil sampling and statistical analysis. Such a process needs to be conducted very rigorously and scientifically to obtain accurate salinity estimates. Numerous calibration procedures have been developed by researchers; however, most remain site-specific and must be conducted after each survey. Apart from calibration of the EM data, an internal calibration of the sensors is also required before each use. Another limitation is the fixed intercoil spacing of certain EM instruments and their single operating frequency, which limit the depths of exploration. To address this problem, newer instruments are being developed with two or three receiver coils placed at different distances from the transmitter coil or with multifrequency sensors. In addition, the EM instruments are subject to electronic drift with changing ambient conditions, such as air temperature and humidity (Sudduth et al., 2003). Their measurements are also affected by metallic or EM objects present within the zone of influence of the transmitter. Such objects can include buried pipelines, culverts, and utilities, as well as above-ground metal fences, debris, tools, power lines, and any metallic items worn by the operator or found on the tow vehicle. Therefore, the EM instrument should always be positioned at least three meters from any metallic object to avoid erroneous EC_a readings. Finally, the cost of the EM instruments remains high and varies from about \$9,000 to \$40,000, depending mostly on the depth of exploration and the number of receiver coils. Table 11.1 summarizes the main benefits and disadvantages of the EM methodology.

11.3.4 EM SENSORS

Several noninvasive survey instruments have been developed over the past 50 years to obtain soil information through EM induction measurements. Each instrument operates at one or several specific frequencies, which determine the depth of investigation. Instruments range from small handheld units to larger devices that can be mounted on vehicles or aircraft. Handheld units provide near-surface measurements of soil salinity, useful in frequent monitoring and mapping. These instruments contain a transmitter and receiver spaced at a fixed distance (1–4 m). Large ground-based and airborne units are more adapted for mapping deep subsurface salinity and identifying variability within the landscape. These instruments can be operated at varying intercoil spacing selected by the operator.

The most commonly used handheld instruments for salinity appraisal at the field scale include the EM31, EM34, and EM38 (Geonics Ltd., Mississauga, ON Canada), and the newer GEM2 (Geophex, Raleigh, NC) (Figure 11.5). The Geonics instruments operate at a fixed frequency,

TABLE 11.1

Advantages and Limitations of Ground EM Induction Technique

Advantages	Limitations
Strong correlation with salinity	Subject to changes in moisture and clay contents
Precise, rapid, and cost-effective measurements	Measurements affected by metallic objects and extreme climatic conditions (temperature)
Remote technique (no ground contact)	Calibration at key locations needed to obtain salinity estimates
Useful in mapping near-surface variability and salt loading	Limited range of intercoil spacing (i.e., depth of exploration)
Once calibrated, technique provides real-time salinity data	Data interpretation needs to be conducted carefully
Effectiveness not limited by the presence of highly resistive layers in the near-surface	Differentiation between primary salinity (naturally occurring in soils and rocks) and secondary salinity (resulting from human activities) is not possible
Portability and ease-of-use of instrument	Expensive instrument

whereas the Geophex device is a multifrequency sensor. All instruments provide rapid measurements with a high degree of precision and accuracy. They are also equipped with calibration controls, digital readout, and real-time (RT) modification for digital output signal.

11.3.4.1 EM38

The EM38 sensor operates at a fixed frequency of 14.6 kHz, which corresponds to $\omega = 91.7 \times 10^3$ rad s^{-1}. With an intercoil spacing of 1 m, this sensor provides EC_a measurements at effective

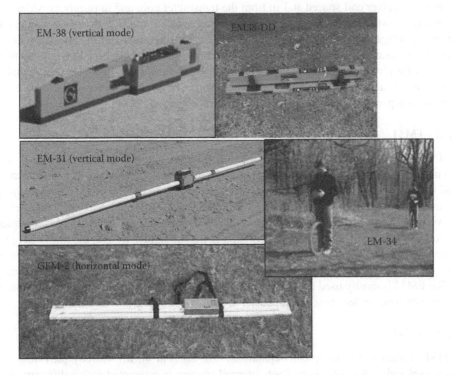

FIGURE 11.5 Various EM instruments commonly used for salinity appraisal and mapping. (From Geonics Ltd., Mississauga, ON Canada and Geophex Ltd, Raleigh, NC)

depths of 0–0.75 m in the horizontal dipole mode and 0–1.5 m in the vertical dipole mode. As indicated above, the horizontal dipole mode response is most sensitive to soil properties near the surface, whereas the vertical response better describes properties of deeper layers, with the greatest sensitivity at about 0.4 m in the case of the EM38. Approximately 80% and 50% of the EM38 response comes from the top 60 cm of the soil profile when placed in the horizontal and vertical dipole mode, respectively. The response is reduced to 8% and 29% for depths greater than 90 cm (Wollenhaupt et al., 1986; McKenzie et al., 1989). The difference between the two measurements can be used to determine the salt profile (regular, inverted) and infer the effectiveness of management practices. Being light-weighted (3 kg), this instrument was originally designed for agricultural applications and is best suited for rapid and shallow root-zone surveys. In addition to bulk EC (quad-phase), the EM38 sensor also provides measurements of magnetic susceptibility (in-phase) useful in detecting buried metallic objects. Obtained separately with the standard EM38 device, these two phase components are provided simultaneously with another instrument, the EM38B. The dual dipole sensor (EM38-DD), consisting of two integrated EM38 units oriented in the horizontal and vertical dipole positions and providing synchronized measurements down to 0.75 and 1.5 m, respectively, is a very popular device among researchers and engineers. With the simultaneous acquisition of horizontal and vertical measurements, the dual dipole instrument greatly reduces the survey time.

However, all EM38 instruments will soon be discontinued and replaced with the new EM38-MK2 models. These later models operate at a fixed frequency of 14.5 kHz, remain very portable (5.4 kg), and are available as two different sensors: EM38-MK2-1 and EM38-MK2-2. The two sensors differ by their number of vertical coils and thus by their exploration depth capability. The EM38-MK2-2 instrument has two receiver coils spaced at 0.5 and 1 m from the transmitter coil, thereby providing simultaneous measurements down to 0.75 and 1.5 m in the vertical dipole orientation, respectively. Additionally, when placed in the horizontal position, the instrument also measures the bulk EC_a at depths of 0–0.375 and 0–0.75 m. Therefore, compared to the EM38-DD sensor, this new device provides additional depths of exploration. The EM38-MK2-1 instrument possesses one receiver coil spaced at 1 m from the transmitter coil, and thus only provides separate measurements down to 0.75 m in the horizontal orientation and 1.5 m in the vertical mode. The main advantages of the new models compared to the previous EM38 series include (1) the simultaneous measurements of the two phase components: conductivity (quad-phase) and magnetic susceptibility (in-phase), (2) a Bluetooth wireless capability in addition to the RS-232 digital option for data communication, and (3) a temperature-compensation circuitry to avoid data drift associated with internal temperature changes.

11.3.4.2 EM31

The EM31 device operates at a fixed frequency of 9.8 kHz and provides measurements down to about 3 m in the horizontal dipole configuration and 6 m in the vertical mode. The instrument has an intercoil spacing of 3.66 m and weighs about 12 kg. Approximately 50% of the instrument response comes from the upper 1.4 and 3 m when placed in the horizontal and vertical orientations, respectively (McNeill, 1980). A short version of the EM31 (EM31-SH) is available for exploration depths down to 4 m. With a smaller coil separation (2 m) and lighter weight, the EM31-SH provides shallower measurements, greater lateral resolution, and easy portability. Both instruments also measure the magnetic susceptibility, i.e., in-phase ratio of the secondary to the primary magnetic fields. The EM31 is mostly used for environmental investigations aimed at appraising soil properties below the root zone or detecting contaminants in subsurface soil layers and aquifers.

11.3.4.3 EM34

The EM34 instrument has three coil separations (10, 20, and 40 m), thereby providing variable depths of exploration down to 60 m. Each intercoil spacing is associated with a different operating frequency: 6.4, 1.6, and 0.4 kHz, respectively. In the horizontal mode, the nominal depth of penetration is about 75% of the coil spacing. In the vertical mode, penetration depth is around 150% of the coil separation distance. The device comes with two meters: one to set the correct intercoil spacing and the second to read the EC_a measurements. Data acquisition remains very slow with the EM34, because (1) two people are required to carry the coils and operate the sensor and (2) the digital data output feature is not available. The EM34 instrument has been used in a few studies for salinity appraisal and delineation at field- and watershed-scales (Brune et al., 1999, 2001), but is mostly intended for very deep salinity mapping, detection of contaminant plumes, and location of groundwater resources (van Lissa et al., 1987; Hazell et al., 1988; Thamke et al., 1999).

11.3.4.4   GEM2

The GEM2 instrument is a digital, broadband, multifrequency sensor. The device operates in a wide range of frequencies, varying from about 300 Hz to 24 kHz, which allow conductivity measurements to be acquired at multiple depths down to 50 m. The GEM2 is very portable, weighing about 4 kg, and contains three coils: transmitter, receiver, and bucking. The transmitter-to-receiver coil separation distance is about 1.7 m. The operation principle, different from that described in Section 11.3.1, is based on the pulse-width modulation technique, whereby the sensor can transmit any arbitrary broadcast waveform of multiple frequencies and can operate either in a frequency or time-domain mode. The transmitter current decreases logarithmically with frequency, and the bucking coil removes the primary magnetic field from the receiver coil. Details on the operation principle of this instrument can be found in Geophex (2007). Although the GEM2 instrument has been used by few researchers for salinity mapping (Won et al., 1996), its applications have mostly focused on imaging and characterizing buried subsurface features, as well as mapping of underground storage tanks, landfills, and contaminant plumes (Huang, 2005).

In summary, each Geonics sensor has a specific intercoil spacing and operates at a fixed frequency. As the intercoil spacing increases and the operating frequency decreases, the depth exploration capability of the sensor is enhanced. Being a multifrequency sensor, the GEM2 provides numerous exploration depths. Table 11.2 summarizes the characteristics of each sensor described above.

11.3.5   APPLICATIONS

Measurements of ground conductivity using EM induction have been applied for more than five decades in the field of geophysical sciences (Belluigi, 1948; Wait, 1954, 1955). In agriculture, the EM technique was introduced in the late 1970s with work conducted by Rhoades et al. (1976), De Jong et al. (1979), and Rhoades and Corwin (1981) for salinity appraisal. It is now commonly employed to assess and map several soil properties, such as salinity, moisture, and clay contents (Cannon et al., 1994). Generally, the EM technique has been most successful in areas where

TABLE 11.2
Specifications of Commonly Used EM Sensors

Intercoil Spacing (m)	Operating Frequency (kHz)	Depth of Exploration (m)		Instrument
		Horizontal	Vertical	
1	14.6	0.75	1.5	EM38
2	9.8	2.0	4.0	EM31-SH
3.66	9.8	3.0	6.0	EM31-MK2
10	6.4	7.5	15	EM34
20	1.6	15	30	EM34
40	0.4	30	60	EM34
1.7	0.3–24	25	50	GEM2

subsurface properties are reasonably homogeneous, the effects of one property dominate over those of the other properties, and variation in the EM response can be related to changes in the dominant property (Rhoades and Corwin, 1990; Cook and Walker, 1992).

Applications of the EM technology are numerous and are listed extensively in Table 11.3. These applications include salinity mapping and monitoring in agricultural lands and natural ecosystems, soil water content measurement, vadose zone characterization, water table depth estimation, soil textural feature identification, contaminant detection in soils and aquifers, deep salt storage location,

TABLE 11.3
Literature Presenting Various Applications of the EM Technology

Properties	Sensor	References
Directly measured soil properties		
Salinity	EM38	Bennett and George (1995), Cameron et al. (1981), Carroll and Oliver (2005), Cassel et al. (2003a, 2006), Ceuppens et al. (1997), Cook and Walker (1992), Corwin and Rhoades (1982, 1990), Diaz and Herrero (1992), Hanson and Kaita (1997), Hendrickx et al. (1992), Herrero et al. (2003), Johnston et al. (1997), Kaffka et al. (2005), Lesch et al. (1992, 1995a, 1995b, 1998, 2005), Mankin and Karthikeyan (2002), McKenzie (2000), McKenzie et al. (1989, 1997), Nettleton et al. (1994), Nogués et al. (2006), Rhoades (1993), Rhoades and Corwin (1981), Rhoades et al. (1990, 1999), Slavish (1990), Slavish and Petterson (1990), Triantafilis (2000), Vaughan et al. (1995), Wollenhaupt et al. (1986)
	EM31	De Jong et al. (1979), Kinal et al. (2006), Sheets et al. (1994)
	EM34	Brune et al. (1999, 2001), Thamke et al. (1999), Williams and Baker (1982)
	GEM	Doolittle et al. (2001), Huang (2005), Won et al. (1996)
Physical and chemical properties (e.g., clay/silt/sand content, depth to clay pan, bulk density)	EM38	Anderson-Cook et al. (2002), Brevik and Fenton (2002), Brus et al. (1992), Corwin and Lesch (2005), Domsch and Giebel (2004), Jung et al. (2005, 2006), Lesch et al. (2005), Rhoades et al. (1999), Sudduth et al. (2003, 2005), Triantafilis et al. (2001a), Triantafilis and Lesch (2005)
	EM31	Inman et al. (2002)
	EM34	Triantafilis and Lesch (2005), Williams and Hoey (1987)
		Boettinger et al. (1997), Doolittle et al. (1994), Kitchen et al. (1996), Sudduth and Kitchen (1993)
Water-related (e.g., water content, water table depth, groundwater exploration, aquifer location, drainage)	EM38	Brevik and Fenton (2002), Hanson and Kaita (1997), Huth and Poulton (2007), Reedy and Scanlon (2003), Scanlon et al. (1999), Schumann and Zaman (2003), Vaughan et al. (1995)
	EM31	Hazell et al. (1988), Scanlon et al. (1999), Sheets and Hendrickx (1995), Kachanoski et al. (1988)
	EM34	van Lissa et al. (1987)
Indirectly measured soil properties		
Organic matter-related (e.g., organic carbon, organic chemical plumes		Brune and Doolittle (1990), Jaynes (1996), Nobes et al. (2000)
CEC	EM31 EM38	McBride et al. (1990), Triantafilis et al. (2002)

TABLE 11.3 (continued)
Literature Presenting Various Applications of the EM Technology

Properties	Sensor	References
Leaching	EM38	Slavish and Yang (1990), Corwin et al. (1999), Rhoades et al. (1999)
Soil map unit boundaries	EM31, EM38	Stroh et al. (2001)
Groundwater recharge		Cook et al. (1989)
Herbicide partition coefficients	EM38	Jaynes et al. (1995)
Nutrient status; contaminants	EM34, EM31, EM38	Drommerhausen et al. (1995), Eigenberg and Nienaber (1998, 2003), Eigenberg et al. (1998), Greenhouse and Slaine (1983, 1986)
Applications in crop production		
Crop productivity	EM38	Kitchen et al. (1999), McKenzie (2000)
Precision agriculture	EM31	Cassel et al. (2003b), Kitchen et al. (2005),
	EM38	Sawchik and Mallarino (2007), Schepers et al. (2004)

Source: Adapted From Corwin, D.L. and Lesch, S.M., *Comput. Electron. Agric.*, 46, 11, 2005.

soil nutrient status, crop productivity assessment, and precision agriculture. The EM technique has also been used to delineate soil mapping units as well as to estimate depth and thickness of horizons. However, as the EM instruments provide direct measurements of ground conductivity, the technology has been predominantly applied to assess spatial and temporal changes in soil salinity (Diaz and Herrero, 1992; Lesch et al., 1998; Cassel et al., 2003a), determine areas at future risk from salinity, review and quantify the extent of salinity problems (McFarlane and George, 1992; Cannon et al., 1994; Triantafilis et al., 2002; Paine et al. 2003; Wittler et al., 2006), evaluate the potential of ecosystem restoration (Sheets et al., 1994), and appraise the impact of agricultural practices, such as irrigation, drainage management, fertilizer, and amendment applications (George and Bennett, 1999; Rhoades et al., 1999; Cassel et al., 2006).

11.4   SOIL SALINITY MAPPING

11.4.1   DATA ACQUISITION

Acquisition of the EM data can be conducted manually or through mobilized survey systems. Manual measurements are obtained by carrying the EM instrument and placing it at or above the soil surface. At each measurement location, the EC_a values are then recorded by hand or downloaded to a data logger or handheld computer. A GPS device can also be carried along with the EM instrument to georeference the measurement locations (Figure 11.6). Although this data acquisition procedure might be the only option in certain survey areas where access to mobilized systems is impossible, it remains relatively slow and is thus not feasible if large areas need to be surveyed. Therefore, to improve the efficiency of EC_a data collection in large survey areas, EM instruments are often mounted on platforms or towed behind vehicles. The georeferencing of the data is easily achieved by mounting a GPS receiver onto the mobilized system. The EM and GPS devices are then connected to a computer through digital interfaces.

The development of mobilized ground systems that integrate EM sensors and GPS receivers with data logging capabilities has considerably improved our capacity to perform extensive salinity mapping. Numerous scientists have built such systems for rapid and detailed characterization of soil salinity variability across large survey areas (Cannon et al., 1994; Rhoades et al., 1999; Triantafilis, 2000; Freeland et al., 2002; Cassel et al., 2003a; Corwin and Lesch, 2005). These systems usually

FIGURE 11.6   Manual salinity surveys with EM sensor, GPS, and handheld computer.

comprise three basic components mounted on a vehicle or platform: (1) one or more EM sensors, (2) a GPS receiver, and (3) a computer with digital interfaces. A soil sampler is sometimes added onto the system for rapid ground-truthing. The EM sensors are placed in a nonconductive PVC carrier-sled attached at the rear of the vehicle or on a front carrier. As stated above, the sensors need to be positioned at least 3 m from any metallic objects to avoid erroneous measurements. The GPS receiver provides guidance and geographical positioning. Differential correction is necessary to ensure submeter accuracy of the GPS data. Digital interfaces connect the EM sensor and GPS receiver to the onboard computer for simultaneous and RT recording of the EM data along with their geographical location. Examples of two mobilized systems developed by the authors in California are presented in Figures 11.7 and 11.8.

In agricultural fields, mobilized EM surveys are conducted by driving the system along field rows or furrows and recording the EM and GPS data on a predetermined time interval. In general, mobilized systems can travel at a speed of 5–8 km h^{-1}, depending on the soil surface conditions. The EM measurements are usually collected along rows and furrows spaced 25–40 m. However, the spacing is mostly dependent on the objectives of the study, the resources (time, labor) available for the survey, and the field row spacing. If we assume a travel speed of 6.5 km h^{-1}, a time record interval of 5 s, and a survey spacing of 35 m, about 2000 measurements can be collected in less than 4 h in a 60 ha field (Figure 11.9). In natural areas such as wetlands, the survey pattern will generally be more irregular and will require additional survey time.

11.4.2   DATA CALIBRATION

The development of remote EM sensors has greatly improved the scale and speed of salinity surveying. However, these instruments only provide EC$_a$ measurements, which require conversion (i.e., calibration) to soil salinity (EC$_e$). This is particularly important as Equation 11.2 becomes nonlinear when the $N \ll 1$ assumption does not hold (i.e., at conductivities above 100 mS m^{-1}). As calibration involving extensive soil sampling is time-consuming and expensive, statistical sampling procedures have been developed to minimize the sampling size. Deterministic and stochastic procedures can be applied for predicting soil salinity (EC$_e$) from survey EC$_a$ data (Rhoades et al., 1999; Lesch et al., 2000; Corwin and Lesch, 2003). Deterministic predictions are performed using either theoretical or empirical models to convert conductivity data (EC$_a$) into

FIGURE 11.7 Mobilized conductivity assessment system used for salinity appraisal in agricultural fields with no or low crop covers.

FIGURE 11.8 Mobilized system developed with an all-terrain vehicle for salinity appraisal in difficult, uneven, or delicate terrains, such as wetlands, golf courses, and cropped fields.

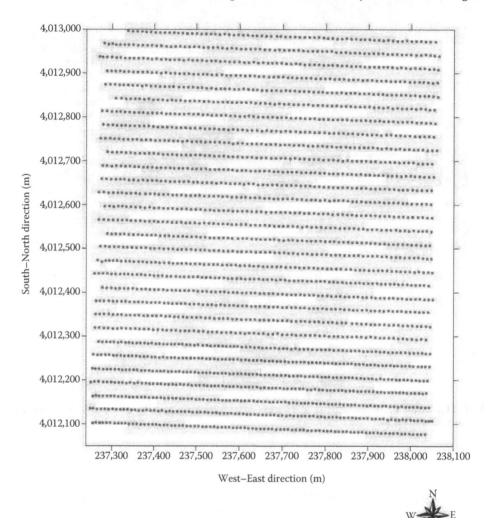

FIGURE 11.9 EM data collection pattern of a typical salinity survey.

salinity (EC_e). Deterministic models are often referred to as static, because the parameters for these models are considered to be known and the acquisition of soil salinity data is not needed. Nonetheless, knowledge of additional soil properties (i.e., soil water content, texture, bulk density, temperature) is required. Stochastic predictions include statistical modeling techniques such as spatial regression or geostatistics (Webster, 1989; Hendrickx et al., 1992; Lesch et al., 1992; Triantafilis et al., 2001b; Carroll and Oliver, 2005). Stochastic predictions are conducted by acquiring a small number of salinity data at selected survey locations to develop a calibration model. Then, using this model and the remaining EC_a data, the soil salinity levels are estimated at all survey locations. Spatial regression techniques are usually more accurate than deterministic models when the secondary soil properties are not known across all survey sites. The geostatistical methods take into account the spatial correlation between random soil variables and, therefore, are preferred for analyzing multivariate survey data. Particularly, cokriging models are often used to improve the estimation accuracy of one variable when information on other variables is available (Yates and

Warrick, 1987; Mulla, 1991). Because stochastic models are dynamic, some soil samples must be acquired during each survey. Additionally, these models also tend to be both time- and location-dependent.

Numerous calibration procedures have been developed over the years (Rhoades and Corwin, 1990; Slavish, 1990; Lesch et al., 1992, 1995b; Wittler et al., 2006). Depth-weighted calibrations were initially developed by Cameron et al. (1981), Wollenhaupt et al. (1986), and McKenzie et al. (1989). These procedures were used for mapping relative differences in soil salinity, but could not accurately assess depth-varying salinity levels within the soil profile. Therefore, Corwin and Rhoades (1982) and Rhoades et al. (1989) developed empirical equations and coefficients for determining EC_a at different soil depth intervals. Their equations were based on the ratio of the vertical and horizontal EM measurements (i.e., salinity profile shape) and were found by Johnston et al. (1997) to be more accurate than the initial depth-weighted relations. Then, Rhoades et al. (1989, 1990) introduced the dual pathway parallel conductance model, which facilitated the estimation of soil salinity from measurements of soil conductivity, texture, bulk density, water content, and temperature. Later, Rhoades (1993) proposed an improved calibration equation that took into account the colinearity between the vertical and horizontal EM data. Lesch et al. (1992, 1995a,b) developed multiple linear regression (MLR) models that were independent of the salinity profile shape and required limited soil information. These calibration models could generate point and conditional estimates of soil salinity as well as field average estimates from EC_a survey data. The main advantage of the MLR methodology lies in the limited sample size required for calibration (Lesch et al., 1995b). A well-accepted and widely used program applying the MLR approach, and particularly designed for EM surveys conducted with mobilized systems, is the estimated salinity assessment program (ESAP) developed by Lesch et al. (2000). This software contains three integrated programs to process the conductivity survey data, generate optimal soil sampling designs from the survey information, develop both deterministic and stochastic conductivity to salinity calibration routines, and generate output maps.

It is important to emphasize that most of the calibration equations developed by investigators are site-specific and cannot be extrapolated to other survey sites. Only recently, inverse procedures with linear (Borchers et al., 1997) and nonlinear (Hendrickx et al., 2002) models have been developed and validated; these models are based on EM physics only and do not require field calibration.

11.4.3 Mapping Soil Salinity from Point Data

Once the EM data have been calibrated and converted into soil salinity estimates, maps are generated using mapping software or geographical information systems (GIS). Since salinity values are continuous over space, they can be represented by continuous surfaces. Apparently very simple, mapping of the salinity estimates requires the application of spatial interpolation methods to create continuous surfaces from data collected at discrete locations (i.e., at points). Selection of the interpolation method and assessment of its accuracy are critical to develop reliable salinity maps.

11.4.3.1 Interpolation

Interpolation is the process of using known data values to estimate values at nonsurveyed locations. It is based on the principle that data points close to one another in space have more similar characteristics than those further away. The two commonly used interpolation procedures for salinity mapping are kriging and inverse-distance weighing (IDW) (Vaughan et al., 1995; Cassel et al., 2003a; Jung et al., 2006). Spline is another interpolation technique available in most mapping software, but is rarely selected. The IDW and spline methods are called deterministic interpolations, because they allocate values at nonsurveyed locations based on the surrounding data. The IDW method calculates the value of a property at an unmeasured point using a distance-weighted average of the designated neighborhood data points. The weight assigned to a particular neighborhood value decreases as the distance from the prediction location increases; thus data closest to the unmeasured point contribute more to the calculated average. The IDW algorithms are well suited for evenly distributed points, but are sensitive to outliers and unevenly distributed data clusters. The spline method creates raster curves by fitting a small number of data points and then joining each curve to obtain continuous surfaces. Spline functions can generate very smooth and sufficiently accurate surfaces from moderately detailed data sets. However, they are also sensitive to outliers and provide poor accuracy if data variance is high.

Kriging is a stochastic interpolation method that takes into account the spatial autocorrelation between surveyed points. Similarly to IDW, kriging uses a weighing factor to assign more influence to the nearest data points. However, it determines the statistical relationship between values at surveyed locations and uses this relationship to make predictions at unvisited points. In general, kriging methods produce better results than IDW and spline, because they account for the spatial interdependency between data and allow for quantification of the prediction error. In addition, kriging can be used with larger data sets than IDW or spline. However, the interpolation procedure requires more modeling and analysis time than the two deterministic methods. Comparison of the three interpolation techniques applied to the same data set is illustrated in Figure 11.10. The kriging method provides a smoother delineation of the different salinity levels observed in the field, while the IDW and spline interpolations provide a more irregular pattern.

With kriging, the spatial structure of the variable is determined through fitted variograms in a two-step procedure, including (1) computation of experimental variograms and (2) selection of theoretical models using the cross-validation procedure (Vieira et al., 1983). Experimental variograms are obtained by performing semivariance analysis. Semivariance is a statistical autocorrelation defined as

$$\gamma(h) = \left[\tfrac{1}{2}N(h)\right] \sum (z_i - z_{i+h})^2 \qquad (11.10)$$

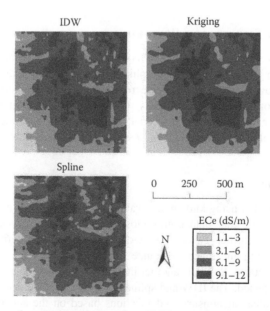

FIGURE 11.10   Salinity surface mapping derived from IDW, ordinary kriging, and spline interpolations.

where

$\gamma(h)$ is the semivariance for interval (separation) distance h

z_i is the measured sample value at point i

z_{i+h} is the measured sample value at point $i + h$

$N(h)$ is the total number of sample couples for the lag interval h

The semivariance is evaluated by calculating $\gamma(h)$ for all possible pairs of points in the data set in an interval distance h. Based on the experimental variograms, theoretical models are chosen using the cross-validation statistical technique in which known data points are evaluated using the fitted model. Spherical, exponential, linear, and Gaussian models most commonly describe the spatial characteristic of salinity data. These models are described based on three parameters: nugget variance (C_0 is the y-intercept of the model), sill ($C_0 + C$ is the model asymptote), and range (A_0 is the distance over which spatial dependence exists), as follows (Robertson, 2000):

$$\gamma(h)_{\text{spherical}} = C_0 + C\left[1.5(h/A_0) - 0.5(h/A_0)^3\right] \quad \text{for } h \leq A_0 \tag{11.11}$$

$$\gamma(h)_{\text{spherical}} = C_0 + C \quad \text{for } h > A_0 \tag{11.12}$$

$$\gamma(h)_{\text{exponential}} = C_0 + C[1 - \exp(-h/A_0)] \tag{11.13}$$

$$\gamma(h)_{\text{linear}} = C_0 + [C(h/A_0)] \tag{11.14}$$

$$\gamma(h)_{\text{Gaussian}} = C_0 + C\left[1 - \exp(-h^2/A_0^2)\right] \tag{11.15}$$

The value C expresses the structural variance. Theoretical models help determine the autocorrelation distance between soil salinity measurements, i.e., the distances between which salinity data are likely to be similar. Figure 11.11 presents kriging models that commonly describe the spatial

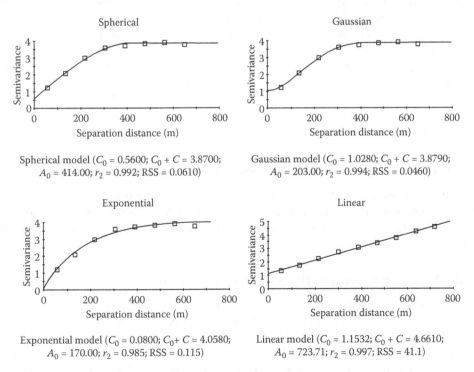

FIGURE 11.11 Examples of kriging models developed from experimental variograms for interpolation of salinity data (denoted as γ).

structure of salinity data. The models were developed based on sample variograms and showed the spatial interdependency of the salinity data. The data described by spherical, exponential, and Gaussian models were spatially dependent up to a distance of about 300 m. The linear model showed that spatial correlation occurred among all salinity data.

11.4.3.2 Mapping Tools

Salinity maps can be developed using mapping software such as Surfer (Golden Software, 2004) or ArcGIS (ESRI, 2005). Each tool includes an interpolation option for developing surface maps. In ArcGIS, different maps can be layered in space and very detailed spatial analyses can be conducted. Maps illustrating different soil and crop properties can be compared spatially and relationships between the properties can be established. For example, salinity maps can be layered with crop yield maps to evaluate the influence of salinity on crop production and understand the cause of within-field yield variability.

11.4.3.3 Interpretation

Interpretation of the salinity maps must take into account all the factors influencing the salinity data. In general, soil salinity patterns do not tend to change significantly over time, unless important changes in management practices occur. Variations in soil moisture content affect the salinity levels, but they remain temporary and seasonal.

Soil salinity maps are often used as a guide to develop management decisions. In irrigated areas affected by salinization, maps are important for implementing irrigation and drainage practices and evaluating the effects of such practices in reducing salt build-up. Salinity maps derived from EM surveys provide reliable information on the levels and distributions of soil salinity within the soil profile. They can be used to (1) determine whether soil salinity falls within acceptable limits for crop production and wetland habitat, (2) evaluate leaching and drainage problems within fields or irrigated areas, (3) infer adequate management practices to reduce salt build-up and increase leaching fraction, and (4) classify soils into categories of low, medium, or high risk of salinization.

There are several agronomic and economic advantages in using soil salinity maps derived from EM measurements for management decisions, including (1) salinity levels and variability can be assessed rapidly and cost-effectively; (2) the effects of irrigation and drainage practices on the salinity profile shape (regular, inverted) and the leaching of salts can be evaluated; and (3) management zones for site-specific application of seeds, fertilizers, and amendments based on the salinity levels can be developed very accurately.

11.5 CASE STUDIES OF SALINITY ASSESSMENT AND MAPPING

This section provides several examples of salinity assessment and mapping studies conducted by the authors in California using the EM technology. These examples encompass different applications, including agricultural drainage management, precision farming, wetland management, and forage evaluation.

11.5.1 Salinity Assessment in Drainage-Water Reuse Systems

Many farmlands in arid and semiarid regions of the world are threatened by excessive soil salinity and inadequate drainage that affect crop yields and soil quality. Therefore, adequate management practices need to be implemented to reduce such problems. In California, the implementation of on-farm drainage practices has been proposed to conserve irrigation water, reduce drainage volume, and, ultimately, lower the amount of salt present in the root zone. Particularly, a system called integrated on-farm drainage management (IFDM) has been developed and implemented in several farmlands. In such a system, drainage water from irrigation applications is collected in tile drains located at 1–2 m below the soil surface and reused several times to irrigate crops of increasing salinity tolerance. Therefore, mapping and monitoring soil salinity on those farmlands are important to evaluate the effectiveness of the IFDM practices.

11.5.1.1 Site Description

The example presented below describes the salinity surveys conducted on one farmland that exhibited very severe salinity problems and implemented an IFDM system to reduce salt build-up. The farmland, located near Five Points in Central California, was divided into four areas for sequential reuse of the saline drainage water. Subsurface tile drains were installed at 1.5–2 m depth and spaced every 80–90 m. A schematic representation of the farm IFDM is presented in Figure 11.12. High-quality canal water was applied to the low-salinity area A (73% of the farm area) where high value salt-sensitive crops, such as vegetables, were grown. Drainage water collected from A was reused to irrigate salt-tolerant commercial crops, such as alfalfa and cotton, produced in the moderately saline area B (20% of the farm area). Drainage water from B was subsequently applied to the relatively high-salinity area C (2% of the farm) where salt-tolerant forages were planted. Finally, the bio-concentrated drainage collected from C was applied to area D (0.8% of the farm) where only highly salt-adapted halophytes were grown. This sequential reuse system utilized about 90% of the drainage water collected within the farm boundaries. The remaining 10% of the drainage water was discharged into a solar evaporation system for rapid water evaporation and salt

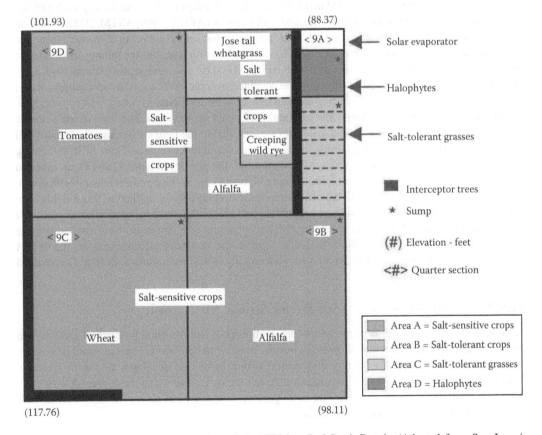

FIGURE 11.12 Schematic representation of the IFDM at Red Rock Ranch. (Adapted from San Joaquin Valley Drainage Implementation Program, 1998. Cervinka, V., Finch, C., Martin, M., Menezes, F., Peters, D., and Buchnoff, K., Drainwater, salt, and selenium management utilizing IFDM/Agroforestry systems. Report to the US Bureau of Reclamation, Sacramento, CA, 2001. With permission.)

crystallization. With these drainage management practices, the owner was able to conserve 20% of good-quality water. Description of IFDM systems can be found in Shannon et al. (1997), Cervinka et al. (2001), and Tanji and Kielen (2002).

11.5.1.2 Data Collection

Soil salinity surveys were conducted in seven fields that successively received reused drainage water. The mobile conductivity assessment system described in Figure 11.7 was used in sections A, B, and C, with a Geonics EM38-DD sensor. In section D, the EM and GPS readings were taken manually due to the soil conditions and various experiments conducted in that field (Figure 11.6). The device was operated in both horizontal and vertical positions at the soil surface to obtain effective measurement depths of 0.75 and 1.5 m, respectively. For each field, the EM and GPS data were collected along transects spaced 25 m apart, and recorded on a 5 s interval.

11.5.1.3 Salinity Assessment

Calibration of the EM data was performed using the ESAP software (Lesch et al., 2000). After an EM survey had been completed, an optimal soil sampling plan was generated. The sampling plan encompassed 6 or 12 locations depending on the number of EM data collected and the size of the survey area. The plan represented the spatial distribution of the EC_a data over the entire survey area. Ground-truth soil sampling was then conducted at each of the selected locations using a GPS for geographical positioning. Soil samples were collected in 0.3 m increments to a depth of 1.2 m and analyzed for EC on 1:1 soil–water extracts, saturation percentage, and water content (APHA, 1999; ASTM, 2000). Based on the EM data and soil determinations, salinity was then estimated for the entire surveyed field using the stochastic methods described by Lesch et al. (2000). Surface maps representing the salinity distribution in each drainage area were generated using ArcGIS (ESRI, 2005) for the average profile depth (0–1.2 m). The GIS maps were created with IDW using the weighted average of 12 measurement points located around a particular point and interpolating the data into 10 m grids (Figure 11.13).

The maps illustrate the drainage management performed on the farm. The lowest salt amounts were observed in area A where fresh canal water was applied to the fields, whereas area D exhibited the highest salinity levels. Salinity levels increased with depth in fields A and part of B, indicating good drainage management and leaching of salts through the profile. In fields C and D, an inverted soil profile shape was observed, with salinity being highest at the surface and decreasing with depth. This was explained by the high salt concentration in the drainage water applied to those fields. The maps also showed that the south and western middle parts of field B were being reclaimed, since salt levels were lower than those observed in the previous year and concentrations were similar to those found in fields A. Thus, the salinity surveys suggest the potential effectiveness of the IFDM practices in reducing salt build-up in area B. Additionally, salt concentrations in field C remained stable, indicating good management.

11.5.2 SALINITY ASSESSMENT FOR SITE-SPECIFIC SEEDING MANAGEMENT

In saline soils, yield variability at the field scale is important, and marginal production areas within the field can lead to considerable decrease in profits. In order to address this spatial variability issue, it is necessary to evaluate the salinity distribution within fields and develop site-specific management practices that will increase yields in salt-affected areas. By following these precision farming practices, fields are not cultivated as one homogeneous soil unit, but as various soil entities with different salinity levels and yield potentials.

11.5.2.1 Site Description

This example presents a salinity assessment study that was conducted in fields grown with cotton on very saline soils of the Westside San Joaquin Valley in California. Cotton (*Gossypium hirsutum* L.)

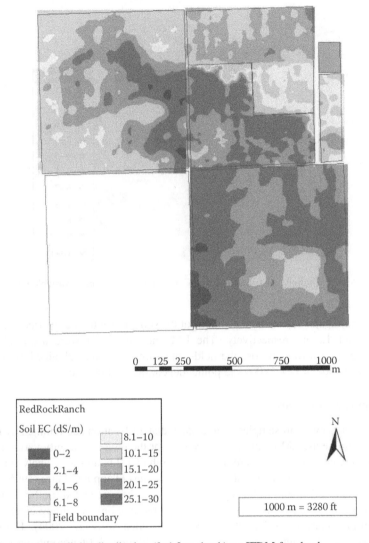

FIGURE 11.13 Average salinity distribution (0–1.2 m depth) on IFDM farmland.

is a major crop in California, where it is grown on more than 220,000 ha with about 95% of the production concentrated in the San Joaquin Valley (USDA National Agricultural Statistics Service, 2007). However, cotton in California is mostly cultivated on salt-affected soils, which reduces seed germination, water infiltration, nutrient uptake, and ultimately yields. In this study, salinity surveys were conducted on a farmland near Lemoore, California, where soil salinity levels commonly range from 1 to 15 dS m^{-1} and exhibit significant within-field variability (Cassel et al., 2003b). The goal of the study was to assess soil salinity in selected cotton fields, where yield variability was important, and to apply variable seeding rates based on the different salinity levels observed across the fields.

11.5.2.2 Data Collection

The salinity surveys were conducted in the cotton fields before planting. The mobile conductivity assessment system, described in Figure 11.7 and equipped with an EM38-DD sensor, was driven along seed rows spaced 35 m apart. As described in the previous example, the EM38 device was

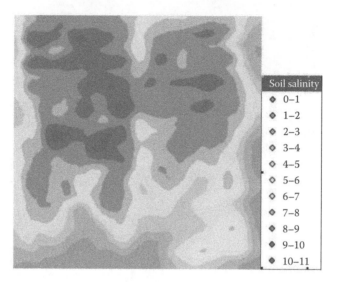

FIGURE 11.14   Predicted salinity distribution (dS m^{-1}) for the application of variable seeding rates.

operated in both horizontal and vertical positions at the soil surface to obtain effective measurement depths of 0.75 and 1.5 m, respectively. The EM and GPS data were acquired every 10 m approximately along each row. Across the fields, measurements were obtained on a 0.1 ha scale, which amounted to about 1500–1800 data points depending on the field size.

11.5.2.3   Salinity Assessment

Following the EM surveys, soil samples were collected at 12 locations determined by ESAP (Lesch et al., 2000) to calibrate the EM readings and account for the range of salinity variability within the fields. Samples were taken in 0.3 m increments to a depth of 1.2 m and were analyzed for saturation percentage, water content, and EC on saturated paste extracts (Topp and Ferré, 1996; Gavlak et al., 2003). Based on the EM data and soil analyses, salinity was then estimated for the entire fields using the stochastic methods available in ESAP. The spatial distributions of salinity levels were mapped for each soil profile depth using the software Surfer (Golden Software, 2004). The salinity maps were created using the kriging interpolation technique. Figure 11.14 presents an example of the salinity distribution in one cotton field.

11.5.2.4   Variable Rate Application

Based on the salinity levels observed in the 0–0.3 m salinity maps, variable seeding rates were applied across the fields. In this particular study, areas exhibiting a soil salinity of <3, 3–7, and >7 dS m^{-1} were planted with 10, 15, and 18 kg seed ha^{-1}, respectively (Figure 11.15). Since salinity reduced the emergence rate, areas with the greatest salinity levels were planted with the highest seeding rate. Site-specific applications of the cotton seeds were accomplished by importing the maps into a GIS program and a computer on board the farm tractor. The seeding applications in the fields were then performed using a controller, a variable rate delivery system, and an auto-guidance system linked to a GPS receiver and the computer.

The maps developed in this study provided precise information on the levels of salinity across the fields. As shown in Figure 11.14, areas in the northern part of the field were more severely affected by soil salinity and, therefore, required a greater seeding rate. By implementing such precision farming practices, the most appropriate seeding rates were applied in each 0.1 ha area of the surveyed fields.

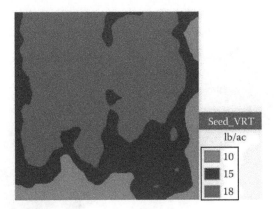

FIGURE 11.15   Prescription map for site-specific application of cotton seeds in surveyed field.

11.5.3   Salinity Assessment in Wetlands

Soil salinity is an important conservation and environmental problem in wetland habitats. Salinity affects plant germination and development, and can lead to significant increases in salt-tolerant species populations, thereby creating imbalances in the wetland ecosystem. Salinity can also influence fauna diversity, such as invertebrate, fish, and bird. Thus, it is important to evaluate the extent and variability of soil salinity in wetlands in order to develop sound planning and management practices for improving long-term habitat health and restoration.

11.5.3.1   Site Description

The selected example presents a salinity mapping project conducted at the San Luis National Wildlife Refuge (SLNWR), one of the very few remaining wetlands in the Central Valley of California. The wetland is a vital source of food and habitat for local birds but also for migratory birds since it falls on the Pacific Flyway (Quinn et al., 2005). Water at the SLNWR wetland is managed through a series of irrigation and drawdown events. Such artificial management affects soil salinity and, therefore, impacts the overall wetland ecosystem. Details of the study can be found in Quinn et al. (2005).

11.5.3.2   Data Collection

The study was conducted at two sites representative of the soil conditions and vegetation population observed at the SLNWR wetland (Cassel, 2004). The salinity surveys were performed using the mobilized system shown in Figure 11.8. The system comprised a GPS receiver and an EM38-DD meter placed in a carrier-sled attached at the rear of an ATV. The EM and GPS data were collected along transects spaced 45–90 m apart, depending on the extent of the vegetation cover, and recorded simultaneously to a laptop computer. Data collection did not follow a regular grid pattern in this study because of the tall vegetation present in the survey areas.

11.5.3.3   Salinity Assessment

After the EM surveys, the data were analyzed for each site using ESAP (Lesch et al., 2000), and a soil sampling plan comprising six locations was developed to calibrate the EM measurements. Soil samples were collected at each location at 0–0.15 and 0.15–0.3 m depths and analyzed for EC_e, moisture and saturation percentage, following standard analytical methods (Topp and Ferré, 1996; Gavlak et al., 2003). Based on the EM data and laboratory analyses, maps of soil salinity were generated for each depth sampled using the IDW interpolation technique in a GIS environment. (ESRI, 2005) A map produced for the 0.15–0.3 m depth is presented in Figure 11.16. The map

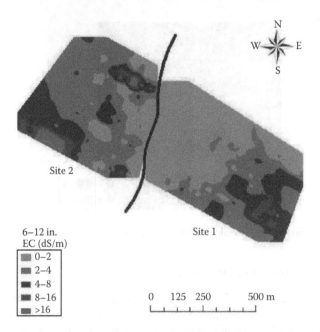

FIGURE 11.16 Soil salinity estimated at 0.15–0.3 m depth for wetland restoration.

shows the areas of high salinity levels that can detrimentally affect the wetland ecosystem. It reveals that management practices need to be implemented to reduce salt build-up, increase flora and fauna diversity, and ameliorate the wetland habitat.

11.5.4 Salinity Management for Forage Production

Land application of food-processing effluent waters is a widely practiced treatment and disposal technique that allows for the beneficial reuse of nutrients and water. However, excessive application of these waters can lead to subsurface water degradation and increase in soil salinity. Such water contamination is encountered across the country, including in the Central Valley of California, where intensive agriculture is practiced and numerous food-processing plants need to dispose of their excess waters.

11.5.4.1 Site Description

The authors have been working with the City of Fresno's Department of Public Utilities to monitor soil and water quality at a disposal site that received winery effluents for several decades. Such monitoring was mandated by the California Regional Water Quality Control Board to determine the organic and nitrogen loadings that occurred on the site. After the disposal of winery effluents ceased, two forage grasses were planted in adjacent plots and irrigated with fresh water. The two grasses studied were Elephant grass (*Pennisetum purpureum*) and Sudan grass (*Sorghum vulgare var. sudanense*). The objective of the study was to test the effectiveness of the two forage grasses in uptaking soil nutrients and their potential for mitigating the impact of soil salinization (Goorahoo and Cassel, 2004; Cassel et al., 2005).

11.5.4.2 Data Collection

The salinity surveys were conducted before planting the two grasses and after a period of 1 year. Since the plots were relatively small, the surveys were performed manually using the EM38-DD sensor and a GPS receiver for georeferencing the measurement locations. The EM and GPS data were collected in a regular 5 × 6 m grid pattern.

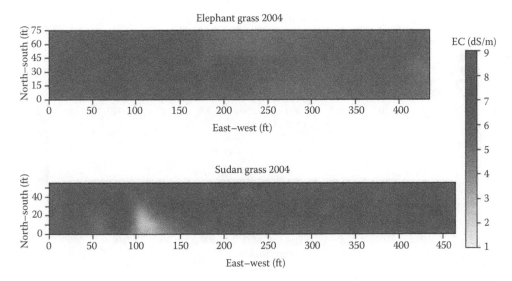

FIGURE 11.17   Soil salinity distribution before planting of Elephant and Sudan grasses (1 ft = 0.3 m).

11.5.4.3   Salinity Assessment

After the EM surveys, ground-truthing was conducted at six locations for calibration of the EM data at each survey site. The sampling locations were selected through statistical analysis performed using ESAP to reflect the spatial variability of the EM data (Lesch et al., 2000). The soil samples were analyzed in the laboratory for EC on saturated paste extracts, as well as for moisture and saturation percentage, following methods described in Topp and Ferré (1996) and in Gavlak et al. (2003). The salinity estimates obtained in the two forage plots were mapped for each survey period (Figures 11.17 and 11.18) using the kriging interpolation method in Surfer (Golden Software, 2004). Figure 11.19 shows the difference in salinity levels observed between the two survey periods for each forage grass. Nutrient analysis of the forages along with the salinity maps revealed that Elephant grass had a better capability than Sudan grass in removing salt throughout the profile

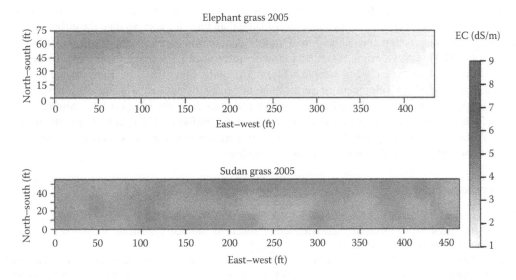

FIGURE 11.18   Soil salinity distribution one year after planting of Elephant and Sudan grasses (1 ft = 0.3 m).

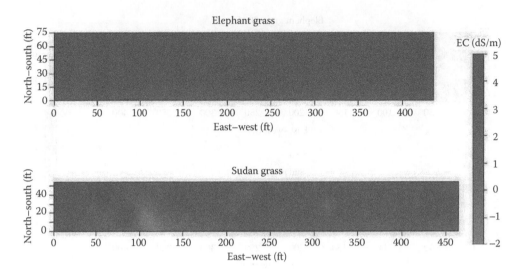

FIGURE 11.19 Comparison of the spatial and temporal differences in salt levels between Elephant and Sudan forage grasses (1 ft = 0.3 m).

depth; this was partly explained by the deep rooting system of Elephant grass. Sudan grass was mostly effective in reducing salt levels in the top 0.3 m.

11.6 CONCLUSIONS

Mapping and monitoring of soil salinity are important for the management and sustainability of agricultural fields and natural ecosystems. Traditional measurements of soil salinity involving in situ sampling and laboratory determinations are slow, expensive, and do not account for the spatial variability of salinity. The EM induction technique offers an alternative for rapid and cost-effective assessment of soil salinity at field, watershed, and regional scales. Based on the principle of induced current flows, the EM technology provides measurements of depth-weighted soil EC_a, which can be converted to salinity estimates through deterministic and stochastic calibration procedures. In this chapter, we provided an overview of the EM technology with its principle, applications, advantages, and limitations. We listed the most commonly used EM sensors and described the benefits of mobilized EM systems for the rapid collection of large data sets. We also detailed the different steps involved in the acquisition and calibration of the EM data for the development of soil salinity maps, including (1) EM survey design, (2) collection of georeferenced EC_a data, (3) soil sampling design, (4) soil sample collection, (5) soil sample analysis, (6) calibration of EC_a measurements using soil sample data, and (7) development of surface salinity maps using spatial statistical analysis. We emphasized the importance of calibration of the EM data and interpolation into salinity maps. Finally, we presented different case studies conducted in California, USA, to highlight the useful-ness of the EM technique for the management of drainage-water reuse systems, wetlands and forages, as well as its application in precision agriculture.

REFERENCES

Anderson-Cook, C.M., Alley, M.M., Royguard, J.K.F., Khosla, R., Noble, R.B., and Doolittle, J.A. 2002. Differentiating soil types using electromagnetic conductivity and crop yield maps. *Soil Science Society of America Journal* 66: 1562–1570.

APHA (American Public Health Association). 1999. *Standard Methods for the Examination of Water and Wastewater*, 20th ed., L.S. Clesceri, A.E. Greenberg, and A.D. Eaton (Eds.), American Public Health Association–American Water Works Association–Water Pollution Control Federation, Washington, DC.

ASTM (American Society for Testing and Materials). 2000. *Annual Book of ASTM Standards*. American Society for Testing and Materials International, West Conshohocken, PA.

Belluigi, A. 1948. Coupling of a vertical coil with a homogeneous earth. *Geophysics* 14: 501–507.

Bennett, D.L. and George, R.J. 1995. Using the EM38 to measure the effect of soil salinity on eucalyptus globules in south-western Australia. *Agricultural Water Management* 27: 69–86.

Boettinger, J.L., Doolittle, J.A., West, N.E., Bork, E.W., and Schupp, E.W. 1997. Nondestructive assessment of rangeland soil depth to petrocalcic horizon using electromagnetic induction. *Arid Soil Research and Rehabilitation* 11: 372–390.

Borchers, B., Uram, T., and Hendrickx, J.M.H. 1997. Tikhonov regularization for determination of depth profiles of electrical conductivity using non-invasive electromagnetic induction measurements. *Soil Science Society of America Journal* 61: 1004–1009.

Brevik, E.C. and Fenton, T.E. 2002. The relative influence of soil water, clay, temperature, and carbonate minerals on soil electrical conductivity readings taken with an EM-38 along a mollisol catena in central Iowa. *Soil Survey Horizons* 43: 9–13.

Brune, D.E. and Doolittle, J. 1990. Locating lagoon seepage with radar and electromagnetic survey. *Environmental Geology and Water Science* 16: 195–207.

Brune, D.E., Drapcho, C.M., Radcliff, R.E., Harter, T., and Zhang, R. 1999. Electromagnetic survey to rapidly assess water quality in agricultural watersheds. ASAE Paper No. 992176, American Society of Agricultural Engineers, St. Joseph, MI.

Brune, D.E., Buchnoff, K., Zhang, R., and Jenkins, B. 2001. Electromagnetic terrain conductivity to delineate and quantify soil salinization in the Central Valley of California. ASAE Paper No. 012221, American Society of Agricultural Engineers, St. Joseph, MI.

Brus, D.J., Knotters, M., van Dooremolen, W.A., van Kernebeek, P., and van Seeters, R.J.M. 1992. The use of electromagnetic measurements of apparent soil electrical conductivity to predict the boulder clay depth. *Geoderma* 55: 79–93.

Cameron, D.R., de Jong, E., Read, D.W.L., and Oorsteveld, M. 1981. Mapping salinity using resistivity and electromagnetic techniques. *Canadian Journal of Soil Science* 61: 67–78.

Cannon, M.E., McKenzie, R.C., and Lachapelle, G. 1994. Soil-salinity mapping with electromagnetic induction and satellite-based navigation methods. *Canadian Journal of Soil Science* 74: 335–343.

Carroll, Z.L. and Oliver, M.A. 2005. Exploring the spatial relations between soil physical properties and apparent electrical conductivity. *Geoderma* 128: 354–374.

Cassel, F., Zoldoske, D., and Spiess, M. 2003a. Assessing spatial and temporal variability of soil salinity on farms implementing drainage management practices. California Agricultural Technology Institute Publication, California State University, Fresno, CA.

Cassel, F., Taylor, B., Roberts, B., and Zoldoske, D. 2003b. Precision farming applications in cotton systems of California. Final Report, Cotton Incorporated, Cary, NC.

Cassel, F. 2004. Estimating soil salinity in wetlands of the San Luis National Wildlife Refuge and Salinas Club. Report submitted to Nigel Quinn, Lawrence Berkeley National Laboratory, Berkeley, CA.

Cassel, F., Adhikari, D., and Goorahoo, D. 2005. Salinity mapping of fields irrigated with winery effluents. Irrigation Association, Falls Church, VI.

Cassel, F., Goorahoo, D., and Senatore, K. 2006. Soil salinity assessment and determination of soil hydraulic properties at Red Rock Ranch. Final report submitted to the United States Bureau of Reclamation, Fresno, CA.

Cervinka, V., Finch, C., Martin, M., Menezes, F., Peters, D., and Buchnoff, K. 2001. Drainwater, salt, and selenium management utilizing IFDM/Agroforestry systems. Report to the US Bureau of Reclamation, Sacramento, CA.

Ceuppens, J., Wopereis, M.C.S., and Miézan, K.M. 1997. Soil salinization in rice irrigation schemes in the Senegal River Delta. *Soil Science Society of America Journal* 61: 1122–1130.

Cook, P.G., Hughes, M.W., Walker, G.R., and Allison, G.B. 1989. The calibration of frequency-domain electromagnetic induction meters and their possible use in recharge studies. *Journal of Hydrology* 107: 251–265.

Cook, P.G. and Walker, G.R. 1992. Depth profiles of electrical conductivity from linear combinations of electromagnetic induction measurements. *Water Resources Research* 28: 2953–2961.

Corwin, D.L. and Lesch, S.M. 2003. Application of soil electrical conductivity to precision agriculture: Theory, principles, and guidelines. *Agronomy Journal* 95: 455–471.

Corwin, D.L. and Lesch, S.M. 2005. Apparent electrical conductivity measurements in agriculture. *Computers and Electronics in Agriculture* 46: 11–43.

Corwin, D.L. and Rhoades, J.D. 1982. An improved technique for determining soil electrical conductivity-depth relations from above-ground electromagnetic measurements. *Soil Science Society of America Journal* 46: 517–520.

Corwin, D.L. and Rhoades, J.D. 1990. Establishing soil electrical conductivity—depth relations from electromagnetic induction measurements. *Communications in Soil Science and Plant Analysis* 21: 861–901.

Corwin, D.L. and Rhoades, J.D. 2005. Apparent electrical conductivity measurements in agriculture. *Computers and Electronics in Agriculture* 46: 11–43.

Corwin, D.L., Carrillo, M.L.K., Vaugan, P.J., Rhoades, J.D., and Cone, D.G. 1999. Evaluation of GIS-linked model of salt loading to groundwater. *Journal of Environmental Quality* 28: 471–480.

Dalton, F.N. 1982. Development of time domain reflectometry for measuring soil water content and bulk soil electrical conductivity. In *Advances in Measurement of Soil Physical Properties: Bringing Theory into Practice*, G.C. Topp, W.D. Reynolds, and R.E. Green (Eds.), Soil Science Society of America Special Publication No. 30, Madison, WI, pp. 143–167.

Dalton, F.N., Herkelrath, W.N., Rawlins, D.S., and Rhoades, J.D. 1984. Time-domain reflectometry: Simultaneous measurement of soil water content and electrical conductivity with a single probe. *Science* 224: 989–990.

De Jong, E., Ballantyne, A.K., Cameron, D.R., and Read, D.W. 1979. Measurement of apparent electrical conductivity of soils by an electromagnetic induction probe to aid salinity surveys. *Soil Science Society of America Journal* 43: 810–812.

Diaz, L. and Herrero, J. 1992. Salinity estimates in irrigated soils using electromagnetic induction. *Soil Science* 154: 151–157.

Domsch, H. and Giebel, A. 2004. Estimation of soil textural features from soil electrical conductivity recorded using the EM-38. *Precision Agriculture* 5: 389–409.

Doolittle, J.A., Sudduth, K.A., Kitchen, N.R., and Indorante, S.J. 1994. Estimating depths to claypans using electromagnetic induction methods. *Journal of Soil and Water Conservation* 49: 572–575.

Doolittle, J., Petersen, M., and Wheeler, T. 2001. Comparison of two electromagnetic induction tools in salinity appraisal. *Journal of Soil and Water Conservation* 56: 257–262.

Drommerhausen, D.J., Radcliffe, D.E., Brune, D.E., and Gunter, H.D. 1995. Electromagnetic conductivity surveys of dairies for groundwater nitrate. *Journal of Environmental Quality* 24: 1083–1091.

Ehsani, R. and Sullivan, M. 2002. Soil electrical conductivity (EC) sensors. Extension Fact Sheet AEX-565-02, Ohio State University, Columbus, OH.

Eigenberg, R.A. and Nienaber, J.A. 1998. Electromagnetic survey of cornfield with repeated manure applications. *Journal of Environmental Quality* 27: 1511–1515.

Eigenberg, R.A. and Nienaber, J.A. 2003. Electromagnetic induction methods applied to an abandoned manure handling site to determine nutrient buildup. *Journal of Environmental Quality* 32: 1837–1843.

Eigenberg, R.A., Korthals, R.L., and Nienaber, J.A. 1998. Geophysical electromagnetic survey methods applied to agricultural waste sites. *Journal of Environmental Quality* 27: 215–219.

(ESRI) Environmental System Research Institute. 2005. *ArcGIS*. Environmental System Research Institute, Cambridge, England.

Freeland, R.S., Yoder, R.E., Ammons, J.T., and Leonard, L.L. 2002. Mobilized surveying of soil conductivity using electromagnetic induction. *Applied Engineering in Agriculture* 18: 121–126.

Gavlak, R.G., Horneck, D.A., Miller, R.O., and Kotuby-Amacher, J. 2003. *Soil, Plant and Water Reference Methods for the Western Region*. A Western Regional Extension Publication. 125, 2nd edition, Oregon State University, Corvallis, OR.

Geophex. 2007. GEM-2 principle of operation. Geophex, Raleigh, NC, Available at www.geophex.com/GEM2/How%20it%20works/Operation%20Principle/operation.htm.

George, R.J. and Bennett, D.L. 1999. Root zone electromagnetic systems explain variations in soil salinity. Paper presented at the Irrigation Association of Australia Conference, Perth, Australia.

Golden Software. 2004. SURFER, version 8.05. Golden Software, Golden, CO.

Goorahoo, D. and Cassel, F. 2004. Vadose zone monitoring for organic and nitrogen loading from winery stillage application. Paper presented at the annual meeting of European Soil Scientists-EUROSOIL, Freiburg, Germany.

Greenhouse, J.P. and Slaine, D.D. 1983. The use of reconnaissance electromagnetic methods to map contaminant migration. *Groundwater Monitoring Review* 3: 47–59.

Greenhouse, J.P. and Slaine, D.D. 1986. Geophysical modeling and mapping of contaminated groundwater around three waste disposal sites in southern Ontario. *Canadian Geotechnical Journal* 23: 372–384.

Hanson, B.R. and Kaita, K. 1997. Response of electromagnetic conductivity meter to soil salinity and soil-water content. *Journal of Irrigation and Drainage Engineering* 123: 141–143.

Hanson, B.R., Grattan, S.R., and Fullon, A. 1999. *Agricultural Salinity and Drainage*. Division of Agriculture and Natural Resources, Publication 3375. University of California, Davis, CA.

Hazell, J.R.T., Cratchley, C.R., and Preston, A.M. 1988. The location of aquifers in crystalline rocks and alluvium in Northern Nigeria using combined electromagnetic and resistivity techniques. *Quarterly Journal of Engineering Geology* 21: 159–175.

Hendrickx, J.M.H., Baerends, B., Raza, S.I., Sadig, M., and Chaudhry, M.A. 1992. Soil salinity assessment by electromagnetic induction of irrigated land. *Soil Science Society of America Journal* 56: 1933–1941.

Hendrickx, J.M.H., Borchers, B., Corwin, D.L., Lesch, S.M., Hilgendorf, A.C., and Schlue, J. 2002. Inversion of soil conductivity profiles from electromagnetic induction measurements: Theory and experimental verification. *Soil Science Society of America Journal* 66: 673–685.

Herrero, J., Ba, A.A., and Aragués, R. 2003. Soil salinity and its distribution determined by soil sampling and electromagnetic technique. *Soil Use and Management* 19: 119–126.

Hillel, D. 2000. *Salinity Management for Sustainable Irrigation: Integrating Science, Environment, and Economics*. The World Bank, Washington, DC.

Hoekstra, P., Lahti, R., Hild, J., Bataes, C.R., and Phillips, D. 1992. Case histories of shallow time domain electromagnetics in environmental site assessment: Case history mapping migration of oil field brines from evaporation pits and ponds. *Ground Water Monitoring Review* 13: 110–117.

Huang, H. 2005. Depth investigation for small broadband electromagnetic sensors. *Geophysics* 70: 135–142.

Huth, N.I. and Poulton, P.L. 2007. An electromagnetic induction method for monitoring variation in soil moisture in agroforestry systems. *Australian Journal of Soil Research* 45: 63–72.

Inman, D.J., Freeland, R.S., Ammons, J.T., and Yoder, R.E. 2002. Soil investigations using electromagnetic induction and ground-penetrating radar in Southwest Tennessee. *Soil Science Society of America Journal* 66: 206–211.

Jaynes, D.B. 1996. Mapping the areal distribution of soil parameters with geophysical techniques. In *Applications of GIS to the Modeling of Non-Point Source Pollutants in the Vadose Zone*, D.L. Corwin and K. Loague (Eds.), Soil Science Society of America Special Publication No. 48, Madison, WI, pp. 205–216.

Jaynes, D.B., Novak, J.M., Moorman, T.B., and Cambardella, C.A. 1995. Estimating herbicide partition coefficients from electromagnetic induction measurements. *Journal of Environmental Quality* 24: 36–41.

Johnston, M.A., Savage, M.J., Moolman, J.H., and du Pleiss, H.M. 1997. Evaluation of calibration methods for interpreting soil salinity from electromagnetic induction measurements. *Soil Science Society of America Journal* 61: 1627–1633.

Jung, W.K., Kitchen, N.R., Sudduth, K.A., Kremer, R.J., and Motavalli, P.P. 2005. Relationship of apparent soil electrical conductivity to claypan soil properties. *Soil Science Society of America Journal* 69: 883–892.

Jung, W.K., Kitchen, N.R., Sudduth, K.A., and Anderson, S.H. 2006. Spatial characteristics of claypan soil properties in an agricultural field. *Soil Science Society of America Journal* 70: 1387–1397.

Kachanoski, R.G., Gregorich, E.G., and VanWesenbeeck, I.J. 1988. Estimating spatial variations of soil water content using non-contacting electromagnetic inductive methods. *Canadian Journal of Soil Science* 68: 715–722.

Kaffka, S.R., Lesch, S.M., Bali, K.M., and Corwin, D.L. 2005. Site-specific management in salt-affected sugar beet fields using electromagnetic induction. *Computers and Electronics in Agriculture* 46: 329–350.

Kinal, J., Stoneman, G.L., and Williams, M.R. 2006. Calibrating and using an EM31 electromagnetic induction meter to estimate and map soil salinity in the jarrah and karri forests of south-western Australia. *Forest Ecology and Management* 233: 78–84.

Kitchen, N.R., Sudduth, K.A., and Drummond, S.T. 1996. Mapping of sand deposition from 1993 Midwest floods with electromagnetic induction measurements. *Journal of Soil and Water Conservation* 51: 336–340.

Kitchen, N.R., Sudduth, K.A., and Drummond, S.T. 1999. Soil electrical conductivity as a crop productivity measure for claypan soils. *Journal of Production Agriculture* 12: 607–617.

Kitchen, N.R., Sudduth, K.A., Myers, D.B., Drummond, S.T., and Hong, S.Y. 2005. Delineating production zones on claypan soil fields using apparent soil electrical conductivity. *Computers and Electronics in Agriculture* 46: 285–308.

Lesch, S.M., Rhoades, J.D., Lund, L.J., and Corwin, D.L. 1992. Mapping soil salinity using calibrated electromagnetic measurements. *Soil Science Society of America Journal* 56: 540–548.

Lesch, S.M., Strauss, D.J., and Rhoades, J.D. 1995a. Spatial prediction of soil salinity using electromagnetic induction techniques: I. Statistical prediction models: A comparison of multiple linear regression and cokriging. *Water Resources Research* 31: 373–386.

Lesch, S.M., Strauss, D.J., and Rhoades, J.D. 1995b. Spatial prediction of soil salinity using electromagnetic induction techniques: II. An efficient spatial sampling algorithm suitable for multiple linear regression model identification and estimation. *Water Resources Research* 31: 387–398.

Lesch, S.M., Herrero, J., and Rhoades, J.D. 1998. Monitoring temporal changes in soil salinity using electromagnetic induction techniques. *Soil Science Society of America Journal* 62: 232–242.

Lesch, S.M., Rhoades, J.D., and Corwin, D.L. 2000. ESAP-95 Version 2.10R: User manual and tutorial guide. Research Report 146, USDA-ARS George E. Brown, Jr. Salinity Laboratory, Riverside, CA.

Lesch, S.M., Corwin, D.L., and Robinson, D.A. 2005. Apparent electrical conductivity mapping as an agricultural management tool in arid zone soils. *Computers and Electronics in Agriculture* 46: 351–378.

Letey, J. 2000. Soil salinity poses challenges for sustainable agriculture and wildlife. *California Agriculture* 54: 43–55.

Mallants, D., Vanclooster, M., Toride, N., Vanderborght, J., van Genuchten, M.T., and Feyen, J. 1996. Comparison of three methods to calibrate TDR for monitoring solute movement in undisturbed soil. *Soil Science Society of America Journal* 60: 747–754.

Mankin, K.R. and Karthikeyan, R. 2002. Field assessment of saline seep remediation using electromagnetic induction. *Transactions of the American Society of Agricultural Engineers* 45: 99–107.

McBride, R.A., Gordon, A.M., and Shrive, S.C. 1990. Estimating forest soil quality from terrain measurements of apparent electrical conductivity. *Soil Science Society of America* 54: 290–293.

McFarlane, D.J. and George, R.J. 1992. Factors affecting dryland salinity in two wheatbelt catchments in Western Australia. *Australian Journal of Soil Research* 30: 85–100.

McKenzie, R.C. 2000. Salinity: Mapping and determining crop tolerance with an electromagnetic induction meter (Canada). EM38 Workshop, 57–68, New Delhi, India.

McKenzie, R.C., Chomistek, U., and Clark, N.F. 1989. Conversion of electromagnetic induction readings to saturated paste extract values in soil for different temperature, texture, and moisture conditions. *Canadian Journal of Soil Science* 69: 25–32.

McKenzie, R.C., George, R.J., Woods, S.A., Cannon, M.E., and Bennett, D.L. 1997. Use of electromagnetic-induction meter (EM38) as a tool in managing salinization. *Hydrogeology Journal* 5: 37–50.

McNeill, J.D. 1980. Electromagnetic terrain conductivity measurements at low induction numbers. Technical note TN-6, Geonics Ltd., Mississauga, ON, Canada.

McNeill, J.D. 1992. Rapid, accurate mapping of soil salinity by electromagnetic ground conductivity meters. In *Advances in Measurement of Soil Physical Properties: Bringing Theory into Practice*, G.C. Topp, W.D. Reynolds, and R.E. Green (Eds.), Soil Science Society of America Special Publication, No. 30, ASA-CSSA-SSSA-Madison-WI, pp. 201–229.

Metternicht, G.I. and Zinck, J.A. 2003. Remote sensing of soil salinity: Potentials and constraints. *Remote Sensing of the Environment* 85: 1–20.

Mulla, D.J. 1991. Using geostatistics and GIS to manage spatial patterns in soil fertility. *Automated Agriculture for the 21st Century*, G. Kranzler (Ed.), American Society of Agricultural Engineers Conference, St Joseph, MI, pp. 336–345.

Nettleton, W.D., Bushue, L., Doolittle, J.A., Endres, T.J., and Indorante, S.J. 1994. Sodium-affected soil identification in South-Central Illinois by electromagnetic induction. *Soil Science Society of America Journal* 58: 1190–1193.

Nobes, D.C., Armstrong, M.J., and Close, M.E. 2000. Delineation of a landfill leachate plumate and flow channels in coastal sands near Christchurch, New Zealand, using a shallow electromagnetic survey method. *Hydrogeology Journal* 8: 328–336.

Nogués, J., Robinson, D.A., and Herrero, J. 2006. Incorporating electromagnetic induction methods into regional soil salinity survey of irrigation districts. *Soil Science Society of America Journal* 70: 2075–2085.

Paine, J.G. 2003. Determining salinization extent, identifying salinity sources, and estimating chloride mass using surface, borehole, and airborne electromagnetic induction methods. *Water Resources Research* 39: 1059.

Paine, J.G., White, W.A., Gibeaut, J.C., Andrews, J.R., and Waldinger, R. 2004. Exploring quantitative wetlands mapping using airborne lidar and electromagnetic induction on Mustang Island, TX. Paper presented at the American Geophysical Union Meeting, Abstract No. B41A-24.

Postel, S. 1999. *Pillar of Sand: Can the Irrigation Miracle Last?* W.W. Norton, New York.

Quinn, N.W.T., Stromayer, K.A.K., Ennis, M.J., and Woolington, D.W. 2005. Real-time water quality monitoring and habitat assessment in the San Luis National Wildlife Refuge. Report LBNL-58813, Earth Sciences Division, Ernest Orlando Lawrence Berkeley National Laboratory, University of California, Berkeley, CA.

Reedy, R.C. and Scanlon, B.R. 2003. Soil water content monitoring using electromagnetic induction. *Journal of Geotechnical and Geoenvironmental Engineering* 129: 1028–1039.

Rhoades, J.D. 1993. Electrical conductivity methods for measuring and mapping soil salinity. In *Advances in Agronomy*, Vol. 49, D.L. Sparks (Ed.), Academic Press, San Diego, CA, pp. 201–251.

Rhoades, J.D. and Corwin, D.L. 1981. Determining soil electrical conductivity-depth relations using an inductive electromagnetic soil conductivity meter. *Soil Science Society of America Journal* 45: 255–260.

Rhoades, J.D. and Corwin, D.L. 1990. Soil electrical conductivity: Effects of soil properties and application to soil salinity appraisal. *Communications in Soil Science and Plant Analysis* 21: 837–860.

Rhoades, J.D. and Loveday, J. 1990. Salinity in irrigated agriculture. In *Irrigation of Agricultural Crops*, B.A. Stewart and D.R. Nielsen (Eds.), Agronomy Monograph 30, American Society of Agronomy, Madison, WI, pp. 1089–1142.

Rhoades, J.D., Raats, P.A.C., and Prather, R.J. 1976. Effects of liquid-phase electrical conductivity, water content and surface conductivity on bulk soil electrical conductivity. *Soil Science Society of America Journal* 40: 651–655.

Rhoades, J.D., Manteghi, N.A., House, P.J., and Alves, W.J. 1989. Soil electrical conductivity and soil salinity: New formulations and calibrations. *Soil Science Society of America Journal* 53: 433–439.

Rhoades, J.D., Shouse, P.J., Alves, W.J., Manteghi, N.A., and Lesch, S.M. 1990. Determining soil salinity from soil electrical conductivity using different models and estimates. *Soil Science Society of America Journal* 54: 46–54.

Rhoades, J.D., Chanduvi, F., and Lesch, S.M. 1999. Soil salinity assessment: Methods and interpretation of electrical conductivity measurements. FAO Irrigation and Drainage Paper 57, Food and Agriculture Organization of the United Nations, Rome, Italy.

Robertson, G.P. 2000. GS$^+$, Geostatistics for the environmental sciences, version 5.1. Gamma Design Software, Plainwell, MI.

San Joaquin Valley Drainage Implementation Program. 1998. *Drainage Management in the San Joaquin Valley: A Status Report*. Department of Water Resources, Sacramento, CA.

Sawchik, J. and Mallarino, A.P. 2007. Evaluation of zone soil sampling approaches for phosphorus and potassium based on corn and soybean response to fertilization. *Agronomy Journal* 99: 1564–1578.

Scanlon, B.R., Paine, J.G., and Goldsmith, R.S. 1999. Evaluation of electromagnetic induction as a reconnaissance technique to characterize unsaturated flow in arid setting. *Groundwater* 37: 296–304.

Schepers, A.R., Shanahan, J.F., Liebig, M.A., Schepers, J.S., Johnson, S.H., and Luchiari, A. Jr. 2004. Appropriateness of management zones for characterizing spatial variability of soil properties and irrigated corn yields across years. *Agronomy Journal* 96: 195–203.

Schumann, A.W. and Zaman, Q.U. 2003. Mapping water table depth by electromagnetic induction. *American Society of Agricultural and Biological Engineers* 19: 675–688.

Shannon, M.C., Cervinka, V., and Daniel, D.A. 1997. Drainage water re-use. In *Management of Agricultural Drainage Water Quality*, C.A. Madramootoo, W.R. Johnston, and L.S. Willardson (Eds.), Water Report 13, International Commission on Irrigation and Drainage, Food and Agriculture Organization of the United Nations, Rome, Italy.

Sheets, K.R. and Hendrickx, J.M.H. 1995. Non-invasive soil water content measurements using electromagnetic induction. *Water Resources Research* 31: 2401–2409.

Sheets, K.R., Taylor, J.P., and Hendrickx, J.M.H. 1994. Rapid salinity mapping by electromagnetic induction for determining riparian restoration potential. *Restoration Ecology* 2: 242–246.

Slavish, P.G. 1990. Determining ECa-depth profiles from electromagnetic induction measurements. *Australian Journal of Soil Research* 28: 443–452.

Slavish, P.G. and Petterson, G.H. 1990. Estimating average rootzone salinity from electromagnetic induction (EM-38) measurements. *Australian Journal of Soil Research* 28: 453–463.

Slavish, P.G. and Yang, J. 1990. Estimation of field-scale leaching rates from chloride mass balance and electromagnetic induction measurements. *Irrigation Science* 11: 7–14.

Stroh, J.C., Archer, S., Doolittle, J.A., and Wilding, L. 2001. Detection of edaphic discontinuities with ground-penetrating radar and electromagnetic induction. *Landscape Ecology* 16: 377–390.

Sudduth, K.A. and Kitchen, N.R. 1993. Electromagnetic induction sensing of claypan depth. ASAE Paper No. 931531, American Society of Agricultural Engineers, St. Joseph, MI.

Sudduth, K.A., Kitchen, N.R., Bollero, G.A., Bullock, D.G., and Wiebold, W.J. 2003. Comparison of electromagnetic induction and direct sensing of soil electrical conductivity. *Agronomy Journal* 95: 472–482.

Sudduth, K.A., Kitchen, N.R., Wiebold, W.J., Batchelor, W.D., Bollero, G.A., Bullock, D.G., Clay, D.E., Palm, H.L., Pierce, F.J., Schuler, R.T., and Thelen, K.D. 2005. Relating apparent electrical conductivity to soil properties across the north-central USA. *Computers and Electronics in Agriculture* 46: 263–283.

Tanji, K.K. and Kielen, N.C. 2002. Agricultural drainage water management in arid and semi-arid areas. FAO Irrigation and Drainage Paper 61, Food and Agricultural Organization of the United Nations, Rome, Italy.

Thamke, J.N., Craigg, S.D., and Mendes, T.M. 1999. Use of terrain electromagnetic geophysical methods to map saline-water contamination, East Poplar oil field, Northeastern Montana. In *Ground Conductivity Meters for Environmental Site Evaluation*. Geonics, Mississauga, ON, Canada.

Topp, G.C. and Davis, J.L. 1981. Detecting infiltration of water through the soil cracks by time-domain reflectrometry. *Geoderma* 26: 13–23.

Topp, G.C. and Ferré, P.A. 1996. Methods for measurement of soil water content. In *Methods of Soil Analysis, Part 4. Physical Methods*, J.H. Dane and G.C. Topp (Eds.), Soil Science Society of America Book Series No. 5, American Society of Agronomy-Soil Science Society of America, Madison, WI.

Topp, G.C., Davis, J.L., and Annan, A.P. 1982. Electromagnetic determination of soil water content using TDR: I. Applications to wetting fronts and steep gradients. *Soil Science Society of America Journal* 46: 672–678.

Triantafilis, J. 2000. Use of EM instruments to describe spatial distribution of soil properties: Experiences in Australian cotton fields. EM38 Workshop, 69–76, New Delhi, India.

Triantafilis, J. and Lesch, S.M. 2005. Mapping clay content using electromagnetic induction techniques. *Computers and Electronics in Agriculture* 46: 203–237.

Triantafilis, J., Huckel, A.I., and Odeh, I.O.A. 2001a. Comparison of statistical prediction-methods for estimating field-scale clay content using different combinations of ancillary variables. *Soil Science* 166: 415–427.

Triantafilis, J., Odeh, I.O.A., and McBratney, A.B. 2001b. Five geostatistical models to predict soil salinity from electromagnetic induction data across irrigated cotton. *Soil Science Society of America Journal* 65: 869–878.

Triantafilis, J., Ahmed, M.F., and Odeh, I.O.A. 2002. Application of a mobile electromagnetic sensing system (MESS) to assess cause and management of soil salinization in an irrigated cotton-growing field. *Soil Use and Management* 18: 330–339.

USDA National Agricultural Statistics Service. 2007. California Agricultural Statistics, 2006 Crop Year. USDA National Agricultural Statistics Service, California Field Office, Sacramento, CA.

van Lissa, R.V., van Maanen, H.R.J., and Odera, F.W. 1987. The use of remote sensing and geophysics for groundwater exploration in Nyanza Province, Kenya. Paper presented at the African Water Technology Conference, Nairobi, Kenya.

Vaughan, P.J., Lesch, S.M., Corwin, D.L., and Cone, D.G. 1995. Water content effect on soil salinity prediction: A geostatistical study using cokriging. *Soil Science Society of America Journal* 59: 1146–1156.

Vieira, S.R., Hatfield, J.L., Nielsen, D.R., and Biggar, J.W. 1983. Geostatistical theory and application to variability of some agronomic properties. *Hilgardia* 51: 1–75.

Wait, J.R. 1954. Mutual coupling of loops lying on the ground. *Geophysics* 19: 290–296.

Wait, J.R. 1955. Mutual electromagnetic coupling of loops over a homogeneous ground. *Geophysics* 20: 630–637.

Webster, R. 1989. Recent achievements in geostatistical analysis of soil. *Agrokemical Talajtan* 38: 519–536.

Williams, B.G. and Baker, G.C. 1982. An electromagnetic induction technique for reconnaissance surveys of soil salinity hazards. *Australian Journal of Soil Research* 20: 107–118.

Williams, B.G. and Hoey, D. 1987. The use of electromagnetic induction to detect the spatial variability of the salt clay contents of soils. *Australian Journal of Soil Research* 25: 21–27.

Wittler, J.M., Cardon, G.E., Gates, T., Cooper, C.A., and Sutherland, P.L. 2006. Calibration of electromagnetic induction for regional assessment of soil water salinity in an irrigated valley. *Journal of Irrigation and Drainage Engineering* 132: 436–444.

Wollenhaupt, N.C., Richardson, J.L., Poss, J.E., and Doll, E.C. 1986. A rapid method for estimating weighted soil salinity from apparent soil electrical conductivity measured with an above-ground electromagnetic induction meter. *Canadian Journal of Soil Science* 66: 315–321.

Won, I.J., Keiswetter, D.A., Fields, G.R.A., and Sutton, L.C. 1996. GEM-2, A new multifrequency electromagnetic sensor. *Journal of Environmental and Engineering Geophysics* 1: 129–137.

Wraith, J.M. 2002. Solute content and concentration-indirect measurements of solute concentration–time domain reflectometry. In *Methods of Soil Analysis, Part 4. Physical Methods*, J.H. Dane and G.C. Topp (Eds.), American Society of Agronomy, Madison, WI.

Wraith, J.M., Robinson, D.A., Jones, S.B., and Long, D.S. 2005. Spatially characterizing apparent electrical conductivity and water content of surface soils with time domain reflectometry. *Computers and Electronics in Agriculture* 46: 239–261.

Yates, S.R. and Warrick, A.W. 1987. Estimating soil water content using cokriging. *Soil Science Society of America Journal* 51: 23–30.

12 Combined Active and Passive Remote Sensing Methods for Assessing Soil Salinity: A Case Study from Jezre'el Valley, Northern Israel

Eyal Ben-Dor, Naftaly Goldshleger, Eshel Mor, Vladmir Mirlas, and Uri Basson

CONTENTS

12.1 INTRODUCTION

In many arid regions, the use of irrigation has caused the loss of formerly productive land through waterlogging, salinization, and increased organic content, leading to the abandonment of cultivated fields. Soil salinity is one of the most common soil degradation processes, particularly in arid and semiarid areas (Rhoades, 1990; Ceuppens et al., 1997). Moreover, the use of effluent and drainage water for irrigation purposes, mostly in clayey soils, leads to increased precipitation of evaporite minerals (Ghabour and Dales, 1993). This is critical in places where the water table is high and evaporation is significant. Often, an increase in soil salinity is followed by an increase in sodicity (Rhoades, 1990). The latter increases the dispersion of soil particles and destruction of the soil structure and causes the formation of crust on the soil surface, which leads to soil erosion and degradation.

Soil salinity problems usually result from water table rise. This can be induced by land clearing alone and is then referred to as dryland salinity, or caused by excessive application of irrigation water and is then referred to as irrigation-induced salinity. Salinity is a dynamic property that can change from one season to another (Ben-Dor et al., 2002), either in its location in the field or in its level of severity. Thus, the salinity status is uncertain to farmers and decision makers. Knowing salinity distribution within a given field, both vertically and horizontally, is an important issue. Traditional methods that use point-by-point soil sampling along a field to generate a salinity hazard map are time-consuming and expensive, and cannot be applied on a routine basis. Comparatively, remote sensing that uses passive or active electromagnetic (EM) radiation provides a more efficient method to map and combat soil salinity in large areas and on a temporal basis. Remote sensing is extensively used in agricultural applications to map and monitor changes in vegetation, soil surface condition, water bodies, and snow cover. Although the remote sensing technology offers many advantages, a crucial limitation is that it cannot view the subsurface properties of the earth's solid surface when passive radiation is used.

Soil salinity is a four-dimensional (4D) issue that varies both in time (1D) and space (3D). Space variation takes place within the soil profile (depth) and along the surface (area), and thus requires advanced tools to remotely sense the spatial salinity domain. Subsoil salt occurrence cannot be detected by passive sensors, whereas surface-affected areas are often viewed only when the salinity problem has reached a terminal stage causing a significant loss of production. On a global scale, the annual loss of 75 million tons of soil costs approximately US$400 billion every year (Eswaran et al., 2001). Detecting and mapping areas that undergo salinization in its early stage, when the process is still invisible to the naked eye, are important for drafting and implementing suitable management plans to avoid irreversible damage to the soil. The purpose is to provide the end-user with a salinity hazard map, based on the soil salinity level and distribution, long before the soil undergoes severe degradation and irremediably loses its productivity.

Since soil salinity problems usually begin in the subsurface layers, as controlled by the drainage system, water table, and soil permeability, there is urgent need to develop and apply a subsurface soil remote sensing approach for imaging the soil profile depth and merge it with surface information in order to generate a 3D representation of the emerging problem. The former can be done by applying an active remote sensing technique, whereas the latter can be done by applying a passive hyperspectral technique. Combining such techniques in an integrated method for mapping soil salinity is important, but apparently lacks practical and scientific applications (Farifteh, 2007). Currently, passive and active remote sensing to detect soil salinity are used separately, each having its own limitations and constraints. Barmes et al. (2003) have reviewed the close and far remote sensing techniques for evaluating soil properties and concluded that the challenge remains in having all techniques integrated to obtain the best from all rather than utilizing each individually.

Chapter 3 of this book postulates the need to integrate available surface and subsurface remote sensing technologies such as field spectroradiometers, frequency-domain electromagnetic methods (FDEM), and ground-penetrating radar (GPR), to enable effective detection and mapping of soil salinity. Although this idea has been stressed in several recent studies, apparently not much work has been done toward its implementation. As a result, studies examining the use of combined spectral and EM induction devices in combating soil salinity are scarce. Farifteh (2007) points out that investigation should focus on further integration of both the active and passive remote sensing means with other approaches such as solute modeling and geohydrological knowledge, all that being projected on a geographical information systems (GIS) interface.

In the light of the above challenge, we describe and discuss the results of a study in which both spectral and EM-based means (FDEM and GPR) were merged with traditional laboratory data (for verification), yielding a final product that provides more information than that generated by using each of them separately. We also analyze a GIS model that displays the results in the spatial domain and provides a salinity hazard map to the end-user, with possible solutions and recommendations on how to solve the salinity problem.

12.2 TEST SITE

The study area is the Jezre'el Valley, located in the semiarid part of Israel. Cultivated soils are prone to salinization because of irrigation with low-quality water, clayey soils, and inefficient drainage (Ravikovich, 1992; Benyamini et al., 2005). Crop yields have decreased because of local salinity problems that change from one year to another. Farmers try to control and mitigate salinity using measures, such as subsoil drainage, which sometimes work and sometimes do not. The Jezre'el Valley is an alluvial valley of nearly 300 km^2, which is extensively cultivated all year around, with an irrigated area of 140 km^2. Annual rainfall is about 450 mm (October–April). Main crops are cotton and wheat. Irrigation water is provided by local reservoirs that store runoff water during the rainy season, and by domestic effluents throughout the year. The irrigation methods are drip, lateral move, and central pivot. Soils are mainly Chromic Vertisols, with high clay content (<65%) and smectites as the dominant clay mineral. There is a semiconfined, shallow aquifer about 10 m below the soil surface that collects the rainwater from the ridges surrounding the valley. High groundwater head in the aquifer causes upward hydraulic pressure in the upper soil layers. In recent years, numerous agricultural plots near the water reservoirs developed severe soil salinity that caused the decline of crop production, together with high groundwater levels and excessive soil moisture. Salinity increased as the water provision to the local reservoirs changed from storage of seasonal (winter) runoff water to the year-round storage of domestic effluents. As a result, the water table in the vicinity of the reservoirs rose, followed by capillary rise of soil moisture to the soil surface. Over the past 30 years, subsurface drainage systems were installed to lower the water table in spring by more than 1 m below the soil surface to prevent soil salinization (Mirlas et al., 2003). Local defects in the drainage systems and the occurrence of sealed underground layers have compounded the salinity issue.

Two irrigated fields that were reported by the farmers to be seasonally affected by soil salinity were selected: Mizra and Genigar (Figure 12.1). The Genigar field, extending over some 30 ha, is confined by two ridges lying at a distance of about 3 km from each other. These ridges are the recharge areas for the local buried riverbeds. A buried riverbed with restricted outlet created an abnormally high hydrodynamic pressure in the upper soil layer and inhibited the lowering of the groundwater by means of subsurface drainage systems. As a result, a saline water table formed at a shallow depth, contributing significantly to increase salinity in the root zone of the crops. The Mizra field, covering around 50 ha, is bounded to the north by the embankment of the large Maale Kishon reservoir. The presence of the reservoir caused the groundwater to rise close to the soil surface around the embankment. Evaporation from the shallow groundwater contributed to the formation of saline soils. Two deep drains with passive relief wells were installed parallel to the embankment of the reservoir to lower the water table and control soil salinization. A saline field strip, 25–30 m wide and located close to the embankment, was selected. This field is situated on a subsurface drainage system with a combination of shallow drains, deep drains, and passive relief wells

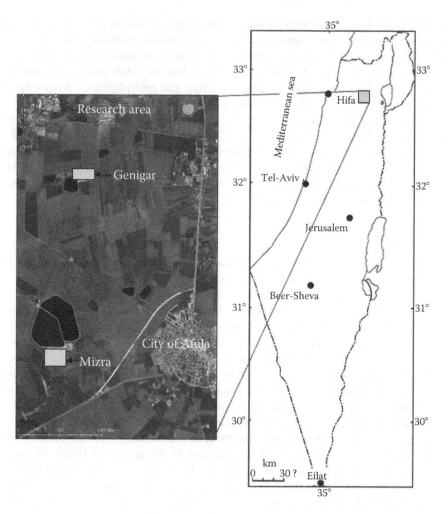

FIGURE 12.1 Geographic location of the two test fields.

(Benyamini et al., 2005). The shallow drains were 0.8–0.9 m deep, with a spacing of about 30 m, whereas the deep drains were about 2–3 m deep with spacing of about 90 m. The passive relief wells, which were filled with gravel and connected to the deep drains, reached a depth of about 8 m (Schwab et al., 1957). They had a diameter of 0.6 m and spacing of about 40 m. These wells reduced the upward flow of the groundwater toward the soil surface, thereby reducing the piezometric head. The soils in both fields are smectite-rich Chromic Vertisols. During the dry summers, preferential flow along wide and deep soil cracks increases deep percolation of irrigation water. During the winter rains or summer irrigations, the soils swell and the cracks close, resulting in a considerable reduction of the saturated conductivity.

12.3 MATERIALS AND METHOD

Active and passive remote sensing techniques, together with laboratory data, were used to track soil salinity in the two aforementioned fields. All information was projected on a GIS interface using a spatial kriging model, to enable spatial interpretation and verification of the results by generating working maps for the end-users. Along each field, sampling lines and points were set for spectral measurements and chemical determinations in boreholes and excavations to various soil depths.

12.3.1 Spectral Measurements

12.3.1.1 Instrument

An ASD field Pro spectrometer covering the VIS–NIR–SWIR regions was used in the research. The ASD field spectrometer FR-Pro model consists of 2151 bands across the 350–2500 nm region. We used a 2 m optic fiber cable with 25° fore-optics. The spectrometer includes three detectors: one in the VIS–NIR region (350–1000 nm) and two in the SWIR region (1000–1790 and 1790–2500 nm). The spectral sampling is 1.4 nm in the VIS–NIR region and 2 nm in the SWIR region. The spectral resolution, as defined by full width half max (FWHM), is 3 nm in the VNIR region and 10 nm in the NIR region.

12.3.1.2 Field Measurements

The spectrometer was programmed to collect 40 samples per second in three modes: (1) a dark current mode, where the electronic noises are (later) subtracted from the original raw data; (2) a white reference mode, where a perfect Lambertian reflector is used as a reference (HALON in this case) at the same configuration; and (3) a sample mode, where the soil sample was measured. The data are computed as reflectance values relative to the standard white reference (HALON) after subtracting the electronic noise in both measurements (dark current). Another set of measurements was done with the contact probe assembly that uses the artificial illumination of a tungsten halogen lamp and is used to measure the samples in both field and laboratory environments.

12.3.1.3 Data Processing

After receiving the spectra and archiving them relative to their exact geographic position (± 1 m) and depth, the spectra of each sample were averaged and examined for irregularities. Subsequently, they were used to determine the minerals associated with each sample from spectral libraries (e.g., USDA, JPL, local database). Next, the spectral information was analyzed in two ways: supervised (where known wavelengths were inspected) and unsupervised (where no previous information was used to correlate the spectral information with the salinity values). For the latter analysis, a multivariate approach was applied using the Unscrambler version 7.0 software. Models were generated to predict salinity indicators (NaCl concentration and electrical conductivity (EC) in the soil solution extract) solely from the spectral information. The results were then put into a GIS environment and interpolated to derive a salinity map at two depths (0–30 and 30–60 cm).

12.3.2 FDEM Measurements

12.3.2.1 The Instrument

The EM measurements, including conductivity and magnetic susceptibility frequency sounding measurements, were acquired using a GEM-2 FDEM system, which is a small, handheld, broadband instrument. The GEM-2 is a fixed-geometry sensor (Slingram) that allows variable multiple-frequency operations suitable for depth sounding. The instrument has a constant transmitter–receiver coil separation of 1.66 m and a standard survey operation height of about 1 m. The sensor operates in 1–10 selected frequencies at a bandwidth ranging from 300 Hz to 48 kHz. Raw data are the in-phase and quadrate (out-of-phase) components of the secondary magnetic field, expressed in parts per million, against the primary field at the receiver coil. From the in-phase and the quadrate profile sounding, lateral and vertical variations in ground conductivity and magnetic susceptibility can be computed and mapped.

12.3.2.2 Field Measurements

The FDEM measurements began by monitoring the environmental EM noise along the bandwidth of the GEM-2 to verify minimal EM noise interference prior to the exact selection of the frequencies. Considering survey goals and EM noise, the measurements were conducted at frequencies (Hz) of 3,025, 6,025, 12,025, 24,025, and 47,975 (odd multiples of 25 Hz, in order to also avoid harmonics of 50 Hz of the electric power line). The data were acquired at a sampling rate of 25 Hz (25 times per second) and three to nine samples were averaged for a better signal-to-noise ratio. Initially, several 2D profiles were conducted at selected locations, following the locations of the boreholes, the GPR profiles, and the surface measurements of the spectral sensors. Subsequently, a 2D "zigzag profile" of about 3400 m length was conducted, using a global positioning system (GPS) with wide area augmentation system (WAAS) corrections to achieve an effective spatial coverage of the area.

12.3.2.3   Data Processing

In addition to analyzing the existing raw in-phase and quadrate data, the first stage of data processing aimed at reducing spikes (due to metal objects and field obstacles) and smoothing the data. The smoothing operation was iterative at different levels to preserve the accuracy of the raw data. In the second stage, in-phase and quadrate values were converted into apparent magnetic susceptibility and apparent conductivity. Girding of the data was done in the third stage in order to produce maps.

12.3.3   GPR

12.3.3.1   Instrument and Field Measurements

The EM properties of the subsurface were measured by using a RAMAC CUII GPR. The GPR system was operated at different frequencies ranging from 100 MHz to 1 GHz, in order to achieve different depths of penetration and resolution.

A variety of reflection profiles was conducted using four center frequencies: 100 MHz (deep penetration—low resolution), 250 and 500 MHz (intermediate penetration—medium resolution), and 1 GHz (shallow penetration—high resolution). The deep profiles were collected to a depth of about 8–10 m using an unshielded 100 MHz antenna configuration. The 100 MHz profiles were manually acquired, with a lateral step (station) interval of 0.25 m and a stack of 64 measurements (at each step it was averaged to increase the signal-to-noise ratio). The intermediate profiles were conducted using shielded antenna configurations to depths of about 3–6 m. The 250 and 500 MHz devices were operated from carts, measurements were automatically acquired at a lateral step interval of 0.05 m, using a survey wheel, and a stack of 8–16 measurements was averaged at each step. The high resolution profiles were conducted using the 1 GHz shielded antenna configuration to a depth of about 1 m. These profiles were automatically acquired at a lateral step interval of 0.01 m, using a 1 GHz cart, and a stack of 4–8 measurements.

12.3.3.2   Data Processing

The profiles were processed in different ways using dewow, band pass, average/median, and average removal filters, with different gains. The raw data (without any filter, besides the display parameters) were presented for comparison. A vertical cross section was generated at problematic sites and the soil profile layers were interrelated.

12.3.4   Laboratory Determinations

The soil samples from the boreholes were sent to the laboratory in sealed plastic bags (to preserve soil moisture) and submitted to several chemical determinations to derive the following attributes: (a) EC (25°C), measured in the soil solution extract (vacuum extraction of saturated paste with distilled water); (b) ions including Na, Ca + Mg, Cl, and SO_4, measured with an atomic absorption spectrophotometer; and (c) soil moisture measured by oven-drying the sample to 105°C for 24 h, and mechanical composition measured by the hydrometer method. All methods are described in detail by Jackson (2005).

Nets of piezometers were placed along the research area at four selected points for monitoring the water table level and water salinity. Each net included three piezometers at different depths: (1) shallow (at 2 m) to characterize the upper soil layer; (2) medium (at 5 m) to characterize the average soil depth; and (3) deep (at 9–12 m) to characterize the water body that creates the artesian effect. Water samples were taken from the piezometers for EC and other measurements.

12.3.5 WORKING SCHEME

In each of the selected fields, measurements were made several times over the year by using simultaneously the three methods (ASD, FDEM, and GPR) mentioned above, in addition to the laboratory determinations for verification. Only the measurements made on July 25, 2005 are used in this study for demonstrating the multisensor approach. The measurements were carried out in subplots of 100–140 m length, crossing each field. The FDEM measurements were made along the GPR lines, with readings recorded every 2 m and every 0.05 m for studying the FDEM density effect. This survey yielded a set of points that were later processed to provide EM susceptibility (FDEM) and backscatter profiles in problematic horizons (GPR). The processed data were projected on a spatial domain based on GPS information.

Along the scanned lines, GPS-located drills were opened every 10 m to a depth of 60 cm (Figure 12.2). Soils were sampled from the soil surface and at 0–30 and 30–60 cm depth to track changes in soil salinity and moisture along the aforementioned control lines, simultaneously with the remote sensing measurements. At each site, measurements were made using a contact probe ASD device, with three replicates for each sample to cover the spectral properties of the soil from all mixtures and directions as obtained in the field. The reflectance of the terrain surface was also measured by the ASD spectrometer with sun illumination, maintaining similar geometry of both the incident and the viewing angle as much as possible. All measurements were processed to obtain reflectance units relative to the HALON white reference, which was measured at the same geometry. In addition, water samples were taken from the piezometers during each field campaign. At several locations along the field, boreholes were opened for detailed soil profile description. All data gathered by the ground-based spectroradiometer, GPR, and FDEM were analyzed and projected on a georeferenced image to visualize and interpret the results on a spatial scale. Based on the resulting information, local farmers were given recommendations regarding how to overcome the local salinity problems.

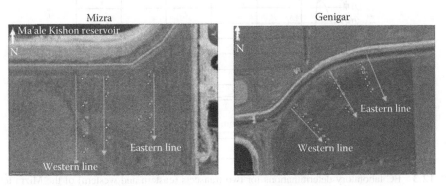

FIGURE 12.2 Location of sampling points over the two test fields. Each point represents soil spectral measurements on the soil surface at a 30–60 cm depth, as well as laboratory determinations. The lines represent the GPR and FDEM survey transects.

12.4 RESULTS AND DISCUSSION

12.4.1 SPECTRAL RESULTS

Two transects, selected in each field on the basis of surface EC values to obtain a first spectral impression, show that the salinity phenomenon is local. In the Mizra field, high salinity (20–90 dS m^{-1}) occurs close to the embankment of the water reservoir, while in the Genigar field relatively high salinity (10–30 dS m^{-1}) occurs in local spots. Subsequently, spectral readings were taken in the Mizra field to identify spectral parameters suitable to track high EC values in the soils. Figure 12.3 shows the EC values along the north-south transect in the Mizra field (the western line), whereas Figure 12.4 shows the spectral readings at each sampling point along the same line. EC values were high at about 10 m from the reservoir embankment but decreased toward the inner field (Figure 12.5). The spectra in Figure 12.7 show significant spectral variation, ranging from high to low EC values. Soil spectra reflect gypsum features in places where EC values are high (Figure 12.6), while a typical Vertisol spectrum corresponds to low EC values (Ben-Dor and Banin, 1995).

Nevertheless, the main spectral features in all samples correspond to absorbed water molecules at around 1400 and 1900 nm. The absorption feature at around 1900 nm shifts toward 1942 nm in high EC samples (some with white salt crust) and to 1911 nm in nonsaline soils (Figure 12.7). The water feature at 1400 nm shows a similar trend: low EC soils have an absorption feature at 1419 nm that shifts toward 1445 nm in high EC soils. In the presence of featureless salt such as halite, the position of the water peaks may be the only indicator helping detect saline areas. Gypsum in our area resulted from a reaction between the SO$_4$ anions coming from the water reservoir and the Ca cations of the soil solution. The plot of the CaSO$_4$ against the NaCl from the soil solution extract shows that both salts are highly correlated (Figure 12.8) and hence, in the case of our area, halite-driven salinity can be identified from spectral features of gypsum or those of the water molecules. This observation is important because, in the case of mapping salinity from airborne hyperspectral data, for instance, the absorption features of the soil water molecules might be overlapped by those of the water vapor, and this relationship would make it possible to assess the existence of spectrally featureless salts. Since gypsum holds also absorption features at around 1700 and 2340 nm and is significantly correlated to halite, it may be used as an indirect indicator to locate salinity-affected areas. Based on the spectral findings obtained in the Mizra field, a similar analysis was applied to the Genigar field (Figure 12.9). The spectral shift of water was significantly dominant in a specific

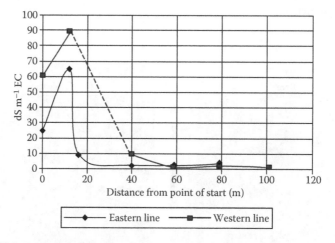

FIGURE 12.3 EC laboratory determinations for two transects (eastern and western) of the Mizra test field. Point 0 refers to the starting point shown in Figure 12.7 (beginning of the arrow of the scanning lines). Distances on the X-axis are in meters.

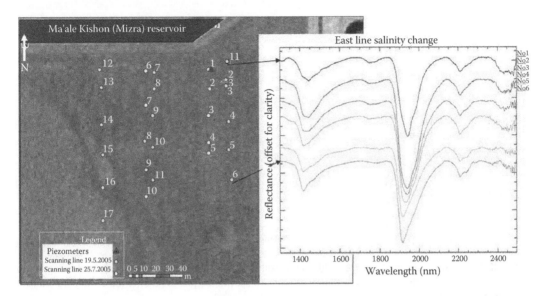

FIGURE 12.4   Sampling points and representative spectra collected from the soil surface at the Mizra test field.

location situated close to points 2, 3, and 11 that have EC values around 30 dS m^{-1}. In the Mizra field, the high EC values appear to correspond to lateral water flow from the reservoir and are located around 10 m from its embankment. However, the relatively high EC values (as detected from the spectroscopy) appear in a more sporadic way, with no plausible explanation. This shows the need to incorporate additional physical remote sensing means, able to sense beneath the soil surface, and spot more light on unexplained high EC values of the Genigar field test.

Visible and infrared reflectance measurements sense only the upper 50 μm of the soil (Ben-Dor et al., 2002), being unable to provide information about the subsoil environment. However, before adopting more sophisticated remote sensing techniques, it is worthwhile to investigate an interesting phenomenon we have found in the field tests studied. Figure 12.10 presents the correlation between the CaSO$_4$ content on the surface (determined in laboratory) and the EC measured at a depth of 30–60 cm. The good correlation obtained between the salinity parameters at the root zone (30–60 cm) and the gypsum content at the soil surface suggests that spectral absorption features of gypsum might be a good indicator for monitoring subsoil salinity without

FIGURE 12.5   Field view of Mizra Field toward the south from the embankment of the Ma'ale Hakishon reservoir.

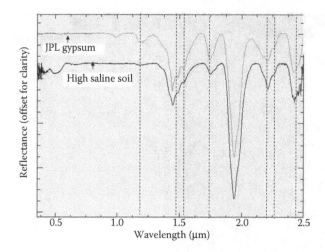

FIGURE 12.6 Spectral signature of a highly saline soil from the Mizra test field, as compared to a gypsum spectral signature from the JPL spectral library (2006). All spectra are presented after the continuum removal algorithm was applied.

using active EM means. Nevertheless, it should be remembered that this kind of relationship is local and may not be valid in other fields.

Based on the spectral information that shows a good relation to the apparent electrical conductivity (ECa) values, a multivariate spectral-based model was generated. Figure 12.11 presents the validation results of this model. Applying the spectral model to each of the surface soil sampling points collected in the field tests generated a data base in which a kriging spatial interpolation process (Oliver and Webster, 1990) was implemented to produced a continuous layer of the area. This spatial layer allocates EC values to areas that have never been sampled or analyzed, providing a regional view of the salinity problem. Validation of the spectral-based model was done on a spatial basis using the laboratory-derived EC values and a similar kriging procedure.

Figures 12.12 and 12.13 show the spatial distribution of EC values as evaluated using the spectral model and the laboratory-derived data for both test fields of Mizra and Genigar. The

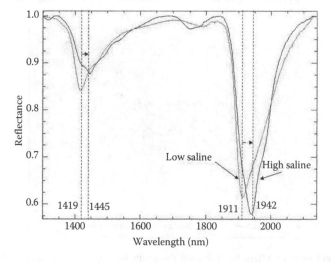

FIGURE 12.7 Spectra of high and low saline soils from the Mizra test field showing water band shifts.

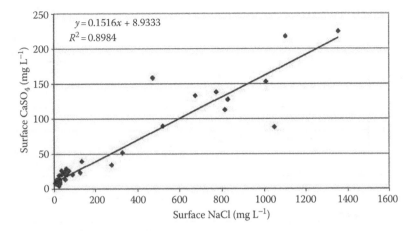

FIGURE 12.8 Statistical correlation between halite (NaCl) and gypsum (CaSO$_4$) in the Mizra test field.

agreement between spectral and laboratory data reaches 92% in Mizra and 81% in Genigar. The model-based image (Figure 12.13a) reveals three highly saline areas (points $1 + 2 + 3$, points $11 + 12$, and points $17 + 18$), while the laboratory-based image (Figure 12.13b) shows only one hot spot (points $1 + 2 + 3$). This information is later checked using FDEM and GPR to find and explain the reasons for these differences in the identification of saline areas.

The high correlation ($>80\%$) found between the spatial distribution of salt-affected areas using the two methods (spectral- and laboratory-derived analyses) suggests that (1) it is possible to replace to some extent the chemical determinations by spectral measurements and (2) the spatial information generated by the spectral-based model provides new information that pinpoints the areas that require more analysis by incorporating additional tools.

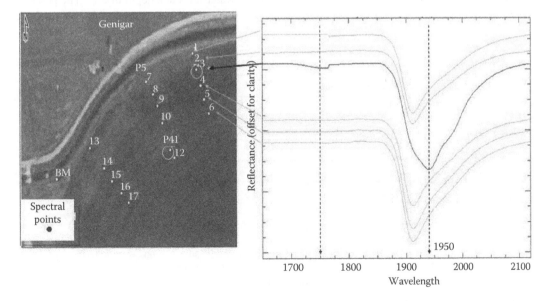

FIGURE 12.9 Sampling points and representative spectra collected from the soil surface of the Genigar test field (wavelength in nm).

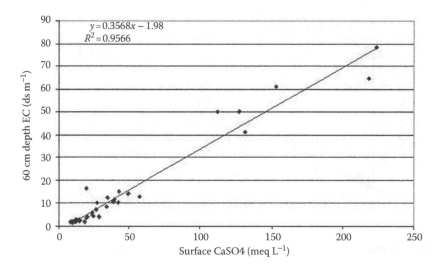

FIGURE 12.10 Statistical correlation between surface gypsum content and EC values measured at 30–60 cm depth.

12.4.2 GPR Results

Based on the spectral results of the field tests, we used EM techniques with a view to assess the type of additional information on soil salinity that can be obtained from active remote sensing. Scanning with a GPR instrument was done at a frequency of 100 MHz to a depth of about 8–10 m in the Genigar test field and at a frequency of 500 MHz to a depth of about 4 m in the Mizra test field.

Figure 12.14 presents the scanning results of the survey conducted in the central part of the Mizra field, sensing the soil subsurface to 3 m depth with about 5 cm range resolution. The subsurface layers can be well observed, showing that the cross section is correlated with drilling borehole observations. The GPR sensed the underground water level at a depth of 2 m. The passive relief wells located in this area were also clearly identified and thus used to verify this information.

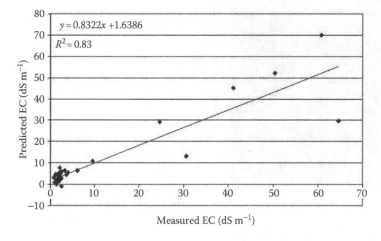

FIGURE 12.11 Spectral model validation results: EC measured in laboratory versus spectrally predicted EC values.

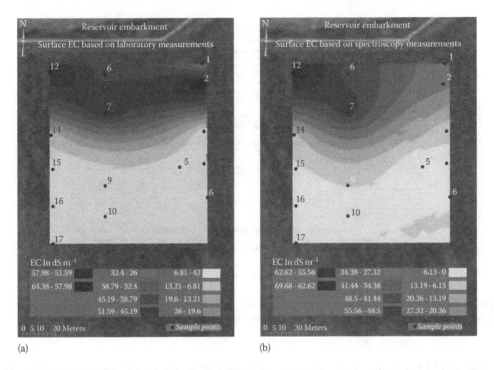

FIGURE 12.12   Kriging interpolation of (a) EC predicted from the model in the Mizra test field and (b) EC measured in laboratory.

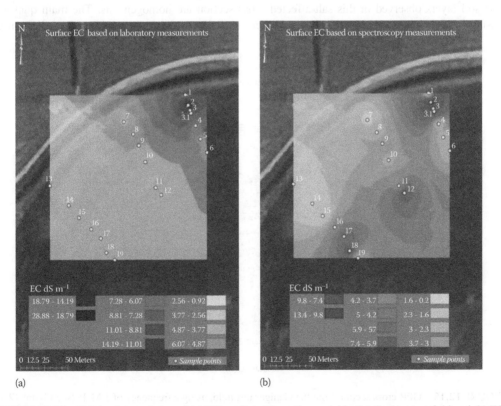

FIGURE 12.13   Kriging interpolation of (a) EC predicted from the model in the Genigar test field and (b) EC measured in laboratory.

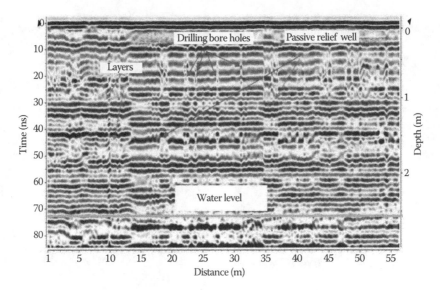

FIGURE 12.14 GPR cross section from the Mizra test field, using a frequency of 1 MHz. The water table, piezometer locations, and several soil layers are clearly seen.

Figure 12.15 presents a cross section located in the western side of the Genigar test field, at 30 m from the eastern transect. The cross section is 12 m long and 2.5 m deep (see Figure 12.16 for its exact location). The resolution of the scanning line was high, with an interval of less than 0.05 m. The soil layers observed in this salt-affected cross section are homogeneous. The main question arising from the measurements relates to whether the shallow soil profile contains physical features

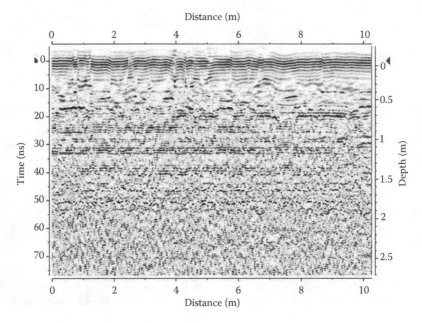

FIGURE 12.15 GPR cross section from the Genigar test field, using a frequency of 1 MHz (see Figure 12.16 for location). The profile is very homogeneous.

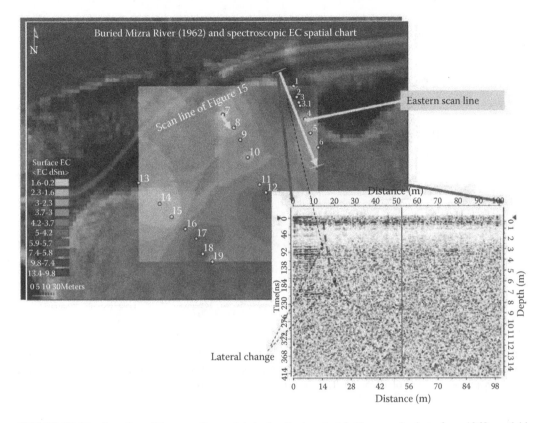

FIGURE 12.16 Location of the sampling points in the Genigar test field on an air photo from 1962 overlaid on a kriging interpolation layer of the EC content as determined by the spectral model. Also shown is a cross section of the 2D GPR profile taken at the eastern line using a frequency of 100 MHz, together with the cross section location of the 2D GPR profile given in Figure 12.15.

that have significant influence on the water regime. From the GPR readings, no significant changes were observed along the cross section, although a discontinuity in the layers was evident at a depth of 0.6–0.7 m. The GPR results at low frequency for points 1–2–3 in the Genigar irrigated test field, depicted by the spectral measurements to be a salt-affected location, confirmed the existence of a buried paleochannel. Figure 12.16 shows the location of these sampling points on an air photo from 1962 and a 2D GPR profile taken at point 3 using a frequency of 100 MHz with the following parameters: an antenna separation of 1 m, 0.25 m step size, a time window of 420 ns $(420 \cdot v)/2 =$ depth, and 64 stacks. The horizontal scale shows the measurement location (m) along the profile and the vertical scales show the time (ns) and the depth (m). The profile revealed a prominent lateral change at a depth of about 2.5 m in the vicinity of point 3 on the east line of the Genigar test field (Figure 12.9), where the aerial photograph shows a buried pebbled paleochannel which favors water infiltration. This explains why points 1, 2, and 3 were found to be salt affected. Thus, it can be concluded that GPR surveys provide important information for better understanding on why a given point holds high EC values and for inferring possibilities to rehabilitate salt-affected soils.

12.4.3 FDEM Results

Whereas the GPR data provide an internal view of the soil profile that suggests reasons for the occurrence of high salinity areas, the FDEM measurements provide information that is strongly related to EC values in given soil layers, depending on the frequencies used. Figure 12.17 shows an

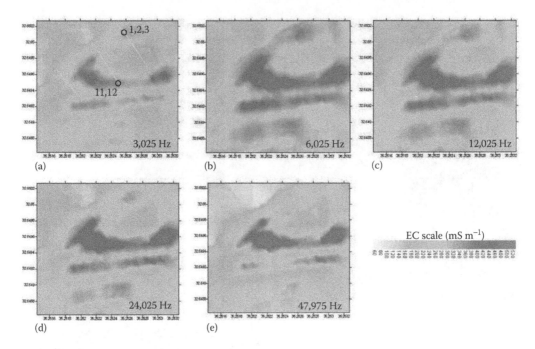

FIGURE 12.17 EC variation maps generated using the FDEM technique. Each map (a–e) represents the EC variations at frequencies of 3,025, 6,025, 12,025, 24,025, and 47,975 Hz, with the corresponding estimated integral depths (m $\pm 25\%$) of 10, 5, 2.5, 1.25, and 0.65 m, respectively.

example of the FDEM findings generated from a set of profiles in the Genigar test field. Five maps of EC variations (mS m^{-1}) were compiled from the measurements conducted at the aforementioned frequencies. The maps represent the integrated subsurface EC values at depths of 10, 5, 2.5, 1.25, 0.65 m ($\pm 20\%$), respectively. A vertical variation from the soil surface to a depth of about 10 m can be observed, as well as a lateral variation along three lines (1, 2, and 3; see Figure 12.2). These lines represent the position of artificial drainage pipes that were constructed years ago. Line 1 is characterized by low EC values, whereas line 2 presents high EC values. Low values indicate healthy and well functioning drainage pipes, whereas high values reveal a problematic situation within the drainage system. Figure 12.17a shows the area where spectral measurements overlay the FDEM survey. Hot spots observed in the spectral model map (Figure 12.13) are also observed in the FDEM map. It appears that the spectral measurement lines were not dense enough to identify saline areas that the FDEM would not be able to detect also, and therefore interpolation of the three reflectance lines underestimated the salinity content. These results show that both methods provide consistent and complementary information. Figure 12.18 highlights an inversion process applied to the FDEM data to reveal local changes as a function of depth. Four maps of EM levels (mS m^{-1}) corresponding to depth ranges of 5–10, 2.5–5, 1.25–2.5, and 0.65–1.25 m are presented. The results show that shallow profiles (0.65–1.25 and 1.25–2.5 m) experience more variations than deeper profiles. In the 2.5–5 m inversion image, a good correlation is obtained with the previous EC images that pinpoints to the position of the drainage pipeline (Figure 12.17), whereas new line features following a north-south direction can be observed in the 5–10 m range. Since the latter features provide no explanation as to the current salinity problem, they were not further examined. It can be concluded that FDEM measurements provide new insights into salinity distribution within the soil profile, but apparently fail to reveal information about the root zone (0–60 cm).

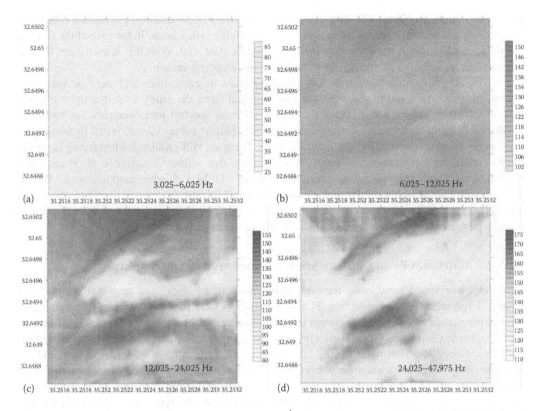

FIGURE 12.18 Inversion maps of EM levels (mS m^{-1}) for depth ranges of 5–10, 2.5–5, 1.25–2.5, and 0.65–1.25 m.

12.5 CONCLUSIONS

12.5.1 CONTRIBUTION OF THE DIFFERENT TECHNIQUES TO MAPPING SOIL SALINITY

Soil spectral sensing by using specific bands (e.g., 1400 and 1900 nm) or by multivariate analysis of the entire spectral region, integrated with laboratory-derived EC values, enable mapping salt-affected areas. However, this study shows that intensive spectral measurements must be conducted in the field. To this end, hyperspectral technology, providing continuous spectral information on a pixel basis, may be a suitable tool for remote salinity mapping. The spectral and laboratory-derived measurements showed the feasibility of using the presence of gypsum at the soil surface as an indirect indicator for assessing the occurrence of (halite) salt at 30–60 cm depth. The strong correlation observed between gypsum and halite in the study area may enable airborne hyperspec-tral-based mapping, because gypsum absorption bands occur in a spectral range different from the water vapor bands of the atmosphere. This finding may be further used to map salinity occurrence in the root zone from airborne spectral sensors. However, the relationship found in the two test fields is limited to our study area, and its extrapolation to other areas should be preceded by local studies. The EC maps generated from the spectral-based models showed consistency with the ones derived from EC determinations in laboratory and helped select potential locations for GPR inspection. On the basis of GPR data, factors controlling the occurrence of highly saline areas, such as the presence of a regional subsurface rock layer that reduces water percolation to deep groundwater, could be scrutinized. The FDEM provided a view of all soil layers, except that of the root zone, pinpointing areas with salinity problems. The images obtained from the FDEM readings provided a subsurface view that helped identify reasons for the occurrence of highly saline areas. In the case of the Genigar test field, malfunction of the drainage pipelines could be observed. With this information, farmers can solve or mitigate the problem before the following cropping season.

All the aforementioned methods are complementary to each other and may be used in a synergistic way. An important conclusion that emerged from the study is that a more efficient method to survey large areas is strongly required. Whereas spectral measurements can be further developed by using airborne hyperspectral technology, FDEM surveys are required in frequencies that can cover relatively shallow depths (0.35–3 m). There are still problems when trying to survey the root zone. In the present case study, it was shown that salinity occurrence at 60 cm depth correlated well with surficial salinity features, making FDEM a suitable mapping tool. Another problem with FDEM technology is that it cannot be operated from airborne platforms to cover wide areas. This case study demonstrated that the synergy between FDEM and spectral measurements offers a powerful approach, whereas GPR data are somehow redundant.

12.5.2 ACTIVE AND PASSIVE REMOTE SENSING TOOLS TO MAP SOIL SALINITY

This chapter reviewed three basic remote sensing methods to assess and map soil salinity. One passive (hyperspectral) and two active (FDEM, GPR) tools were covered. Whereas passive spectral sensors are unable to obtain data beneath the soil surface, FDEM shows the feasibility of sensing from shallow to deep layers (0.4–10 m). GPR can also provide information about subsurface salinity problems, though it is sensitive to many factors including water content, soil compaction, and soil structure. Hyperspectral measurements have been found to hold chemical and physical information about soil properties, and this is why researchers are increasingly interested in its application. Spectral signatures can be obtained for a limited number of salt types (e.g., gypsum), whereas indirect indicators (e.g., soil aggregation and moisture) may provide basic spectral responses to assess soil salinity. In this regard, several significant spectral models to assess soil salinity currently exist. Compared to the hyperspectral tool, FDEM and GPR are extensively used for subsurface soil salinity detection, with emphasis on FDEM. Many studies have shown that these tools are relatively cost-effective in field conditions. The main problem with the active sensing tools is their inability to survey large areas, because they operate on the ground either by close contact (GPR) or 1 m above the surface (FDEM). This prevents mapping salt-affected areas under vegetation cover or in variable terrain conditions. Field-based spectral tools can be replaced by airborne hyperspectral technology that provides continuous spectral information on a pixel basis. Although only a few studies have merged the three methods, there is a strong need to do so. In our case study, we have applied the synergy of these tools to assess and map soil salinity in two agricultural fields from Israel. The results were compared to conventional laboratory data of soil salinity and were found to complement each other significantly. The FDEM technique revealed to be a very accurate tool for mapping salinity at different soil depths, whereas the passive spectral technique was able to generate interesting results on surface salinity and its relation to depths of 30–60 cm. A spectral model to map soil salinity showed significant agreement with the traditional laboratory-derived approach. Nevertheless, the spectral readings would have to be carried out at higher sampling density to be comparable to the FDEM data. Hyperspectral data should be first calibrated against FDEM measurements, and therefore cannot stand alone. A combined airborne technology using both spectral and FDEM means would be the ultimate tool to handle soil salinity problems.

ACKNOWLEDGMENT

This work was supported by the Ministry of Energy and Infrastructure, Earth Sciences Administration, grant 23-2007E and the Ministry of Agriculture, grant 8550053/05.

REFERENCES

Barmes, F., Ricci, M., Zannoni, C., and Cleaver, D.J. 2003. Computer simulations of hard pear-shaped particles. *Physical Reviews E* 68, 021708.

Ben-Dor, E., Patkin, K., Banin, A., and Karnieli, A. 2002. Mapping of several soil properties using DAIS-7915 hyperspectral scanner data. A case study over clayey soils in Israel. *International Journal of Remote Sensing* 23: 1043–1062.

Ben-Dor, E. and Banin, A. 1995. Near-infrared analysis (NIRA) as a simultaneous method to evaluate spectral featureless constituents in soils. *Soil Science* 159: 259–269.

Benyamini, Y., Mirlas, V., Marish, S., Gotesman, M., Fizik, E., and Agassi, M. 2005. A survey of soil salinity and groundwater level control systems in irrigated fields in the Jezre'el Valley, Israel. *Agricultural Water Management* 76: 181–194.

Ceuppens, J., Wopereis, M.C.S., and Miezan, K.M. 1997. Soil salinization processes in rice irrigation schemes in the Senegal river delta. *Soil Science Society of America Journal* 61: 1122–1130.

Eswaran, H., Lal, R., and Reich, P.E. 2001. Land degradation: An overview. In: *Response to Land Degradation*, Eds. E.M. Bridges, I.D. Hannam, L.R. Oldeman, F.W.T. Penning de Vries, S.J. Scherr, and S. Sombatpanit. Science Publishers, Enfield, NH, pp. 20–35.

Farifteh, J. 2007. *Imaging Spectroscopy of Salt-Affected Soils: Model-Based Integrated Method*. PhD dissertation 143, International Institute for Geo-Information Science and Earth Observation (ITC), the Netherlands.

Ghabour, T.K. and Dales, L. 1993. Mapping and monitoring of soil salinity of EL-Fayoum depression by the aid of Landset imagery. *Egyptian Journal of Soil Science* 33(4): 355–370, ISSN 0302–6701.

Jackson, M.L. 2005. *Soil Chemical Analysis: Advanced Course*. UW-Madison Libraries Parallel Press.

JPL spectral library. 2006. ENVI 4.3 program.

Mirlas, V., Benyamini, Y., Marish, S., Gotesman, M., Fizik, E., and Agassi, M. 2003. Method for normalization of soil salinity data. *Journal of Irrigation and Drainage Engineering* 129: 64–66.

Oliver, M.A. and Webster, R. 1990. Kriging: A method of interpolation for geographical information system. *International Journal of Geographical Information Systems* 4: 313–332.

Ravikovich, S. 1992. *The Soils of Israel: Formation, Nature and Properties*. Hakibbutz Hameuchad, Tel Aviv, Israel.

Rhoades, J.D. 1990. Soil salinity—causes and controls. In: *Techniques for Desert Reclamation*, School of Geography, University of Oxford, UK.

Schwab, G.O., Manson, P.W., Luthin, J.N., Reeve, R.C., and Edminster, T.V. 1957. Engineering aspects of land drainage. In *Drainage of Agricultural Lands*, Ed. J.N. Luthin. American Society of Agronomy, Madison, WI, pp. 287–395.

Part III

Diversity of Approaches to Modeling
Soil Salinity and Salinization

13 Mapping Salinity Hazard: An Integrated Application of Remote Sensing and Modeling-Based Techniques

Dhruba Pikha Shrestha and Abbas Farshad

CONTENTS

13.1 INTRODUCTION

Land degradation, a decline in the quality of the land caused by human activities, has become a major global issue (Eswaran et al., 2001). More than half of the dry areas worldwide are affected by land degradation (Dregne and Chou, 1994). Although water erosion is the dominant human-induced soil degradation process, an extent of 0.8 million km^2 suffers from secondary salinization caused by land mismanagement, with 58% of these in irrigated areas alone, and nearly 20% of all irrigated land is salt affected (Ghassemi et al., 1995). The main source of salt in soils comes from the weathering of rocks and primary minerals, either formed in situ or transported by water and/or wind. Main causes of salinity development are irrigation with saline water; disturbance of the water balance between rainfall, on the one hand, and stream flow, groundwater level and evapotranspiration, on the other; groundwater rise caused by forest clearance, overgrazing, and cutting bushes; water percolation through saline materials; and intrusion of seawater. If the groundwater traverses salt-bearing rocks, then there is a risk that the overlying formations be salt-contaminated. These factors cause and/or enhance salinization, sometimes in association with alkalinization. The latter leads to the formation of sodic soils, although these can also result from the desalinization of salt-affected soils that are poor in divalent cations. Following the slogan that "prevention is better than cure," it is argued that mapping salinity hazard might be more useful than mapping salinity itself, hazard being defined as "potential threat to humans and their welfare" (Smith, 2000).

The objective of this study is to introduce an approach to map salinity hazard, the application of which should help track down the occurrence of salt in the subsoil at an early stage, before the salt has reached the surface. A case study carried out in Thailand illustrates the method.

13.2 METHOD AND TECHNIQUES FOR MAPPING AND MONITORING SOIL SALINITY

13.2.1 REMOTE SENSING TECHNIQUES

Salt-affected soils can be detected on aerial photographs and other remote sensing imagery often in the advanced stage of the salinization/alkalinization process, that is, when the soil surface is already affected. In visual image interpretation, the presence of salt is inferred from photographic elements and landscape features such as position in the landscape, graytone, drainage condition, lithology of the surroundings, vegetation, and land use. A poor crop stand on a salt-affected soil, even without surficial salt appearance, shows a different surface reflectance than that of a healthy vegetation cover on a salt-free soil. Salt detection is often satisfactory when aerial photographic interpretation is combined with satellite data analysis. Difference in surface reflectance helps separate salt-affected soils from nonaffected ones. A variety of remote sensing data, including aerial photos, video images, infrared thermography, visible and infrared multispectral, microwave and airborne geophysical data, is available for salinity mapping and monitoring. Many researchers have applied several techniques together to map and monitor saline soils (Metternicht and Zinck, 2003). Howari (2003) used supervised classification, spectral extraction, and matching techniques to investigate types and occurrences of salts in the semiarid regions of the United States–Mexico border areas. In China, Peng (1998) applied an approach combining Landsat TM data transforms with the depth and mineralization rate of groundwater to map soil salinity. In Egypt, Masoud and Koike (2006) have detected and monitored salinization from changes in surface characteristics, expressed from vegetation indices and tasseled cap transformations, and from changes in radiometric thermal temperature, and linked its acceleration to a specific year. In Mexico, an adjusted normalized difference vegetation index derived from Landsat TM data, called combined spectral response index, is reported to have high negative correlation with salt contents (Fernández-Buces et al., 2006). In Israel, hyperspectral airborne sensor data were processed to yield quantitative maps of soil salinity (Ben-Dor et al., 2002). A representative set of soil samples was first taken for laboratory determinations, including organic matter content, pH, and electrical conductivity, and this was followed by the selection of the most correlated spectral bands with the measured soil properties. In this study, multiple regression analysis was used to predict soil salinity level. When hyperspectral data are available, wavelength positions of absorption features related to salinity can be identified, and parameters such as depth, width, area, and asymmetry of the absorption features can be correlated with salinity. A study in southern Spain shows that soil salinity is highly correlated with the depth of the absorption features at 615–625 and 800–810 nm, while it is negatively correlated with the width and asymmetry of the absorption features at 800–810 nm (Shrestha et al., 2005a). In addition to optical remote sensing, synthetic-aperture radar data are used for mapping soil salinity especially in tropical areas (Bell et al., 2001). Similarly, time-domain reflectometry is reported to be an effective method for the assessment of salinity hazard in flooded paddy soils (Grunberger et al., 2005). Once the factors determining the salinity hazard of an area are recognized, a salinity hazard map can be produced. An area with high salinity hazard becomes saline only if there is a change in management practices that affects the water balance and mobilizes salt in the landscape (Gordon et al., 2006).

In summary, salinity is a dynamic process and may vary seasonally. A variety of remote sensing data and techniques has been used to map salt-affected soils (Mougenot et al., 1993; Douaoui et al., 2006; Farifteh et al., 2006). The detection of saline soils can be hindered or inaccurate for various reasons, including the lack of specific absorption bands of some salt types, limited availability of satellite sensor data with high spectral resolution, and variability of saline soils in time and space (Metternicht and Zinck, 2003). Detecting and mapping saline soils become easier when a salt-related symptom (e.g., crusts) is visible on the surface, in which case salinity is already in an advanced stage. The salinization process often starts in the subsoil and slowly develops upward. Salinity mapping can be more complicated when soil moisture affects the surface reflectance, especially in the visible portion of the spectrum. Assessing salinization is time-demanding and laborious, as changes in salinity must be monitored over a long time. Based on laboratory data from soil samples collected over a period of 24 years, Herrero and Pérez-Coveta (2005) were able to diagnose a decrease of soil salinity in the upper meter of soil in southern Spain. Apart from identifying salt-affected soils and monitoring soil salinity, it is important to assess the trend in salinity development and map the salinity hazard that might threaten a given area.

13.2.2 STEPWISE METHODOLOGICAL APPROACH

Considering the difficulty of mapping salinity, a stepwise approach was designed to map salinity hazard, instead of merely mapping salt-affected soils (Figure 13.1), and this approach was implemented in a case study in Thailand.

In the first step, the areas exposed to salinity hazard because of their climatic conditions are set apart, since salinity usually develops in comparatively drier environments, such as the arid and semiarid regions where evapotranspiration is often higher than rainfall and groundwater tends to rise. To check whether the annual evapotranspiration rate is higher than rainfall, an empirical equation (Blaney and Criddle, 1962) can be applied:

$$U = kf \tag{13.1}$$

where
  U is the monthly consumptive use of crops or monthly evapotranspiration rate
  k is an empirical crop coefficient
  f is a climatic factor

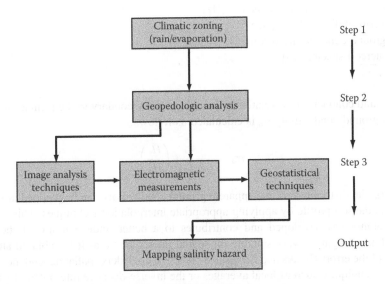

FIGURE 13.1 Stepwise approach to mapping salinity hazard.

The climatic factor can be computed as follows:

$$f = p(0.46t + 8.13)$$

(13.2)

where

  p is the monthly percentage of daylight hours in relation to the annual total of daylight hours
  t is the mean monthly air temperature

Obviously, this first step is applicable only in large areas showing climatic diversity.

In the second step, the geo(hydro)pedological setting of the landscape is introduced to help model salinity hazard. Remote sensing data are used in combination with a digital elevation model (DEM) for stereoscopic interpretation of the landscape. The output shows areas prone to salinization.

The third step includes site examination and application of various image transformation techniques, such as enhancement of soil surface features by applying band rotation (Shrestha, 2005) or deriving salinity-related indices (Douaoui et al., 2006). In this step, attention is also given to the study of the subsoil using techniques of near-surface geophysical survey, followed by geostatistical analysis, and rule-based modeling in geographic information system (GIS) environment. Geophysical data help detect areas of high electrical conductivity where no surface expression of salt is evident.

The near-surface electromagnetic (EM) induction technique helps in mapping terrain (bulk) conductivity practically as fast as the operator can walk (McNeill, 1980). The EM technique works with a transmitter coil and a receiver coil. The transmitter coil is energized with an alternating current that creates a time-varying magnetic field, which induces a very small current in the ground. This generates a secondary magnetic field H_s, which is sensed together with the primary field H_p by the receiver coil (McNeill, 1980). The relationship between the secondary and the primary magnetic fields at the receiver coil is defined as follows:

$$\frac{H_s}{H_p} \approx \frac{i\omega\mu\sigma(s)^2}{4}$$

(13.3)

where

  H_s is the secondary magnetic field at the receiver coil
  H_p is the primary magnetic field at the receiver coil
  ω equals $2\pi f$, where f is the frequency (Hz)
  μ is the permeability of free space
  σ is the ground conductivity (dS m^{-1})
  s is the intercoil spacing (m)
  i equals $\sqrt{-1}$

Since ground conductivity is related to the ratio of the secondary to the primary magnetic field, the apparent ground conductivity σ_a is calculated as follows:

$$\sigma_a = \frac{4}{\omega\mu(s)^2}\left(\frac{H_s}{H_p}\right)$$

(13.4)

Finally, the spatial distribution of the apparent ground conductivity can be mapped at various depth classes within the soil profile by applying appropriate interpolation techniques. This shows at what soil depth salinity has developed and contributes to a better understanding of the salinization process. Ordinary kriging is very suitable for spatial interpolation, as it is unbiased and minimizes the variance of the errors (Issaks and Srivastava, 1989). It considers spatial dependency unlike other interpolation techniques, such as local averages or the inverse-distance interpolation where weights are function of the distances to known points. To apply ordinary kriging, spatial dependency needs to be assessed by computing semivariances and fitting semivariograms so that the nugget effect, the sill variance and the range are known. Semivariances are derived from

$$\gamma(h) = \frac{1}{2m} \sum [Z(xi) - Z(xi + h)]^2 \tag{13.5}$$

where
  $\gamma(h)$ is the semivariance at lag h
  $Z(xi)$ is the value at a given location (xi)
  h is the lag (both distance and direction)
  m is the pair of observation points separated by h

In ordinary kriging, estimation is done by linear combinations of the available data as follows:

$$Z'(xo) = \sum_{i=1}^{n} wi \, Z(xi) \tag{13.6}$$

where
  $Z'(xo)$ is the unknown value
  $Z(xi)$ are the known values at different locations
  wi are the weights with the condition that $\sum_{i=1}^{n} wi = 1$

The weights used in the interpolation are generated using distances between all possible pairs of data locations. Finally, the reliability of the prediction is evaluated by computing the root-mean-square error (RMSE) as follows:

$$\text{RMSE} = \sqrt{\left[\frac{1}{n} \sum_{i=1}^{n} (Z(Xi) - Z'(Xi))^2\right]} \tag{13.7}$$

where
  $Z(Xi)$ is the actual EC value
  $Z'(Xi)$ is the predicted EC value
  n is the number of validation points

13.3 CASE STUDY IN THAILAND

13.3.1 STUDY AREA

A study area of 7400 ha was selected in the Nong Suang district, Nakhon Ratchasima province of Thailand, between 101°45′ and 102°E and between 15° and 15°15′N (Figure 13.2). The area is relatively dry with an average annual rainfall of 1035 mm computed from meteorological data of a 30-year period (1971–2000). Eighty percent of the rainfall occurs from May to October, with the highest rainfall of 226 mm in the month of September. Except for the months of September and October, monthly evaporation is usually higher than precipitation with an average annual evaporation of 1817 mm. As evaporation largely exceeds rainfall, the area has potential for salinity development from the climatic point of view. A major part of the native Dipterocarp forest has been converted into agricultural land (Sukchan and Yamamoto, 2003). Maize and kenaf were introduced in the 1960s, cassava in the 1970s and sugarcane in the 1980s.

Salinity distribution in the study area is clearly related to the geopedologic setting (Soliman, 2004). Salt-bearing rocks occur at about 80 m depth with effect on the groundwater (Imaizumi et al., 2002). The saline groundwater reaches the surface through natural channels (faults, fractures) or openings created by salt mining activities. In the soil mantle, salts also move upward because of

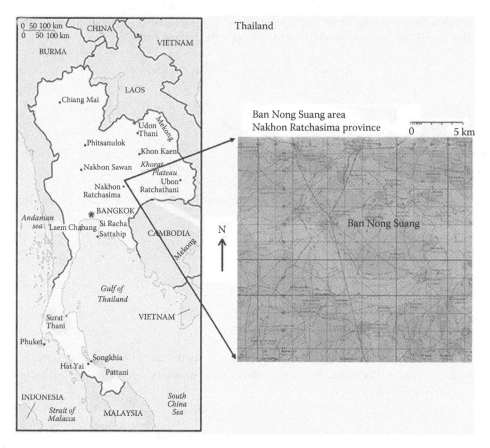

FIGURE 13.2   Study area.

high evapotranspiration (Farshad et al., 2005). Maize yields have declined because of increasing soil salinity (Yadav, 2005). Salt spreading at the surface of paddy fields has been related to the occurrence of a densipan at around 50–70 cm depth. This impervious layer has been locally broken, letting the saline groundwater reach the surface. This has resulted in saline spots and lateral salt spreading forming surface crusts (Figure 13.3).

13.3.2   GEOPEDOLOGIC ANALYSIS AND MASKING AREAS PRONE TO SALINITY DEVELOPMENT

Landsat TM data, acquired in February 2003, were georeferenced to the UTM coordinate system (zone 47) with Indian 1975 datum. A topographic map at scale 1:50,000 (sheet no. 5339II) was scanned and the contour lines were digitized on screen. Using linear interpolation, a DEM was generated with 30 m spatial resolution in order to match with the spatial resolution of the Landsat TM data. Finally, a stereoimage pair was created from the DEM and the Landsat TM band 4 data (Figure 13.4), on which geomorphic interpretation was carried out following a geopedologic approach (Zinck, 1988/89). Delineations were traced directly on the stereoimage on the screen. The advantage of on-screen digitizing is that the interpretation lines are geocoded at once and do not need posterior transformation. Fieldwork was carried out in August–September 2003 and 2004. Two landscape units were recognized, namely peneplain (Pe) and valley (Va) with several sub-units. A peneplain consists of isolated hills, glacis, vales, and depressions, whereas a Valley consists of floodplain and terraces (Table 13.1).

   The basement rocks, belonging to the Korat group, play an important role in the geopedologic setting of the region. They include claystones and shales, interbedded with two or three layers of

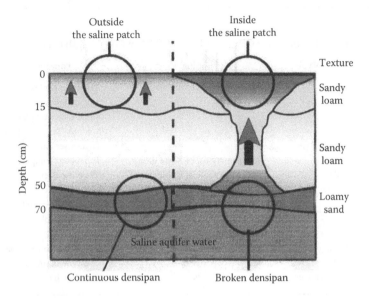

FIGURE 13.3   Broken densipan and its effect on groundwater and salt movement.

evaporites (halite, gypsum, anhydrite). Five soil order classes were distinguished using the USDA soil taxonomy (USDA, 1998). Ultisols occur on ridges, while Alfisols (Ustalfs and Aqualfs) are mainly in sloping areas adjacent to the ridges (Figure 13.5). Vertisols occur in the northern part of the area, along rivers and channels, where vertic horizons form due to the presence of swelling clays. Two suborders, namely Aquerts and Usterts, were distinguished based on the soil moisture regime. Inceptisols are common in the lower part of the lateral valleys that dissect the peneplain lobes. Inceptisols are mainly Aquepts due to poor drainage conditions that lead to the development

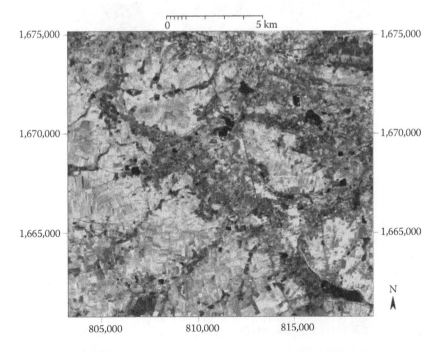

FIGURE 13.4   Stereo-anaglyph of the study area generated using Landsat TM4 and DEM.

TABLE 13.1

Image Interpretation Legend for Geopedologic Mapping (see Figure 13.5)

Landscape	Relief Type	Lithology	Landform	GP Code
Peneplain (Pe)	Ridge	Sedimentary rocks retainasis	Summit complex	Pe111
		Korat group	Side complex	Pe112
			Slope-facet complex	Pe113
			Summit	Pe114
			Tread-riser complex	Pe115
	Glacis	Sedimentary rocks Korat group	Tread-riser complex	Pe211
	Vale	Sedimentary rocks Korat group	Slope complex	Pe311
	Lateral vale	Sedimentary rocks Korat group	Side complex	Pe411
			Bottom-side complex	Pe412
			Bottom complex	Pe413
	Depression	Sedimentary rocks Korat group	Basin	Pe511
Valley (Va)	Floodplain	Alluvial deposits	Levee-overflow complex	Va111
	Old terraces	Alluvial deposits	Overflow-basin complex	Va211
	New terraces	Alluvial deposits	Overflow-basin complex	Va311

of gleyic color, with no abrupt textural change. Wet Psamments, classified as Gleysols according to the FAO World Reference Base for Soil Resources (FAO, 1998), occur in a few sloping spots on residual material derived from sandstone.

It is of general understanding that soil salinity develops in the lower parts of the landscape, where the water table is close to the terrain surface, and that salt moves upward in the soil profile.

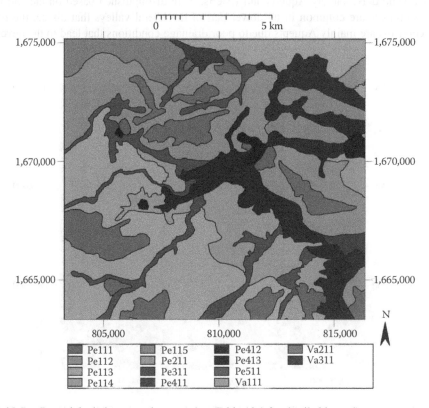

FIGURE 13.5 Geopedologic interpretation map (see Table 13.1 for detailed legend).

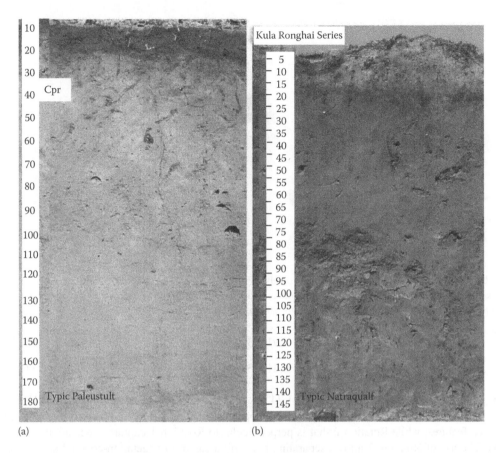

FIGURE 13.6 Soils occurring in two landforms within the peneplain (a) Typic Paleustult on undulating terrain; (b) Typic Natraqualf in a vale. (From Farshad, A., Udomsri, S., Yadav, R.D., Shrestha, D.P., and Sukchan, S., Understanding geopedologic setting is a clue for improving the management of salt-affected soils in Non Suang district, Nakhon Ratchasima, Thailand. LDD-ITC Research Project, Conference Paper, Hua Hin, Thailand, 2005. With permission.)

Thus, the geomorphic units occurring in the lower parts of the landscape, such as floodplains, terraces, and vales, have high potential for salinity development. They were masked out using spatial analysis in GIS and displayed on a map that shows only areas prone to soil salinity. Typical soils of these low-lying areas are shown in Figure 13.6.

13.3.3 BAND ROTATION TO MAXIMIZE SOIL SURFACE FEATURES

Spectral features can be enhanced by rotating red and near-infrared bands, which maximizes information from soil surface features (Shrestha, 2005). On satellite image, saline areas show higher reflectance values than those of nonsaline areas due to the concentration of salt on the terrain surface and the formation of salt crusts. Such saline surface features can be maximized by applying a band rotation technique (Shrestha et al., 2005b). In order to estimate a suitable angle for rotation, training samples were collected from bare soil surfaces. A linear least square model was fitted to obtain the best fitting line ($r = 0.95$) (Figure 13.7) as shown below:

$$y = 0.45x + 19.55 \tag{13.8}$$

The slope of the line (0.45) gives the rotation angle, which is equal to approximately 24° (arc tangent of 0.45 is 24.23°). Equations for anticlockwise rotation of bands are then given by

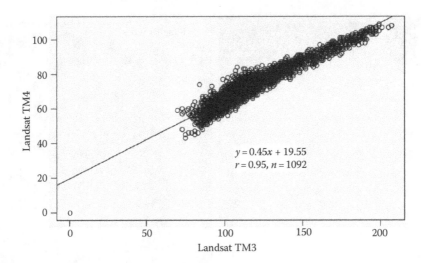

FIGURE 13.7 Best-fitting line to bare soil pixels. (From Shrestha, D.P., Soliman, A.S., Farshad, A., and Yadav, R.D., Salinity mapping using geopedologic and soil line approach. In *Proceedings of the 26th Asian Conference of Remote Sensing*, Hanoi, Vietnam, 2005b. With permission.)

$$\text{Rotation } 1 = 0.91 tm3 + 0.41 tm4 - 19.55 \tag{13.9}$$

$$\text{Rotation } 2 = 0.41 tm3 + 0.91 tm4 - 19.55 \tag{13.10}$$

The coefficient 0.91 is the cosine value of 24°; 0.41 is the corresponding sine value; and 19.55 is the constant to shift the line for starting at the origin. Rotation 1 maximizes information of soil surface features, while Rotation 2 that is perpendicular to Rotation 1 captures vegetation information. The use of Rotation 2 allows separating bare soil areas from vegetation-covered areas where salinity is not the main issue. Also sloping areas were excluded, using a slope map computed from DEM. Level slicing was then applied to the masked areas (i.e., floodplain, terraces, vales, and depressions) to generate salinity level classes.

13.3.4 Use of Geophysical Technique for Subsurface Salinity Mapping

To collect data for subsurface salinity mapping, EM induction measurements were carried out in an area of 100×300 m ($30,000$ m^2), using an EM38 instrument. A GPS receiver (Garmin 12X) was used to determine the coordinates of the four corners of the plot. EM readings were taken every 2 m along 20 transects 300 m long and 5 m apart (Figure 13.8). At each location, two EM measurements, one in vertical mode and one in horizontal mode, were made to correlate with actual soil salinity values at different depths. In addition, soil samples were collected for electrical conductivity determination (EC_e) at three depths (0–30, 30–60, and 60–90 cm).

The EC_e data were correlated with the EM data to assess the reliability of predicting soil salinity from EM data. A linear regression of the EC_e data against the EM data at both horizontal and vertical modes was established. Finally, ordinary kriging was applied to map the spatial distribution of salinity at the three above-mentioned depths.

13.3.5 Results and Discussion

The geopedologic survey in step 2, combining digital terrain and image analysis, helped segregate areas prone to salinity development. Further processing of the image was carried out by means of band rotation, which enhances soil surface features. This was used to map various levels of soil salinity. The results show that nearly one-third of the area (30%) is salt affected (Figure 13.9).

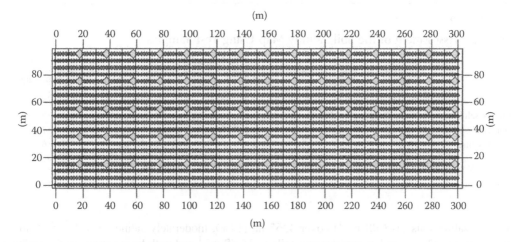

FIGURE 13.8 Sample plot design for geophysical survey (small dots represent EM readings; larger dots represent soil sample locations).

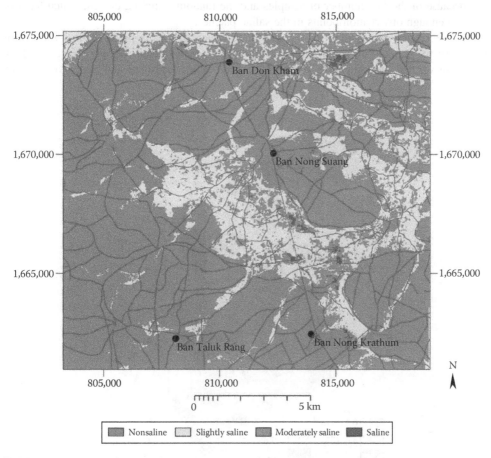

FIGURE 13.9 Classification of saline soils. (From Shrestha, D.P., Soliman, A.S., Farshad, A., and Yadav, R.D., Salinity mapping using geopedologic and soil line approach. In *Proceedings of the 26th Asian Conference of Remote Sensing*, Hanoi, Vietnam, 2005b. With permission.)

TABLE 13.2

Accuracy Assessment of the Soil Salinity Map Using Test Samples

	Test Samples				
	Nonsaline	Slightly Saline	Moderately Saline	Highly Saline	Number of Samples
Nonsaline	26 (68%)	8	4	0	38
Slightly saline	1	4 (80%)	0	0	5
Moderately saline	0	2	5 (71%)	0	7
Highly saline	0	0	0	6 (100%)	6

Notes: Average accuracy, 80%; overall accuracy, 72%; total number of test samples, 56.

Slightly saline soils (4–8 dS m^{-1}) cover 4365 ha (19%), moderately saline soils (8–15 dS m^{-1}) cover 2460 ha (11%), and strongly saline soils (>15 dS m^{-1} and with bright spots on the surface) cover 65 ha (<1%). Areas presently not affected by salinity but located close to saline areas are potentially exposed to salinization, especially if they are occupying low-lying positions in the landscape. Accuracy was assessed by using the results of 56 randomly sampled sites (Table 13.2). The mean accuracy is 80%. The estimated overall accuracy is somewhat lower (72%) because of the low number of samples and the random sampling design, which has led to not having enough observation points in the saline patches.

Ordinary kriging was applied to the EM data, which served as predictor for soil electrical conductivity, to generate soil salinity maps for the three selected depth classes (Figure 13.10). The results were verified by comparing with conductivity values measured at 75 locations. There is

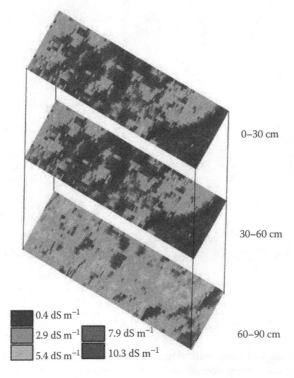

FIGURE 13.10   Soil salinity prediction at three depths.

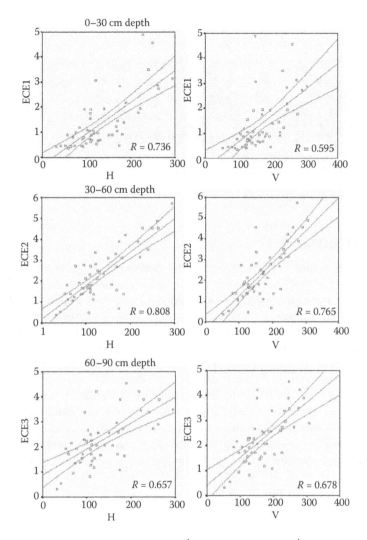

FIGURE 13.11   Correlation between EC_e data (dS m^{-1}) and EM data (mS m^{-1}) in horizontal (H) and vertical (V) modes (middle line represents linear regression; outer lines represent 95% confidence limits).

high positive correlation between EM and EC_e data in the subsoil at 30–60 cm depth ($r = 0.81$ for the horizontal mode and $r = 0.76$ for the vertical mode) (Figure 13.11). In the case of the surface layer (0–30 cm), the EM data seem to predict better in the horizontal mode ($r = 0.74$) than in the vertical mode ($r = 0.59$). Finally, the root mean square error was computed to check the reliability of the prediction. A set of 24 sandy soil points and a set of 21 clayey soil points were used for reliability assessment. The prediction resulted to be better for the sandy soils (RMSE less than 1.0) than for the clayey soils (RMSE slightly higher than 1.0). Slight to moderate salinity problems show up as scattered patches on the soil surface and in the subsurface layer (0–30 and 30–60 cm), but larger areas with relatively higher electrical conductivity values were identified in the deeper layer (60–90 cm). In the course of time, salinization could move upward by capillary action, and this needs monitoring. Finally, a hazard map was prepared by combining the soil salinity maps of the three depth classes (Figure 13.12), using an image classification technique. Three hazard classes (low, moderate, and high) show potential areas for salinity development.

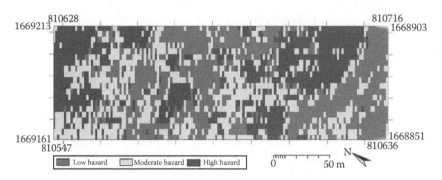

FIGURE 13.12   Soil salinity hazard map.

13.4 CONCLUSION

The case study demonstrates a practical method and technique to map salinity hazard. It is argued that mapping hazard should be preferred, simply because areas with high salinity hazard will become saline if no appropriate management is applied. The stepwise approach allows identifying salt-prone areas in the landscape where detailed studies may be carried out. It also shows that geopedologic mapping helps understand soil–landscape relations, which play a vital role in salt movement and salinity development. Timely detection and monitoring of the process are essential to track the salt movement so that necessary mitigation measures can be adopted. Although advances in EM techniques speed up subsoil salinity detection, the results should not be accepted blindly. For instance, in this study it is shown that the EM readings are influenced by soil texture.

For salinity studies, proper fieldwork timing is crucial. In this study, fieldwork was carried out in August–September (rainy season), constrained by academic calendar. Usually, a dry season may be preferred to carry out salinity studies in the field. This may cause problems when soil survey interpretation has to be done, for instance, in the case where the soil salinity map forms the basis for landuse suitability evaluation. As salt occurrence in the soil is season-specific, a rainfall event may dissolve the surface salt and cause the salinity level in the topsoil to decrease. Such a situation can be anticipated through salinity hazard mapping.

ACKNOWLEDGMENTS

Materials used in the case study were derived from a joint research project of ITC, the Netherlands, with the Land Development Department, Ministry of Agriculture and Cooperatives, Bangkok, Thailand. Work done by ITC students, especially Aiman Soliman from Egypt and Ram Dular Yadav from Nepal, is duly acknowledged.

REFERENCES

Bell, D., Menges, C., Ahmad, W., and van Zyl, J.J. 2001. The application of dielectric retrieval algorithms for mapping soil salinity in a tropical coastal environment using airborne polarimetric SAR. *Remote Sensing of Environment* 75(3): 375–384.

Ben-Dor, E., Patkin, K., Banin, A., and Karnieli, A. 2002. Mapping of several soil properties using DAIS-7915 hyperspectral scanner data: A case study over clayey soils in Israel. *International Journal of Remote Sensing* 23(6): 1043–1062.

Blaney, H.F. and Criddle, W.D. 1962. Determining consumptive use and irrigation water requirements. *USDA Technical Bulletin*, Beltsville, MD.

Douaoui, A.E.K., Nicolas, H., and Walter, C. 2006. Detecting salinity hazards within a semiarid context by means of combining soil and remote-sensing data. *Geoderma* 134(1–2): 217–230.

Dregne, H.E. and Chou, N.T. 1994. Global desertification dimensions and costs. In *Degradation and Restoration of Arid Lands*, H.E. Dregne (Ed.), Texas Technical University, Lubbock, TX.

Eswaran, H., Lal, R., and Reich, P.F. 2001. Land degradation: An overview. In *Response to Land Degradation*, E.M. Bridges, I.D. Hannam, L.R. Oldeman, F.W.T.P. de Vries, S.J. Scherr, and S. Sombatpanit (Eds.), Oxford and IBH Publishing, New Delhi/Calcutta, India.

FAO. 1998. *World Reference Base for Soil Resources*. World Resources Report 84, Food and Agriculture Organization of the United Nations, Rome, Italy.

Farifteh, J., Farshad, A., and George, R.J. 2006. Assessing salt-affected soils using remote sensing, solute modelling, and geophysics. *Geoderma* 130(3–4): 191–206.

Farshad, A., Udomsri, S., Yadav, R.D., Shrestha, D.P., and Sukchan, S. 2005. Understanding geopedologic setting is a clue for improving the management of salt-affected soils in Non Suang district, Nakhon Ratchasima, Thailand. LDD-ITC Research Project, Conference Paper, Hua Hin, Thailand.

Fernández-Buces, N., Siebe, C., Cram, S., and Palacio, J.L. 2006. Mapping soil salinity using a combined spectral response index for bare soil and vegetation: A case study in the former lake Texcoco, Mexico. *Journal of Arid Environments* 65(4): 644–667.

Ghassemi, F., Jakeman, A.J., and Nix, H.A. 1995. *Salinisation of Land and Water Resources: Human Causes, Extent, Management and Case Studies*. The Australian National University, Canberra, Australia, and CAB International, Wallingford, Oxon, UK.

Gordon, I., Pearce, B., Heiner, I., and Biggs, A. 2006. Land series: Salinity hazard mapping in Qld. Department of Natural Resources and Water, The State of Queensland, Australia. Available at http://www.nrw.qld.gov.au/factsheets/pdf/land/l57.pdf (accessed December 4, 2007).

Grunberger, O., Maeght, J.L., Montoroi, J.P., Rattana-Anupap, S., Wiengwongnam, J., and Hammecker, C. 2005. Assessment of salinity hazard by time domain reflectometry in flooded sandy paddy soils. In *Proceedings of Conference on Management of Tropical Sandy Soils for Sustainable Agriculture*, Khon Kaen, Thailand.

Herrero, J. and Pérez-Coveta, O. 2005. Soil salinity changes over 24 years in a Mediterranean irrigated district. *Geoderma* 125(3–4): 287–308.

Howari, F.M. 2003. The use of remote sensing data to extract information from agricultural land with emphasis on soil salinity. *Australian Journal of Soil Research* 41(7): 1243–1253.

Imaizumi, K., Sukchan, S., Wichaidit, P., Srisuk, K., and Kaneko, F. 2002. Hydrological and geochemical behavior of saline groundwater in Phra Yun, northeast Thailand. Land Development Department, Khon Kaen, Thailand.

Issaks, E.H. and Srivastava, R.M. 1989. *An Introduction to Applied Geostatistics*. Oxford University Press, New York.

Masoud, A.A. and Koike, K. 2006. Arid land salinization detected by remotely-sensed landcover changes: A case study in the Siwa region, NW Egypt. *Journal of Arid Environments* 66(1): 151–167.

McNeill, J.D. 1980. Electromagnetic terrain conductivity measurement at low induction numbers. Geonics, ON, Canada.

Metternicht, G.I. and Zinck, J.A. 2003. Remote sensing of soil salinity: Potentials and constraints. *Remote Sensing of Environment* 85(1): 1–20.

Mougenot, B., Pouget, M., and Epema, G.F. 1993. Remote sensing of salt affected soils. *Remote Sensing Reviews* 7(3–4): 241–259.

Peng, W. 1998. Synthetic analysis for extracting information on soil salinity using remote sensing and GIS: A case study of Yanggao basin in China. *Environmental Management* 22(1): 153–159.

Shrestha, D.P. 2005. Image transformation and geo-statistical techniques to assess sedimentation in southern Nepal. *Asian Journal of Geoinformatics* 5(3): 24–31.

Shrestha, D.P., Margate, D.E., van der Meer, F., and Anh, H.V. 2005a. Analysis and classification of hyperspectral data for mapping land degradation: An application in southern Spain. *International Journal of Applied Earth Observation and Geoinformation* 7(2): 85–96.

Shrestha, D.P., Soliman, A.S., Farshad, A., and Yadav, R.D. 2005b. Salinity mapping using geopedologic and soil line approach. In *Proceedings of the 26th Asian Conference of Remote Sensing*, Hanoi, Vietnam.

Smith, K. 2000. *Environmental Hazards, Assessing Risk and Reducing Disaster*, 3rd ed., Taylor & Francis Group, London and New York.

Soliman, A.S. 2004. Detecting salinity in early stages using electromagnetic survey and multivariate geostatistical techniques: A case study of Nong Suang district, Nakhon Ratchasima, Thailand. MSc thesis, International Institute for Geo-Information Science and Earth Observation (ITC), Enschede, the Netherlands.

Sukchan, S. and Yamamoto, Y. 2003. Classification of salt affected areas using remote sensing and GIS. Land Development Department, Khon Kaen, Thailand.

USDA. 1998. *Keys to Soil Taxonomy*, 8th ed. US Department of Agriculture, Washington, DC.

Yadav, R.D. 2005. Modelling salinity effects in relation to soil fertility and crop yield: A case study of Nong Suang district, Nakhon Ratchasima, Thailand. MSc thesis, International Institute for Geo-Information Science and Earth Observation (ITC), Enschede, the Netherlands.

Zinck, J.A. 1988/89. Physiography and soils. ITC lecture notes. International Institute for Geo-Information Science and Earth Observation (ITC), Enschede, the Netherlands.

14 Stochastic Approaches for Space–Time Modeling and Interpolation of Soil Salinity

Ahmed Douaik, Marc Van Meirvenne, and Tibor Tóth

CONTENTS

14.1 INTRODUCTION

The concentration of soluble salts in the soil is one of the most important factors of land degradation that affects land utilization and reclamation. The effective control of soil salinity and waterlogging requires, among others, knowledge of the magnitude, extent and distribution of root zone salinity (inventory), knowledge of the changes and trends in soil salinity over time (monitoring), and the ability to determine the impact of management changes upon saline conditions (Rhoades et al., 1999).

The spatial variability and dynamic nature of soil salinity are controlled by a variety of factors, including soil factors (permeability, water table depth, salinity of the groundwater, topography and parent material), management-related factors (irrigation, drainage, and tillage), and climatic factors (rainfall, wind, and relative humidity).

Soil salinity is determined conventionally by measuring the electrical conductivity of a saturated soil paste extract (EC_e) in laboratory (Soil and Plant Analysis Council, 1992). This procedure is expensive and time-consuming so that only a limited number of samples is usually processed.

However, because of the spatiotemporal variability of soil salinity, numerous samples need to be taken and the measurements need to be repeated as conditions change. There are two options to solve this issue: (1) evaluating soil salinity in the field by measuring the bulk soil or apparent electrical conductivity (EC_a) with electrode probes or electromagnetic induction techniques or (2) determining salinity indices from remote sensing (RS) images. These two procedures are cheaper and easier than the laboratory analysis of soil samples, even if they both require some supporting laboratory determinations. Therefore, they are now widely used for the inventory and monitoring of soil salinity in space and time.

Geospatial technology tools such as global positioning system (GPS), geographic information system (GIS), and RS can help much in mapping and monitoring soil properties, in particular soil salinity. GPS is used to locate precisely the sites where samples have been taken; remotely sensed data can be used as a GIS layer to derive salinity indices like the normalized difference salinity index (Shrestha, 2006); and GIS allows operations on different kinds of data such as overlay and buffering. Soil samples are collected on point locations, but property values must be interpolated to get a layer that can be used within GIS. Geostatistics (Goovaerts, 1997) and Bayesian maximum entropy (BME) (Christakos, 2000) offer a convenient way to interpolate soil properties and get an idea about the uncertainty attached to the interpolated values.

The application of RS to quantitative evaluation of soil properties was reviewed by Ben-Dor (2002), while Barnes et al. (2003) considered the opportunities of combining remote and proximal sensed data. For the specific case of soil salinity and salt-affected soils, an earlier review was done by Mougenot et al. (1993), while a more recent literature review was reported by Metternicht and Zinck (2003).

It is also possible to infer soil property values at unknown space locations and time instants from the values determined at sampled locations and time instants. There are many space–time interpolation methods, some of them being able to integrate data from different sources. A review of digital or predictive soil mapping, with description of sources of ancillary data (including ground and remote sensing) and statistical and geostatistical methods for data analysis, can be found in McBratney et al. (2000, 2003) and Scull et al. (2003).

The usefulness of spatial statistics in general and geostatistics in particular for the analysis of RS data was addressed by Curran and Atkinson (1998) and Stein et al. (1998a). Similarly, Dobermann and Ping (2004) described geostatistical approaches for the integration of RS and crop yield data, which can be easily applied to improving soil salinity mapping. Douaoui et al. (2006) illustrated the integration of electrical conductivity (EC) data and satellite images using geostatistical methods for better detection of salinity hazards. Shrestha (2006) related EC to RS data using a classical statistical method (multiple regression with forward stepwise selection), without taking into account the likely spatial autocorrelation and cross-correlation between EC and RS data. All these research works were limited to the spatial domain only, without repeated measurements over time.

Space–time variability of soil salinity has been appraised using a variety of approaches. Lesch et al. (1998) used a classical statistical method to monitor the temporal change of soil salinity between two time periods. This approach can be applied easily to a few measurement times but is of limited practical use when dealing with series of time periods, as the procedure must be repeated for each pair of times and does not account of any possible temporal correlation between two or more successive measurements.

Douaik et al. (2004) proposed an alternative approach. Using calibration equations based on regression models, EC_a measurements were scaled into $EC_{2.5}$ values that correspond to the EC determined in laboratory from 1:2.5 soil–water suspensions, taken as a surrogate of the EC of the water-saturated soil-paste extract EC_e (Carter and Pearen, 1985). This was followed by spatiotemporal kriging to predict soil salinity at unknown places and times. The approach takes into account the spatial and temporal correlations between soil salinity measurements. However, the $EC_{2.5}$ values resulting from the calibration equations are only estimates of the actual soil salinity, thus suffering from some degree of uncertainty that needs to be considered in the analysis.

The method of BME (Christakos, 2000) enables a rigorous data analysis by distinguishing formally between the accuracy of laboratory EC determinations, on the one hand, and that of field EC measurements or RS-based salinity indices, on the other. The former are hard data provided by direct and accurate measurements of the soil salinity. The latter are soft data derived from uncertain estimates of soil salinity through indirect measurements. Uncertainty is attached to soil salinity estimated from EC_a values, because these values depend not only on the salt content but also on variables such as soil moisture content, temperature, and particle size distribution (see Chapter 11). Similarly, the soil surface reflectance, expressed by the RS digital numbers (DNs), is influenced by many factors such as vegetation cover, nature and amount of salts, organic matter, soil moisture, and soil color.

The objective of this chapter is to compare the predictive performance of two space–time interpolation methods, kriging and BME, in building continuous surface maps of soil salinity from a limited number of $EC_{2.5}$ values and more intensively sampled ancillary measurements. Each of the two methods has two variants. Kriging can be implemented (1) with observed $EC_{2.5}$ data alone or (2) with observed and estimated $EC_{2.5}$ data together. In turn, BME can be applied with two types of soft, or uncertain, data (1) in the form of intervals or (2) in the form of probability distribution functions (pdf).

14.2 THEORETICAL CONCEPTS

There are two main approaches for fitting models to data: the deterministic approach and the probabilistic or stochastic approach. In the deterministic approach, the observed data are modeled as being exactly predictable from a small set of parameters. This means that the values of the parameters and the variables in the model are known with certainty, as there are no random fluctuations in the data. The probabilistic or stochastic modeling takes into account that there may be some element of variability or randomness (uncertainty) in at least one of the parameters or variables. Thus, predictions from this kind of model are probability distributions of the possible values.

Deterministic modeling is usually easy to implement and understand. However, its conceptual status is somewhat unclear, given the inevitable discrepancies between model and data. Also, overfiting can cause error rates to be understated. In addition, deterministic modeling does not provide information regarding the accuracy of the estimates and does not take into consideration the structure of the natural process being studied. In contrast, probabilistic modeling accounts for modeling errors, and this allows quantifying the uncertainty and estimating more accurately the error probabilities. It is also suitable in risk analysis and assessment, as well as for the appraisal of investment options. Probabilistic or stochastic modeling is particularly appropriate for monitoring soil salinity, as our knowledge about the processes and factors that act at different spatial and temporal scales is imperfect.

Soil salinity modeling has followed different approaches, focusing mainly on the spatial domain. These approaches include artificial neural network (Patel et al., 2002), classification and regression tree (Tóth et al., 2002), fuzzy logic (Metternicht, 2001), generalized Bayesian analysis (Peng, 1998), geostatistics (Triantafilis et al., 2001), and wavelet transform (Lark et al., 2003). In this chapter, focus is on the geostatistical approach.

14.2.1 SPACE–TIME RANDOM FIELD

The distribution of the EC or DN values in space and time is represented adequately by a space–time random field (STRF), $X(\boldsymbol{p})$, which takes values at points $\boldsymbol{p} = (s, t)$ in a space–time domain, where $s = (s_1, s_2)$ represents a 2D spatial location and t is the time (Christakos, 1992). The random field $X(\boldsymbol{p})$ at $n + 1$ points $\boldsymbol{p}_i$ ($i = 1, \ldots, n, k$; n observation points and one estimation point) is denoted by the random variables $\boldsymbol{x}_{\text{map}} = [x_1, \ldots, x_n, x_k]'$ and their realization is $\boldsymbol{\chi}_{\text{map}} = [\chi_1, \ldots, \chi_n, \chi_k]'$; the prime represents the transpose of the vector.

The STRF assigns a probability that a realization χ in n dimensions will occur following the multivariate pdf of $X(p)$:

$$\text{Prob}(\chi) = P_x[\chi_1 \leq x_1 \leq \chi_1 + d\chi_1, \ldots, \chi_n \leq x_n \leq \chi_n + d\chi_n] = f_x(\chi)d\chi \qquad (14.1)$$

The above definition represents a complete characterization of the STRF. A partial, but sufficient, characterization of the STRF is provided by its space–time statistical moments.

The first-order moment, which is the mean function, expresses trends or systematic structures in space–time:

$$m_x(p) = E[X(p)] = \int \chi f_x(\chi) d\chi \qquad (14.2)$$

The moment of second-order, the covariance function, expresses correlations and dependencies between two different points p and p':

$$c_x(p, p') = E\left\{[X(p) - m_x(p)]\,[X(p') - m_x(p')]\right\}$$
$$= \iint [\chi - m_x(p)]\,[\chi' - m_x(p')]f_x(\chi,\chi')d\chi d\chi' \qquad (14.3)$$

with $f_x(\chi, \chi')$ being the bivariate pdf of the random function taken between points p and p'.

It is a function that relates the variance to the space–time separation and provides a concise description of the scale and pattern of the space–time variability.

In addition to these two moments, a STRF can be also characterized by its variogram:

$$\gamma_x(p, p') = \frac{1}{2}E\left\{[X(p) - X(p')]^2\right\}$$
$$= \frac{1}{2}\iint (\chi - \chi')^2 f_x(\chi,\chi'; p,p')d\chi d\chi' \qquad (14.4)$$

For spatially homogeneous and temporally stationary random fields, the mean function is constant:

$$m_x(p) = m_x \qquad (14.5)$$

and the covariance function depends only on the spatial lag $h = s - s'$ and the temporal lag $\tau = t - t'$ between any two points $p = (s, t)$ and $p' = (s', t') = (s + h, t + \tau)$:

$$c_x(p, p') = c_x(s - s', t - t') = c_x(h, \tau) \qquad (14.6)$$

The space–time covariance function can be written

$$c_x(h, \tau) = E\{[X(s, t) - m_x]\,[X(s + h, t + \tau) - m_x]\} = E[X(s, t)X(s + h, t + \tau)] - m_x^2 \qquad (14.7)$$

If the mean function is known (in fact, it is estimated from data), m, the moments estimator of the space–time covariance function is

$$\hat{c}_x(h, \tau) = \frac{1}{N(h, \tau)} \sum_{i,j=1}^{N(h,\tau)} \left\{[x(s_i, t_j) - m]\,[x(s_i + h, t_j + \tau) - m]\right\} \qquad (14.8)$$

with $N(h, \tau)$ being the number of pairs of data separated by the spatial lag h and the temporal lag τ.

If the increments $X(s+h, t+\tau) - X(s, t)$ are second-order stationary, the space–time variogram is defined as

$$\gamma_x(h, \tau) = \tfrac{1}{2}\text{var}[X(s+h, t+\tau) - X(s, t)] = \tfrac{1}{2}E\{[X(s+h, t+\tau) - X(s, t)]^2\} \qquad (14.9)$$

Its moment-based estimator is given by Stein et al. (1998b):

$$\hat{\gamma}_x(h, \tau) = \frac{1}{2N(h, \tau)} \sum_{i,j=1}^{N(h,\tau)} [x(s_i, t_j) - x(s_i + h, t_j + \tau)]^2 \qquad (14.10)$$

with $x(s_i, t_j)$ and $x(s_i + h, t_j + \tau)$ being pairs of observations with a spatial distance equal to h and a temporal distance equal to τ; the number of such pairs is $N(h, \tau)$.

For a second-order stationary STRF, the space–time variogram and space–time covariogram are related to each other:

$$\gamma_x(h, \tau) = c_X(0, 0) - c_X(h, \tau) \qquad (14.11)$$

with $c_x(0, 0) = \sigma_X^2$ being the variance of the STRF.

To introduce space–time anisotropy, the spatial and temporal distances are combined to give a space–time distance, d, as follows:

$$d = \sqrt{h^2 + \varphi\tau^2} \qquad (14.12)$$

with φ being the space–time anisotropy ratio.

14.2.2 KRIGING

When a theoretical model is fitted to the experimental space–time covariance function or variogram, it becomes possible to tackle the problem of predicting attributes at unsampled space locations and/or time instants. Consider the problem of predicting the value of a continuous attribute x at any unsampled space location s_0 and time instant t_0, $x(s_0, t_0)$, using only the x-data available over the space region S and time domain T, say, $n = \sum_{j=1}^{n_t} n_j$ data $\{x(s_{ij}, t_j), j = 1, \ldots, n_t; i = 1, \ldots, n_j\}$.

Kriging is a family of generalized least squares regression algorithms that allow to predict $x(s_0, t_0)$ (Goovaerts, 1997). In geostatistics, the latter is considered as a realization of the STRF $X(s_0, t_0)$. The problem of predicting $x(s_0, t_0)$ is defined as (Rouhani and Myers, 1990):

$$\hat{X}(s_0, t_0) = \sum_j \sum_i \lambda_{ij} X(s_{ij}, t_j) \qquad (14.13)$$

with λ_{ij} being the kriging weights for the time instant t_j and the space location s_{ij}. These weights are determined from the variogram or the covariance function such that the predicted value is unbiased and optimal (its variance is minimal). These two properties confer to kriging the denomination of best linear unbiased predictor.

Two kriging algorithms can be distinguished: the two-step space–time kriging and the anisotropic space–time kriging. The latter was used in our case study. For this algorithm, the prediction is done in one unique step, capitalizing on a space–time covariance function model which incorporates the spatial and temporal dependencies with a space–time anisotropy ratio (Rouhani and Myers, 1990; Stein et al., 1994, 1998b). The advantage of the anisotropic over the two-step space–time kriging is that it allows predictions to be made at every point in space and time, while also avoiding the uncertainties related to the predictors determined at the first step for the two-step space–time kriging.

14.2.3 BAYESIAN MAXIMUM ENTROPY

In most of the classical geostatistical methods of interpolation (i.e., kriging), prediction is based solely on hard data. However, for space locations and time instants where hard data are missing, other data sources can be useful to improve the accuracy of the predictions. Such data include different types of soft data (interval, probabilistic, etc.), physical laws, or moments of higher order.

BME is an approach recently developed for the spatiotemporal mapping of natural processes using uncertain information. It offers the flexibility to incorporate various sources of physical knowledge that enable global prediction features and the adoption of probability distributions without need for any assumptions like, for example, the Gaussianity.

14.2.3.1 Knowledge Bases

In the BME framework, the total knowledge (K) available about a natural process is considered to be formed from two main bases: the general knowledge (G) and the specificatory knowledge (S) such that $K = G \cup S$. The general knowledge represents the knowledge that one has about the distribution of a natural variable to be mapped, before any specific data are used. It encompasses physical laws, statistical moments of any order (including the mean and variogram or covariance function), multipoint statistics, etc. It is said to be general because it does not depend on the specific random field realization at hand.

The pdf defined by Equation 14.1 forms the prior pdf, which can be derived by an estimation process that considers physical constraints provided by prior information or knowledge (G). These physical constraints are given by

$$E[g_\alpha] = \int g_\alpha(\chi_{map}) f_G(\chi_{map}) d\chi_{map}, \quad \alpha = 0, \ldots, N_c \tag{14.14}$$

where $f_G(\chi_{map})$ has the same definition as Equation 14.1, except that x is replaced by G. It is, in fact, the unknown multivariate pdf associated with the general knowledge. The functions g_α are chosen such that the general knowledge base, G, is taken account of in full in the prediction process, and their expectations, $E[g_\alpha]$, provide the space–time statistical moments of interest (means, variances, covariances, etc.).

The specific knowledge includes the data specific to a given experiment or situation. It refers to a particular occurrence of the natural variable at a particular spatial location and time instant. It is divided into two main categories, the hard and soft data, depending on the accuracy of the data. The hard data

$$\chi_{hard} = [\chi_1, \ldots, \chi_h]' \tag{14.15}$$

are exact measurements of the natural process and considered to be error-free. They encompass accurate measurements obtained from real-time observation devices, computational algorithms, and simulation processes.

The soft data

$$\chi_{soft} = [\chi_{h+1}, \ldots, \chi_n]' \tag{14.16}$$

include uncertain observations, empirical charts, and assessments by experts. These are incomplete or qualitative data, linked to opinions, intuition, or expert knowledge. Different kinds of soft data are available, often in the form of interval domain data or of probabilistic nature. The interval soft data are given by

$$\chi_{\text{soft}} = \{[\chi_{h+1}, \dots, \chi_n]' : \chi_i \in I_i = [l_i, u_i], \quad i = h+1, \dots, n\} \tag{14.17}$$

with l_i and u_i representing the lower and upper limits of the interval I_i, respectively.

The probabilistic soft data are given by

$$\chi_{\text{soft}} : P_s(\chi_{\text{soft}} \leq \xi) = \int_{-\infty}^{\xi} f_s(\chi_{\text{soft}}) d\chi_{\text{soft}} \tag{14.18}$$

with $f_s(\chi_{\text{soft}})$ being the specificatory pdf.

The total data available for mapping is then the union of both hard and soft data:

$$\chi_{\text{data}} = [\chi_{\text{hard}}, \chi_{\text{soft}}]' \tag{14.19}$$

14.2.3.2 Three Steps of the BME Analysis

The BME analysis is done in three main steps of knowledge acquisition, integration and processing: the prior step, the meta-prior step, and the integration or posterior step. In the structural or prior step, the goal is to maximize the information content considering only the general knowledge before any use of the data, which corresponds to the first goal of the BME. A random field is completely defined by its multivariate pdf, which forms the prior pdf. The latter should be derived by means of an estimation process that takes into consideration physical constraints under the form of prior information or knowledge. In the BME context, this information is measured using the Shannon's entropy function (Shannon, 1948). Thus, the E in BME expresses the given information in the random vector x_{map} as

$$\text{Info}(x_{\text{map}}) = -\ln[f_G(\chi_{\text{map}})] \tag{14.20}$$

This equation represents the uncertainty, in the form of the pdf $f_G(\chi_{\text{map}})$, regarding the random vector x_{map}: the higher the probability, the lower the uncertainty about x_{map} and the lesser the amount of information provided by the pdf about x_{map}.

The expected information is given by the following entropy function:

$$E[\text{Info}(x_{\text{map}})] = -\int \ln[f_G(\chi_{\text{map}})] f_G(\chi_{\text{map}}) d\chi_{\text{map}} \tag{14.21}$$

This function needs to be maximized subject to the physical constraints provided by prior information or knowledge G (Equation 14.14), which justifies the M in the BME acronym. This maximization requires the use of Lagrange multipliers. At this stage, the prior or G-based multivariate pdf is

$$f_G(\chi_{\text{map}}) = Z^{-1} \exp\left[\sum_{\alpha=1}^{N_c} \mu_\alpha g_\alpha(\chi_{\text{map}})\right] \tag{14.22}$$

$$\text{where } Z = \int \exp\left[\sum_{\alpha=1}^{N_c} \mu_\alpha g_\alpha(\pmb{\chi}_{\text{map}})\right] d\pmb{\chi}_{\text{map}} = \exp(-\mu_0) \tag{14.23}$$

is a normalization constant and μ_0 is the first Lagrange multiplier.

During the meta-prior step, the specificatory knowledge is collected and organized in appropriate quantitative forms that can be easily incorporated in the BME framework. The available data are divided into hard and soft data and the latter can be either interval or probabilistic.

In the integration or posterior step, the two knowledge bases (G and S) are integrated. The goal is to maximize the posterior pdf given the total knowledge K, which is the second goal of BME. The G-based pdf is updated with site-specific data using a Bayesian conditionalization, corresponding thus to the B in BME:

$$f_K(\chi_k|\pmb{\chi}_{\text{data}}) = f_G(\pmb{\chi}_{\text{map}})/f(\pmb{\chi}_{\text{data}}) \tag{14.24}$$

where $f_K(\chi_k|\pmb{\chi}_{\text{data}})$ and $f_G(\pmb{\chi}_{\text{map}})$ are the posterior and the prior pdfs, respectively. The posterior pdf should be maximized with respect to χ_k. This stage yields the K-based pdf, $f_K(\chi_k)$.

In the case of interval soft data, the posterior pdf has the following form:

$$f_K(\chi_k) = \frac{\int_l^u f_G(\pmb{\chi}_{\text{map}}) d\pmb{\chi}_{\text{soft}}}{\int_l^u f_G(\pmb{\chi}_{\text{data}}) d\pmb{\chi}_{\text{soft}}} \tag{14.25}$$

$$\text{where } f_G(\pmb{\chi}_{\text{data}}) = \int f_G(\pmb{\chi}_{\text{map}}) d\chi_k \tag{14.26}$$

For probabilistic soft data, the posterior pdf is defined by

$$f_K(\chi_k) = \frac{\int f_S(\pmb{\chi}_{\text{soft}}) f_G(\pmb{\chi}_{\text{map}}) d\pmb{\chi}_{\text{soft}}}{\int f_S(\pmb{\chi}_{\text{soft}}) f_G(\pmb{\chi}_{\text{data}}) d\pmb{\chi}_{\text{soft}}} \tag{14.27}$$

14.2.3.3 Some BME Estimators

The posterior pdf, which is not limited to the Gaussian type, fully describes the random field at the estimation point. It provides a complete picture of the mapping situation as well as different estimators and their associated estimation uncertainty. The prediction points lie, in general, on a regular grid and the predictions are used to create space–time maps. The different estimates presented in the following case study are specific to a general knowledge involving the first two statistical moments and a specificatory knowledge encompassing hard data as well as interval and probabilistic soft data.

The conditional mean estimate, which is usually a nonlinear function of the data, is suitable for mapping situations where one is interested in minimizing the mean square estimation error:

$$\bar{\chi}_k = \int f_K(\chi_k) \chi_k d\chi_k \tag{14.28}$$

A measure of the uncertainty associated with the estimated values is provided by the variance of the estimation error:

$$\hat{\sigma}_k^2 = \int (\chi_k - \bar{\chi}_k)^2 f_K(\chi_k) \mathrm{d}\chi_k \tag{14.29}$$

This variance is data-dependent here, whereas it is data-free in kriging. Using this uncertainty estimate and the conditional mean, the confidence intervals can be computed assuming a Gaussian distribution. For example, for an error of type I of 5%, the confidence limits are

$$[\bar{\chi}_k - 1.96\hat{\sigma}_k, \bar{\chi}_k + 1.96\hat{\sigma}_k] \tag{14.30}$$

In addition, the confidence intervals can be computed directly from the posterior pdf (Serre and Christakos, 1999), providing a more realistic assessment of the estimation error than the error variance or the confidence limits derived from Equation 14.30.

14.2.3.4 Kriging as a Special Case of BME

When the general knowledge is limited to the mean and covariance functions and the specificatory knowledge is restricted to the hard data only, the BME posterior pdf is Gaussian (the mean and the mode are equal) and the BME mean estimate is equivalent to the kriging estimate (best minimum mean-squared error, MMSE) (Lee and Ellis, 1997), which is the conditional mean:

$$\hat{\chi}_{k,\mathrm{MMSE}} = E[X(\boldsymbol{p}_k)|\boldsymbol{\chi}_{\mathrm{hard}}] \tag{14.31}$$

When the STRF is Gaussian, Equation 14.31 becomes linear and optimal among all MMSE estimators, and it is expressed as

$$\hat{\chi}_{k,\mathrm{MMSE}} = \boldsymbol{\lambda}'\boldsymbol{\chi}_{\mathrm{hard}} \tag{14.32}$$

where $\boldsymbol{\lambda}$ is a vector of weights associated with the data points and involving the space–time mean and covariance functions. Thus, BME is a more general interpolation approach and kriging is a special case in limiting situations.

14.3 CASE STUDY: SOIL SALINITY AND APPARENT ELECTRICAL CONDUCTIVITY

14.3.1 DATA DESCRIPTION

The study site was selected from the most characteristic native sodic grassland of Hungary in the Hortobagy National Park, east of the country. This park forms a subregion of the Great Hungarian Plain where more than 95% of salt-affected soils of the country are located. The study area (Figure 14.1) covers about 25 ha with central coordinates 47°30′ N and 21°30′ E.

Two data sets were collected. The first data set, termed "data set to be calibrated," includes the measurement of the EC_a in the field at 413 locations in a pseudo-regular grid of 25×25 m^2. The EC_a was measured using an EC probe equipped with four electrodes (Rhoades and Miyamoto, 1990). The spacing between the inner sensing electrodes is 90 cm and there is 10 cm distance on both sides between the sensing/receiver (outer pair) and sensing (inner pair) electrodes. The electrodes are inserted in the soil at 8 and 13 cm depth. The corresponding EC_a measures (dS m^{-1}) are characteristic for the 0–20 and 0–40 cm soil depths, respectively. At each calibration point, there were always three parallel measures of EC_a, but only one at the other points.

A spatial site selection algorithm based on response surface design (Lesch et al., 1995) was followed to determine a minimal number of calibration sites. The selected sites were based on the spatial configuration of locations for which EC_a was measured and on the values of the measurements. These were spatially representative of the study area and allowed accurate estimation of the calibration parameters. These selected sites constitute the second data set called calibration

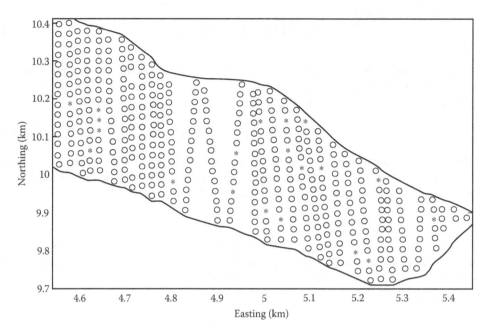

FIGURE 14.1 Extent of the study area (circles represent locations where EC_a was measured; asterisks represent locations where both EC_a and $EC_{2.5}$ were measured). (From Douaik, A., Van Meirvenne, M., and Tóth, T., *GeoEnv IV: Geostatistics for Environmental Applications*, Kluwer Academic, Dordrecht, the Netherlands, 413–424. With permission.)

data set. Soil samples were collected from all sites, air-dried, and crushed to pass through a 2 mm sieve. The 1:2.5 soil–water suspensions were prepared and shacked. After 16 h, pH and EC were measured. The $EC_{2.5}$ values were adjusted to a standard 25°C temperature. In addition to the EC_a field values, measurements of EC from the 1:2.5 suspension ($EC_{2.5}$ in dS m^{-1}), gravimetric moisture (%) and pH were available. Gravimetric moisture was determined by drying soil samples in airtight containers at 105°C until the mass stabilized. The $EC_{2.5}$ is a simple proxy of the water-saturated soil-paste extract (EC_e), which is the conventional measure of soil salinity.

At the calibration points, soil samples were collected to 40 cm depth, by 10 cm increments. Bulked samples were taken from two augerings located between the electrode pairs of the conductivity probe at a distance of 50 cm. The sampling of the data set to be calibrated and the calibration data set was repeated at 21 time instances: November 1994; March, June, September, and December 1995; March, June, September, and October 1996; March, June, September and December 1997; September 1998; April, July, and September 1999; April and December 2000; and March and June 2001. For some dates (e.g., September and October 1996), EC_a data are missing because of instrumental malfunction and vandalism. Thus, the usable multitemporal sampling consisted of 19 time observations. The average temporal lag was 3 months but ranged from 2 to 9 months. Initially, 20 calibration sites were selected and kept for future sampling campaigns, but at a few time instants some of the calibration sites could not be measured due to standing water conditions. Thus, the number of calibration sites varied between 13 and 20 for the different sampling campaigns.

14.3.2 Data Analysis

The histograms of EC_a and $EC_{2.5}$ showed skewed distributions that became less asymmetric after logarithmic transformation. The linearity of their relationship was also improved. All the analyses were based on the transformed data. The mean trend and the covariance function were used as general knowledge for both space–time interpolation methods presented in this study.

The observed $EC_{2.5}$ values were taken as hard data, available at $h = 20$ locations in space. Interval and probabilistic data were used as soft data in the BME framework, whereas the midpoint of interval data was considered in the second approach of kriging. The calibration data set was used to calculate the interval data. The pairs of EC_a and $EC_{2.5}$ data values were used to determine the calibration equations, one for each time instant, by calculating simple ordinary least squares regression models:

$$\ln(EC_{2.5}) = a + b \ln(EC_a) \tag{14.33}$$

where
   a is the intercept
   b is the slope of the regression model

These calibration equations were applied to the "data set to be calibrated" to obtain the expected values and their standard deviations for all 413 locations and 17 time periods. Two periods corresponding to the sampling campaigns 18 and 19 in March and June 2001, respectively, were kept for validation. These expected values and standard deviations were used to determine the 95% confidence intervals whose lower and upper limits form the interval soft data.

The interval midpoint data used in kriging were calculated as the average of the lower and upper limits, which are in fact the expected values. The same expected values and their corresponding standard errors were used to determine the soft pdfs, $f_s(\xi)$, for each of the 413*17 points, assuming a Gaussian distribution. These soft pdfs were used as probabilistic soft data.

Two methods of space–time prediction were compared, each with two variants and differing in the way the soft data are processed:

1. Ordinary kriging using only hard data (HK), which provides no direct way of integrating soft data and ignores them
2. Ordinary kriging using hard data and the midpoint of the interval soft data, which considers the latter as if they were hard data (HSK) and disregards their uncertainty. The difference between (1) and (2) lies in the number of data considered during the analysis. Ordinary kriging (HK) is limited to the hard data (observed $EC_{2.5}$) only, whereas HSK treats essentially both sets of data (observed and predicted $EC_{2.5}$) as hard data
3. BME using the hard and interval soft data (BMEI), which integrates the interval soft data in the prediction and maintains the difference in the degree of uncertainty between hard and soft data
4. BME using the hard and probabilistic soft data (BMEP), which integrates the full distribution of the soft data in the prediction

14.3.3 VALIDATION AND COMPARISON CRITERIA

The methods were compared by cross-validation on observed $EC_{2.5}$ data of the sampling campaigns 18 and 19. Soil salinity was predicted at each of the sites (19 for March 2001 and 20 for June 2001), which were excluded from all previous computations and for which measurements were available, by deleting in turn the value of each location where the prediction was being made. This generated pairs of estimated–observed soil salinity values for the two time periods, using the first three methods (HK, HSK, and BMEI) but not BMEP. Three quantitative criteria were computed from these pairs of values: the Pearson correlation coefficient (r), the mean error (ME) or bias, and the root-mean-squared error (RMSE).

Let x_i and y_i $(i = 1, \ldots, n)$ represent the estimated and the observed soil salinity values, respectively. The ME or bias is defined by

$$\text{ME} = \frac{1}{n} \sum_{i=1}^{n} (x_i - y_i) = \bar{x} - \bar{y}, \qquad (14.34)$$

where
   $\bar{x}$ and $\bar{y}$ represent the means of the estimated and the observed values, respectively
   n is the number of locations for which observations are available

   The RMSE is

$$\text{RMSE} = \sqrt{\frac{1}{n} \sum_{i=1}^{n} (x_i - y_i)^2} \qquad (14.35)$$

The Pearson correlation coefficient, r, measures the strength of the linear relation between the estimated and the observed soil salinity values, and should be close to one for an accurate prediction. The ME should be close to zero and the RMSE should be as small as possible. The distribution of the estimation errors is represented graphically for visual comparison.

TABLE 14.1
Summary Statistics for the Hard Data EC$_{2.5}$ (dS m^{-1})

EC$_{2.5}$	N	Mean	SD	Range	r
Calibration data					
November 1994	13	1.42	0.34	0.96	0.85
March 1995	20	2.29	0.70	2.23	0.91
June 1995	20	2.02	1.10	3.66	0.88
September 1995	20	2.01	0.99	3.73	0.94
December 1995	20	1.84	0.90	2.59	0.92
March 1996	16	2.07	0.81	2.57	0.87
June 1996	20	1.83	0.90	2.86	0.87
March 1997	20	1.61	0.63	2.20	0.89
June 1997	15	1.77	0.97	3.12	0.83
September 1997	20	1.63	1.24	4.25	0.94
December 1997	20	1.60	1.04	3.61	0.90
September 1998	20	3.30	2.17	6.95	0.85
April 1999	20	1.84	1.74	6.41	0.93
July 1999	13	2.27	1.57	4.86	0.91
September 1999	20	2.29	1.89	6.52	0.91
April 2000	18	2.11	1.63	6.32	0.94
December 2000	20	2.32	1.91	7.38	0.93
Validation data					
March 2001	19	1.80	1.21	4.42	0.97
June 2001	20	1.99	1.75	5.83	0.86

Notes: N, number of observations; SD, standard deviation; r, Pearson correlation coefficient.

14.3.4 RESULTS AND DISCUSSION

14.3.4.1 Exploratory Data Analysis

The main statistical parameters of the hard data ($EC_{2.5}$) are given in Table 14.1 for both the calibration and validation data sets. The mean values range between 1.42 dS m^{-1} (November 1994) and 3.30 dS m^{-1} (September 1998), which indicates strong temporal variability in soil salinity. The range values vary between 0.96 dS m^{-1} (November 1994) and 7.38 dS m^{-1} (December 2000), suggesting considerable spatial variation. The Pearson correlation coefficients between EC_a and $EC_{2.5}$ vary from 0.83 to 0.97, indicating a strong relationship between EC_a and $EC_{2.5}$.

14.3.4.2 Structural Analysis

The spatial, temporal, and spatiotemporal dependencies in the salinity data were described and modeled using covariance functions after removal of the mean space–time trend. The spatial covariance function was fitted with a nested exponential model as illustrated in Figure 14.2a. The small-scale variation has a range of 250 m with a sill of 0.27 (dS m^{-1})^2 that represents 79.4% of the total variance. The large-scale range lies beyond the dimensions of the study area (1500 m):

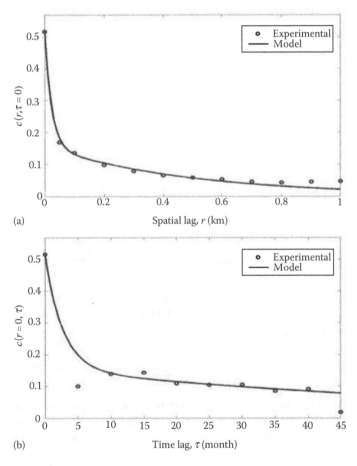

FIGURE 14.2 Covariance functions (a) spatial, (b) temporal (circles represent experimental covariance function; curve represents fitted model). (From Douaik, A., Van Meirvenne, M., Tóth, T., and Serre, M., *Stochast. Environ. Res. Risk Assess.* 18, 219, 2004. With permission.)

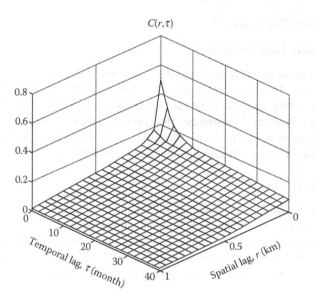

FIGURE 14.3 Space–time covariance function of the residual data $R(\mathbf{s}, t)$. (From Douaik, A., Van Meirvenne, M., Tóth, T., and Serre, M., *Stochast. Environ. Res. Risk Assess.* 18, 219, 2004. With permission.)

$$c(\boldsymbol{h}, 0) = c_{01} \exp\left(-3\boldsymbol{h}/as_1\right) + c_{02} \exp\left(-3\boldsymbol{h}/as_2\right) \qquad (14.36)$$

where
  c_{01} and c_{02} are the sills of the nested models
  as_1 and as_2 are their corresponding ranges

The temporal covariance function (Figure 14.2b) was fitted with the same nested model as mentioned above. The small-scale range is 8 months, while the large-scale range lies far beyond the time period covered (200 months):

$$c(0, \tau) = c_{01} \exp\left(-3\tau/at_1\right) + c_{02} \exp\left(-3\tau/at_2\right) \qquad (14.37)$$

with at_1 and at_2 being the small-scale and large-scale ranges, respectively.

The space–time covariance function (Figure 14.3) was modeled as a nested structure of two space–time separable covariance models:

$$c(\boldsymbol{h}, \tau) = c_{01} \exp\left(-3\boldsymbol{h}/as_1\right) \exp\left(-3\tau/at_1\right) + c_{02} \exp\left(-3\boldsymbol{h}/as_2\right) \exp\left(-3\tau/at_2\right) \qquad (14.38)$$

The non separable space–time covariance function in Figure 14.3 provides a more accurate and realistic representation of the correlation structure of salinity in both space and time than that described by a purely spatial covariance model, or a covariance model where time is taken as an additional spatial coordinate.

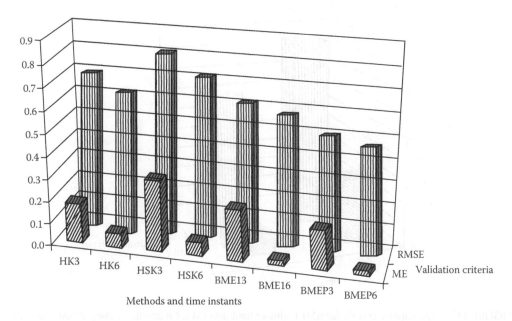

FIGURE 14.4   Quantitative criteria for the comparison of the four approaches (ME, mean error (in absolute value); RMSE, root mean squared error; HK, ordinary kriging using only hard data; HSK, ordinary kriging using hard and soft data; BMEI, Bayesian maximum entropy with interval soft data; and BMEP, Bayesian maximum entropy with probabilistic soft data; 3 and 6 refer to March and June, 2001, respectively).

14.3.4.3   Cross-Validation

Using the four approaches discussed above, soil salinity was predicted for two time periods, March and June 2001. Figure 14.4 gives the cross-validation criteria (ME and RMSE) for both times. The distribution of the errors for March 2001 is given in Figure 14.5.

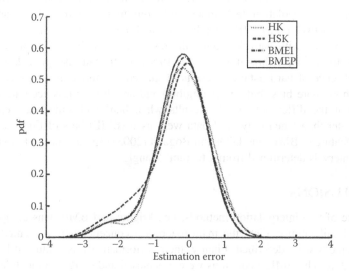

FIGURE 14.5   Distribution of the estimation errors for March 2001 (solid line represents BMEP; dash-dotted line represents BMEI; dotted line represents HK; and dashed line represents HSK).

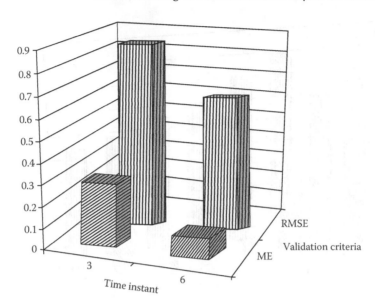

FIGURE 14.6   Quantitative criteria for BMEI without hard data (ME, [in absolute value]; RMSE, root mean squared error; 3 and 6 refer to March and June, 2001, respectively).

The HSK results are the poorest as they have the largest bias (ME) compared to the three other approaches, although it is still not significantly different from zero, and the largest RMSE for both time periods. HSK has the broadest error distribution (Figure 14.5), mostly on the negative side of the curve, which implies that this method is likely to produce larger errors than the others. The errors for BME have a higher mode and a narrower distribution compared with both kriging techniques (Figure 14.5). This fact is confirmed by the smallest RMSE values for BME (Figure 14.4). HK provides more accurate estimates than HSK, but less accurate ones than both BME techniques since its RMSE lies between those of BME and HSK. However, HK gives estimates that are less biased than those of BME (mainly for March 2001). The estimates are strongly correlated with the observations for all interpolation techniques and both time periods. The Pearson correlation coefficients ranged between 0.87 (HK, March 2001) and 0.95 (BMEI and BMEP, June 2001).

When BME is used with hard data only, the results are strictly equivalent to HK. This is in accordance with the theory (Christakos and Li, 1998). BME can also provide estimates of soil salinity in the absence of hard data; the results of this cross-validation are given in Figure 14.6. The estimates show more bias (but still negligible) and are slightly less accurate (particularly for March 2001), but the differences in the results when hard and soft data were used are not significantly greater than when only hard data used (BMEI bars of Figures 14.4 and 14.6). This is a useful feature of BME, and D'Or and Bogaert (2003) used this property to map soil texture using only the intervals determined from a textural triangle.

14.4   CONCLUSIONS

The performance of two interpolation methods, i.e., kriging and BME, was compared for space–time mapping of soil salinity. First, the main theoretical concepts were presented. Then, the two interpolation methods were described. Finally, their performance was evaluated based on a cross-validation procedure. The BME predictions were less biased and more accurate than those obtained by kriging. They gave, also, predictions, which were better correlated with the observed values than in the case of the two kriging techniques. Of the two kriging techniques, the one using soft data resulted in more bias and less accuracy in the predictions than kriging using only hard data. The results showed, also, that BME using probabilistic soft data improved the accuracy of the predictions compared to that using interval soft data. This work shows that BME can incorporate and process soft data rigorously, leading to more accurate predictions.

Ancillary data are usually cheap and readily available, sometimes for the whole study area. This (exhaustive) secondary information can be used in an efficient way to complement the scarcity of direct measurements of a soil property. Cheap, dense, and easily obtained data, such as the EC_a or RS images, can be used in the BME framework to estimate, with less bias and more accuracy, a scarce, time-consuming and expensive soil property of interest such as $EC_{2.5}$ or EC_e. The resulting surface map can be used in a GIS, together with other soil layers, to delineate, for example, different classes of salt-affected soils for subsequent reclamation work.

REFERENCES

Barnes, E.M., Sudduth, K.A., Hummel, J.W., Lesch, S.M., Corwin, D.L., Yang, C., Daughtry, C.S.T., and Bausch, W.C. 2003. Remote- and ground-based sensor techniques to map soil properties. *Photogrammetric Engineering and Remote Sensing* 69: 619–630.

Ben-Dor, E. 2002. Quantitative remote sensing of soil properties. *Advances in Agronomy* 75: 173–243.

Carter, M.R. and Pearen, J.R. 1985. General and spatial variability of Solonetzic soils in north central Alberta. *Canadian Journal of Soil Science* 65: 157–167.

Christakos, G. 1992. *Random Field Models in Earth Sciences*. Academic Press, San Diego, CA.

Christakos, G. 2000. *Modern Spatiotemporal Geostatistics*. Oxford University Press, New York.

Christakos, G. and Li, X. 1998. Bayesian maximum entropy analysis and mapping: A farewell to kriging estimators. *Mathematical Geology* 30: 435–462.

Curran, P.J. and Atkinson, P.M. 1998. Geostatistics and remote sensing. *Progress in Physical Geography* 22: 61–78.

Dobermann, A. and Ping, J.L. 2004. Geostatistical integration of yield monitor data and remote sensing improves yield maps. *Agronomy Journal* 96: 285–297.

D'Or, D. and Bogaert, P. 2003. Continuous-valued map reconstruction with the Bayesian maximum entropy. *Geoderma* 112: 169–178.

Douaik, A., Van Meirvenne, M., and Tóth, T. 2004. Spatio-temporal kriging of soil salinity rescaled from bulk soil electrical conductivity. In *GeoEnv IV: Geostatistics for Environmental Applications*, X. Sanchez-Vila, J. Carrera, and J. Gomez-Hernandez (Eds.), Kluwer Academic, Dordrecht, the Netherlands, pp. 413–424.

Douaik, A., Van Meirvenne, M., Tóth, T., and Serre, M. 2004. Space-time mapping of soil salinity using probabilistic Bayesian maximum entropy. Stochastic Environmental Research and Risk Assessment 18: 219–227.

Douaoui, A., Nicolas, H., and Walter, C. 2006. Detecting salinity hazards within a semiarid context by means of combining soil and remote sensing data. *Geoderma* 134: 217–230.

Goovaerts, P. 1997. *Geostatistics for Natural Resources Evaluation*. Oxford University Press, New York.

Lark, R.M., Kafka, S.R., and Corwin, D.L. 2003. Multiresolution analysis of data on electrical conductivity of soil using wavelets. *Hydrology* 272: 276–290.

Lee, Y.-M. and Ellis, J.H. 1997. On the equivalence of kriging and maximum entropy estimators. *Mathematical Geology* 29: 131–151.

Lesch, S.M., Strauss, D.J., and Rhoades, J.D. 1995. Spatial prediction of soil salinity using electromagnetic induction techniques 2. An efficient spatial sampling algorithm suitable for multiple linear regression model identification and estimation. *Water Resources Research* 31: 387–398.

Lesch, S.M., Herrero, J., and Rhoades, J.D. 1998. Monitoring for temporal changes in soil salinity using electromagnetic induction techniques. *Soil Science Society of America Journal* 62: 232–242.

McBratney, A.B., Odeh, I.O.A., Bishop, T.F.A., Dunbar, M.S., and Shatar, T.M. 2000. An overview of pedometric techniques for use in soil survey. *Geoderma* 97: 293–327.

McBratney, A.B., Mendonça Santos, M.L., and Minasny, B. 2003. On digital soil mapping. *Geoderma* 117: 3–52.

Metternicht, G.I. 2001. Assessing temporal and spatial changes of salinity using fuzzy logic, remote sensing and GIS: Foundations for an expert system. *Ecological Modelling* 144: 163–179.

Metternicht, G.I. and Zinck, J.A. 2003. Remote sensing of soil salinity: Potentials and constraints. *Remote Sensing of Environment* 85: 1–20.

Mougenot, B., Pouget, M., and Epema, G. 1993. Remote sensing of salt-affected soils. *Remote Sensing Reviews* 7: 241–259.

Patel, R.M., Prasher, S.O., Goel, P.K., and Bassi, R. 2002. Soil salinity prediction using artificial neural networks. *Journal of American Water Resources Association* 38: 91–100.

Peng, W. 1998. Synthetic analysis for extracting information on soil salinity using remote sensing and GIS: A case study of Yanggao basin in China. *Environmental Management* 22: 153–159.

Rhoades, J.D. and Miyamoto, S. 1990. Testing soils for salinity and sodicity. In *Soil Testing and Plant Analysis*, R. Westerman (Ed.), ASA Publication Series 3, Madison, WI, pp. 299–336.

Rhoades, J.D., Chanduvi, F., and Lesch, S.M. 1999. Soil salinity assessment: Methods and interpretation of electrical conductivity measurements. *FAO Irrigation and Drainage Paper no 57*, Rome, Italy.

Rouhani, S. and Myers, D.E. 1990. Problems in space–time kriging of geohydrological data. *Mathematical Geology* 22: 611–623.

Scull, P., Franklin, J., Chadwick, O.A., and McArthur, D. 2003. Predictive soil mapping: A review. *Progress in Physical Geography* 27: 171–197.

Serre, M.L. and Christakos, G. 1999. Modern geostatistics: Computational BME in the light of uncertain physical knowledge—The Equus Beds study. *Stochastic Environmental Research and Risk Assessment* 13: 1–26.

Shannon, C.E. 1948. A mathematical theory of communication. *Bell System Technical Journal* 27: 379–423.

Shrestha, R.P. 2006. Relating soil electrical conductivity to remote sensing and other soil properties for assessing soil salinity in Northeast Thailand. *Land Degradation and Development* 17: 677–689.

Soil and Plant Analysis Council. 1992. *Handbook on Reference Methods for Soil Analysis*. Georgia University Station, Athens, GA.

Stein, A., Kocks, C.G., Zadocks, J.C., Frinking, H.D., Ruissen, M.A., and Myers, D.E. 1994. A geostatistical analysis of the spatiotemporal development of downy mildew epidemics in cabbage. *Phytopathology* 84: 1227–1239.

Stein, A., Bastiaanssen, W.G.M., De Bruin, S., Cracknell, A.P., Curran, P.J., Fabbri, A.G., Gorte, B.G.H., Van Groenigen, J.W., Van Der Meer, F.D., and Saldaña, A. 1998a. Integrating spatial statistics and remote sensing. *International Journal of Remote Sensing* 19: 1793–1814.

Stein, A., Van Groenigen, J.W., Jeger, M.J., and Hoosbeek, M.R. 1998b. Space–time statistics for environmental and agricultural related phenomena. *Environmental and Ecological Statistics* 5: 155–172.

Tóth, T., Kabos, S., Pasztor, L., and Kuti, L. 2002. Statistical prediction of the presence of salt-affected soils by using digitized hydrogeological maps. *Arid Land Research and Management* 16: 55–68.

Triantafilis, J., Odeh, I.O.A., and McBratney, A.B. 2001. Five geostatistical models to predict soil salinity from electromagnetic induction data across irrigated cotton. *Soil Science Society of America Journal* 65: 869–878.

15 Mapping Soil Salinity from Sample Data and Remote Sensing in the Former Lake Texcoco, Central Mexico

Norma Fernández Buces, Christina Siebe,
José Luis Palacio Prieto, and Richard Webster

CONTENTS

15.1 INTRODUCTION

Soil salinity affects more than 76 million ha worldwide (Eger et al., 1996). Its causes are both natural and anthropogenic. Seriously affected areas have little if any plant cover and are therefore exposed to erosion by wind and water. In dry environments, wind readily picks up fine particles from the bare soil, thereby causing poor air quality, which is detrimental to human health, and sedimentation on surrounding agricultural land. Wind erosion needs to be controlled by vegetative cover. Unfortunately, salt-affected soils have large osmotic potentials, and this prevents plants from absorbing enough water in the rooting zone. As a consequence, plants suffer from water stress; they may also suffer from toxic concentrations of sodium and chloride ions or from the caustic effect of strongly alkaline soil pH (Szabolcs, 1989). Only a few salt-tolerant species can survive in these conditions, and some of them have been successfully introduced to control erosion. An alternative option is to rehabilitate or improve the soil condition itself.

There are numerous options for improving saline soils, including mechanical, agronomic, and biological techniques. They are all costly and labor-intensive. Drains may be laid to lower the groundwater table. Deep ploughing and addition of organic matter improve the soil's physical properties and increase the hydraulic conductivity and the water storage capacity. Supplementary irrigation can contribute by washing out excessive salts, and the application of gypsum promotes the exchange of sodium for calcium. Species may be planted according to their salt tolerance in either furrows or ridges (Tanji, 1996). The selection of appropriate techniques depends on the characteristics of the terrain and the origin and nature of the salts. Periodic determination of salinity indicators, such as electrical conductivity (EC), pH in saturation extract and sodium adsorption ratio (SAR), helps monitor and assess the effectiveness of the various techniques.

As salinity varies from place to place, salinity indicators must be mapped to determine where to attempt to improve the soil, and also to evaluate the results afterwards. Geographic variation of salinity might be substantial and can occur at a wide range of spatial scales. Traditional soil survey based on intensive sampling and measurement is unaffordable for large areas, especially as most of the spatial variation occurs within a few tens of meters. In these circumstances, remote sensing might be a satisfactory alternative. In fact, remote sensing techniques have been successfully used for mapping soil salinity, especially where efflorescences of salt on bare ground have characteristic reflectance, easy to detect by remote sensors. Mapping is usually based on the correlation between reflectance and salt content in the soil. Mapping scales vary according to the spectral and spatial resolutions of the sensors and the type of products generated (i.e., satellite images, aerial photographs, video imagery, field and laboratory radiometric data), as reported by Milton (1987), Long and Nielsen (1987), Everitt et al. (1988), Tóth et al. (1991), Csillag et al. (1993), Verma et al. (1994), and Valeriano et al. (1995). Satisfactory results have been obtained for bare soil with large salt contents. However, the reflectance of the vegetation cover is rarely considered, although indicators such as diversity, cover density, plant growth and dominance are correlated with soil salinity and alkalinity in accordance with the tolerance of the various species (Kertéz and Tóth, 1994).

In an earlier study, we developed a combined spectral response index (COSRI) that takes into account both the reflectance of bare soils with a variety of colors and the reflectance of several types of vegetation cover (Fernández Buces et al., 2006). We found it to be cost-effective for monitoring changes in soil salinity in the plain of the former Lake Texcoco. At the beginning of the twentieth century, this naturally saline lake was allowed to dry up to avoid floods; the result was a flat bare surface of 60 km^2 with saline soils and a groundwater table at 0.3–2 m depth. Native grass (*Distichlis spicata*) and other halophytes started colonizing the former lake bed, but most of the soils were too salty, and so large areas remained bare. Once dry, the exposed soils were highly susceptible to erosion by wind. The soil particles were and still are borne by the wind into the Valley of Mexico during the dry season. They are the main cause of poor air quality in metropolitan Mexico City during the dry season, where they are responsible for severe public health problems (Cruickshank, 1995).

In the early 1970s, government agencies started implementing several strategies to encourage vegetation in this area so as to control wind erosion (Cruickshank, 1995). Drainage channels were dug to lower the groundwater table; wastewater and drip irrigation were applied to wash out the salts and keep the soil moist; several halophyte species were introduced; and soil amendments were tested in small areas, including the application of gypsum and sulfuric acid. In 1971 and 1998, surveys of soil salinity indicators were carried out, with determination of EC of the saturation extracts, SAR, soil pH, and soluble salts. In the 1971 survey, 800 soil samples were taken from four depths (0–30, 30–60, 60–90, and 90–120 cm) and, in the 1998 survey, 446 soil samples were taken from two depths (0–30 and 30–60 cm). Despite this substantial sampling effort, the data had not previously been analyzed to track changes in salt content that might have occurred in the intervening 27 years.

In this study we compare the two sets of data recorded in 1971 and 1998. For both, we mapped salinity by kriging and compared the results to assess changes in the soil's EC in the interval. Likewise, traditional soil sampling and laboratory data were compared with remote sensing data by the COSRI index. We discuss the advantages and limitations of each procedure.

15.2 MATERIALS AND METHODS

15.2.1 STUDY AREA

The study area is a flat surface covering about 60 km^2 in the lowermost portion of the Mexico basin at 2220–2225 m above sea level (Figure 15.1). It was once occupied by the Lake Texcoco. The climate is temperate semiarid with warm summers and cold winters. The maximum average temperatures range between 30°C and 32°C during April to June and the minimum temperatures between −2°C and −5°C during October to March. Annual rainfall is 460–600 mm, mainly from June to September; skies are usually clear from November to February. Potential evapotranspiration is large from March to May, when there is little precipitation.

Soils are saline–alkaline and belong to the Solonchak and Solonetz classes (FAO, ISRIC and ISSS, 1999). They have developed from lake sediments, mainly volcanic ashes that have been

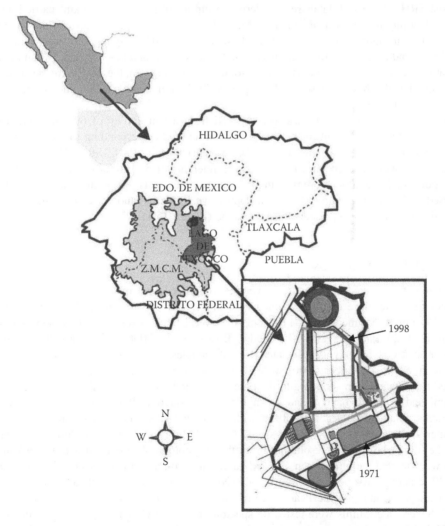

FIGURE 15.1 Location of the study area and areas covered by ancillary soil information (SRH, 1971 and CNA, 1998). The survey area covered by CNA (1998) is common to both surveys and covered by Landsat ETM image and iar photographs.

transformed into overhydrated thixotropic clay material, locally known as "jaboncillo" (soap). The surface horizon is clay loam to silty clay loam, and its thickness varies from 0 to 50 cm. The main source of salts is a subsurface aquifer, frequently at about 30 cm depth, with large EC values (6,000–10,000 dS m^{-1}). Salts move to the surface by capillarity and precipitate as salt efflorescence and crusts.

The vegetation is a halophytic community with predominance of a grass species (*D. spicata* (L.) Greene) and a Chenopodiacea (*Suaeda torreyana* (S.) Watson). These species, together with other Chenopodiaceae, Compositae, and Poaceae, have gradually colonized and improved originally bare soil (Fernández Buces, 2006).

15.2.2 Available Soil Data

We analyzed two soil databases provided by government agencies (SRH, 1971; CNA, 1998). They are as follows.

- The SRH (1971) soil database was derived from a soil survey of 430 km^2 carried out in 1971 during the dry season (January–May). It includes not only the former lake area but also the surrounding agricultural land. In this survey, 200 sampling sites were randomly chosen and sampled at four depths (0–30, 30–60, 60–90, and 90–120 cm). EC and pH in saturation extracts and exchangeable sodium percentage in the soil were determined. From this database, we selected 54 points (0–30 cm depth) within the former Lake Texcoco (Figure 15.1).
- The CNA (1998) soil database was derived from a soil survey of 69 km^2 carried out in May 1998, at the end of the dry season, in an area corresponding to the former lake (Figure 15.1). This area lies within the one surveyed in 1971 (SRH, 1971). A set of 223 sites was sampled on a regular grid at intervals of 500 m, at 0–30, and 30–60 cm depth. The EC and pH were measured in saturation extracts. From this database, we have used 222 sampling points (0–30 cm depth); we had to exclude one sampling point for which the database reported a missing value (Figure 15.1).

15.2.3 COSRI Calculation

In a previous study, spectral responses (average ground radiance measurements recorded in four bands equivalent to those of Landsat imagery: band 1, 430–525 nm; band 2, 510–600 nm; band 3, 600–700 nm; and band 4, 780–1100 nm) of several plant species and bare soil colors were determined at 86 sampling sites (Fernández Buces et al., 2006). The normalized difference vegetation index (NDVI) was calculated for all samples with a georeferenced Landsat ETM image acquired in December 2000, with 30 m × 30 m pixel size (path 26 and row 46). We extracted three bands corresponding to the blue, green, and red (bands 1, 2, and 3) portions of the spectrum, and a fourth band (band 4) in the near-infrared (NIR). Additionally, we used air photographs obtained in December 1999 with a Nikon F90× camera and with a Kodak DCS 420 camera, with a much smaller pixel size (2.6 m × 2.6 m), to test the procedure in smaller areas within the former lake at fine resolution. From the air photographs obtained with the Nikon F90× camera we extracted the blue, green, and red bands, and from the images obtained with the Kodak DCS 420 camera we extracted the fourth band (NIR, 850–1100 nm). We then computed, for both types of images separately, the Pearson correlation coefficients r for the relation between NDVI and EC. Several algorithms were tested on spectral bands 1–4 of the images in combination with the NDVI, in order to incorporate the reflectance of the bare soil and thereby find the best correlation with soil EC. The resulting integrated algorithm was named COSRI to highlight the combination of the spectral responses of bare soil and vegetation (Fernández Buces et al., 2006).

Once a significant relation was noted between the COSRI and soil EC, we regressed the soil's EC on COSRI to obtain an equation that would allow us to predict EC. The relation between salinity and COSRI seemed to be well described by exponential models, and so we fitted these to predict salinity. The analyses were done with GenStat 5 (Payne, 2006).

15.2.4 SOIL SALINITY MAPPING

EC maps were produced by ordinary kriging from measured values. To compute the experimental variograms from the data, we used Matheron's method of moments (Webster and Oliver, 2007):

$$\hat{\gamma}(\mathbf{h}) = \frac{1}{2m(\mathbf{h})} \sum_{j=1}^{m(\mathbf{h})} \left\{ z(\mathbf{x}_j) - z(\mathbf{x}_j + \mathbf{h}) \right\}^2 \tag{15.1}$$

In this equation, $z(\mathbf{x}_j)$ and $z(\mathbf{x}_j+\mathbf{h})$ are the measured values at places $\mathbf{x}_j$ and $\mathbf{x}_j+\mathbf{h}$ separated by the lag $\mathbf{h}$, and $m(\mathbf{h})$ is the number of paired comparisons at that lag. As variation appeared to be isotropic, the lag was treated as a scalar in distance only (i.e., $h = \|\mathbf{h}\|$). We then fitted models to the experimental variograms by weighted least squares. Figure 15.2 shows the results with the experimental values as points and the best fitting models as curves. The exponential model (Equation 15.2) fitted the year 1971 variogram best, while the Gaussian model (Equation 15.3) was the best for the 1998 variogram.

$$\gamma(h) = c_0 + c \left\{ 1 - \exp\left(-\frac{h}{a} \right) \right\} \tag{15.2}$$

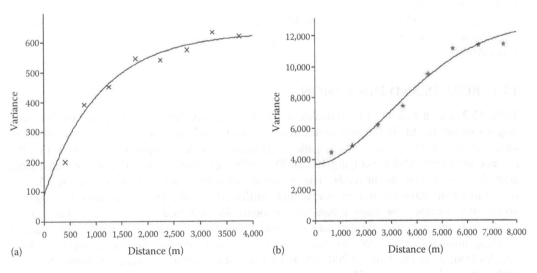

(a) Distance (m) (b) Distance (m)

FIGURE 15.2 Experimental variograms and fitted models of soil EC data from the sample surveys in (a) 1971 and (b) 1998.

TABLE 15.1

Parameter Values of the Models Fitted to the Experimental Variograms

Parameter	1971 Data	1998 Data
a/m	1086 (standard error 328)	4516 (standard error 592)
c	546.4	9111
c_0 (nugget)	88.44	3668

$$\gamma(h) = c_0 + c\left\{1 - \exp\left(-\frac{h^2}{a^2}\right)\right\}$$ (15.3)

In these equations, c_0 is the nugget variance, c is the sill of the correlated variance, and a is a distance parameter. The values of the parameters are listed in Table 15.1. We then interpolated the EC at intervals of 100 m on a raster by ordinary punctual kriging with local neighborhoods containing between 7 and 20 points. Changes in EC between 1971 and 1998 were computed from the differences in predicted EC at the raster points. Maps of predicted EC, the respective kriging variances and the changes between 1971 and 1998 were drawn in a geographical information system framework (ILWIS, 2005).

Additionally, maps of the predicted EC were produced from the COSRI values obtained from the Landsat ETM image and the airborne photographs. The COSRI values were calculated pixel by pixel from the spectral bands extracted from the images. Subsequently, the calculated COSRI values were converted into EC values using the previously fitted exponential model (Fernández Buces et al., 2006) for the relation:

$$EC = 348.1 \exp(-18.372 \times COSRI)$$ (15.4)

for which $R^2 = 0.826$.

Using this equation, raster maps of EC as predicted by COSRI were made. Map slicing and confidence intervals were used to print maps for arbitrarily defined classes (Fernández Buces et al., 2006).

15.3 RESULTS AND DISCUSSION

Table 15.2 lists the descriptive statistics of the two sets of data. As the mean, median, and ranges indicate, the EC in the saturation extracts seems to have increased from 1971 to 1998. The variogram for the 1971 data shows a spatial correlation extending to approximately 3 km ($3 \times$ the distance parameter 1086 m) (Figure 15.2a). The 1998 variogram (Figure 15.2b), computed from gridded data at 500 m intervals, has a longer effective range: $4516 \text{ m} \times \sqrt{3} \approx 7.5 \text{ km}$. Both variograms have positive intercepts and, particularly in the 1998 variogram, the spatially uncorrelated variance (or nugget variance) accounts for one-third of the total variance, indicating that an important portion of the spatial variation is occurring within the shortest sampling interval (500 m). We confirmed this by a nested sampling design, in which we took samples from 0–30 cm depth at 200, 50, and 10 m distances from ten selected points chosen at random from the grid set at 500 m intervals (Fernández Buces, 2006). Figure 15.3 shows the resulting variogram, which indicates that nearly 24% of the EC variation occurs at distances of less than 10 m and 12% at distances of 10–50 m, while the rest occurs at distances between 100 and 500 m.

TABLE 15.2
Descriptive Statistics of the Data

Data Set	SRH (1971)	CNA (1998)
Date	February–May 1971	May 1998
N	54	222
Depth	0–30 cm	0–30 cm
Grid interval	Irregular (>1000 m)	Regular (500 m)

EC Measurement	Saturation Extracts	
Mean (dS m^{-1})	45.0	105.8
Median (dS m^{-1})	46.9	84.0
Variance	589.36	8269
Standard deviation (dS m^{-1})	24.3	90.9
Skewness	0.16	0.70
Range (dS m^{-1})	0.6–99	2.0–410

The kriged map made from the 1971 data (Figure 15.4a) shows that the smallest EC values occur in the eastern part of the area (EC: 10–25 dS m^{-1}), followed by those in the southern part, while the largest EC values are in the middle and in the northern part of the area (EC: 70–90 dS m^{-1}). The kriging variances (Figure 15.4b) range between 200 and 800, and show an irregular patchy pattern. This is because the sampling points were chosen at random. The largest kriging variances occur in the east and at the edges of the area because that is where sampling was sparsest.

The kriged map made from the 1998 data (Figure 15.5a) shows the smallest EC values in the south-eastern part of the area, followed by the south, while the largest EC values occur again in

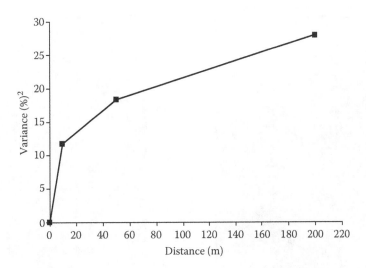

FIGURE 15.3 Variogram of soil EC generated by accumulating the components of variance obtained from a nested sampling design. Departing from 10 nodes chosen at random on a 500 m square grid, additional points were placed with random orientations at 200, 50, and 10 m distance. (Reprinted from Fernández Buces, N. Variabilidad espacial de la salinidad y su efecto en la vegetación en el ex-lago de Texcoco: implicaciones para su monitoreo por percepción remota. Tesis doctoral. Instituto de Geología, Universidad Nacional Autónoma de México, México D.F., 2006. With permission.)

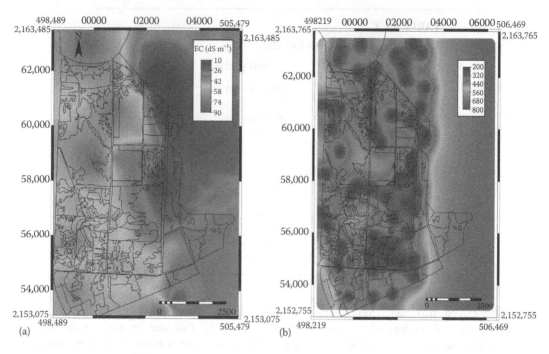

FIGURE 15.4 Maps of soil EC made by ordinary punctual kriging of data from (a) 1971 and (b) the corresponding kriging variances.

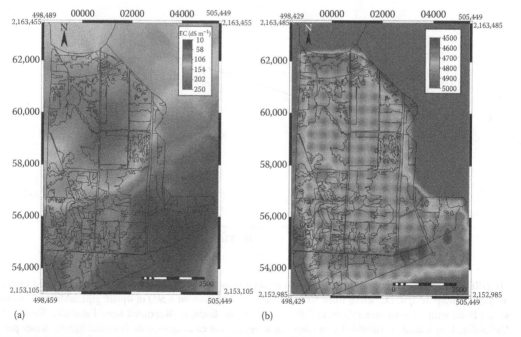

FIGURE 15.5 Maps of soil EC made by ordinary punctual kriging of data from (a) 1998 and (b) the corresponding kriging variances.

the middle but covering a larger extent than in 1971. The kriging variances again have a patchy pattern, but now the pattern is regular with circles around the sampling points at 500 m intervals (Figure 15.5b). As for the 1971 survey, the variances are large at the margins of the area because there are no data beyond them. One missing value in the middle west of the area caused the kriging variances to increase up to values similar to those at the edges. The absolute variances increased by one order of magnitude between 1971 and 1998. Apparently, the soil measures taken to rehabilitate the area increased the spatial variation locally. For example, salts moved from the furrows to the crest of the ridges and from areas bordering drainage pipes to the intervals between pipes.

After 27 years of ameliorative management, it seems that salt contents diminished slightly or remained constant in the south-eastern part of the area, while they increased in the middle to values twice as large as before (from 70–90 dS m^{-1} to values exceeding 200 dS m^{-1}). The calculated changes are mapped in Figure 15.6.

The Landsat ETM image taken in December 2000 (Figure 15.7) shows that the southern part is almost covered by vegetation, whereas large areas remain bare in the north. In the mid-northwest, water covers the surface. Figure 15.8 shows the map of EC calculated by Equation 15.4 to convert COSRI into EC. This equation provides EC values that are similar to those obtained by the traditional soil survey of 1998, but only for the vegetated areas. In the other areas, the remote sensing technique predicts EC values larger by up to one order of magnitude. We attribute this to the fact that COSRI registers only the reflectance of the soil surface, with salt concentrating often in thin crusts a few mm thick. The samples that we took in the previous study to correlate COSRI with EC came from the top 10 cm of the soil. The existence of thin surface crusts might explain part of the

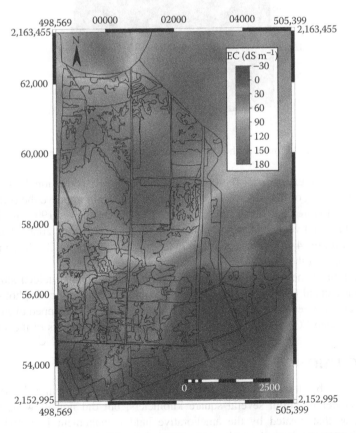

FIGURE 15.6 Map of changes in soil EC estimated from the sample surveys done in 1971 and 1998, after 27 years of ameliorative management.

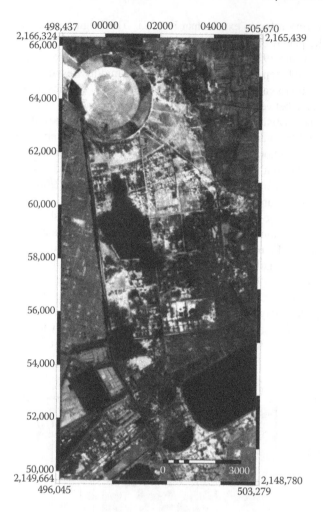

FIGURE 15.7   Landsat ETM image of the study area acquired in December 2000.

variation for which the exponential regression of EC on COSRI (Equation 15.4) did not account (almost 18%). Sampling of the uppermost 10 mm of the soil might improve the correlation between COSRI and EC. In Figure 15.9, depth functions of soil EC determined in 1999 are shown for typical soil profiles on bare and vegetated terrain in the area (Fernández Buces, 2006). Bare surfaces have often salt crusts with EC 4–10 times larger than in the soil below, while EC values in vegetated areas are generally smaller at the soil surface and increase with depth.

The application of the COSRI index to the airborne photographs yielded similar results, but with much greater detail (Figure 15.10). The visual comparison with the air photograph shown in Figure 15.11 demonstrates that the variation in EC can be accurately mapped even within furrows, although the predicted EC values are restricted to the top few millimeters of the soil.

15.4 CONCLUSION

EC data obtained by traditional soil survey at fairly large sampling intervals (>500 m) detected general trends within several square kilometers, but did not allow us to map detailed variation such as that created by the ameliorative land management in individual fields of a few ha. The latter would need much more detailed sampling as the nested sampling design demonstrated.

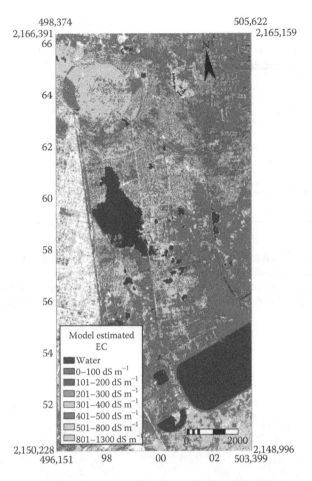

FIGURE 15.8 Distribution of EC values at the soil surface estimated by COSRI applied to the Landsat ETM image. (Reprinted from Fernández Buces, N., Siebe, C., Cram, S., and Palacio-Prieto, J.L., *J. Arid Environ.*, 65, 644, 2006. With permission.)

To account for the large-scale variation, sampling intervals should be smaller than 50 m, and this means that salinity mapping in the study area would need a much larger sampling effort. It might not be necessary to sample at regular grid intervals of 50 m or less in the whole area, but a nested sampling design, with observations at distances smaller than the general 500 m grid interval (e.g., at 200, 50, and 10 m distance from each 500 m grid point), would significantly improve the mapping (Webster et al., 2006). This would lead to a 4-fold increase in the sampling intensity, with a correspondingly large increase in the survey cost. To monitor EC changes in time using traditional survey, the staff charged with soil amelioration at Texcoco will have to accept that fieldwork and laboratory processing will need several months. Additionally, they will have to learn to compute, model and interpret the variograms and to produce the final maps by kriging.

Remote sensing can detect, with reasonable accuracy, the presence of salt in the surface layer of the topsoil. Results can be obtained in a few days, once an algorithm that expresses the relation between reflectance and soil salinity has been established for a given area. This will need traditional sampling and laboratory analyses to obtain the ground data, but should not be necessary on every occasion for monitoring the change. Costs of imagery are diminishing, and resolution has improved substantially in the last few years. Remote sensing techniques allow us to monitor

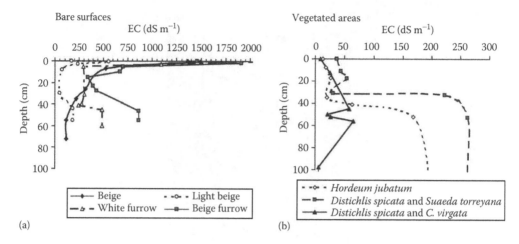

FIGURE 15.9 Depth functions of EC in typical soil profiles within the study area. (From Fernández Buces, N., *Variabilidad espacial de la salinidad y su efecto en la vegetación en el ex-lago de Texcoco: implicaciones para su monitoreo por percepción remota*. Tesis doctoral. Instituto de Geología, Universidad Nacional Autónoma de México, México D.F., 2006. With permission.) Examples of bare surfaces (a) and vegetated areas (b). Colors refer to soil surface patterns.

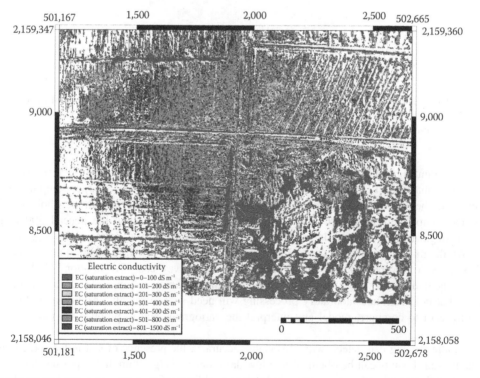

FIGURE 15.10 EC values at the soil surface estimated by calculating COSRI pixel by pixel on an air photograph taken in 1999 and by converting COSRI data into EC data using the exponential model obtained in a previous study. (From Fernández Buces, N, Siebe, C., Cram, S., and Palacio-Prieto, J.L., *J. Arid Environ.*, 65, 644, 2006.)

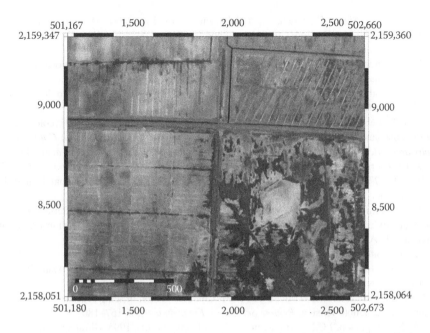

FIGURE 15.11 Original panchromatic air photograph (2.6 m × 2.6 m pixel size) used to calculate COSRI.

changes in salt content of the soil in the short term and with a desired resolution. The particular sensitivity to soil surface characteristics could even be of much larger interest for planning corrective land management. However, the investment in training and image processing equipment and software has also to be considered, if this technique is chosen for monitoring changes.

Combining remote sensing and traditional survey is recommended for assessing medium-term changes. Field data are indispensable to validate the remote sensing data, particularly if the salt crust is sampled separately from the rest of the topsoil. Samples of the first 20–30 cm should provide valuable information for revegetation strategies, which remote sensing alone cannot provide.

REFERENCES

CNA. 1998. *Estudio agrológico detallado de 3000 ha en la zona federal del ex-lago de Texcoco*. Comisión Nacional del Agua, Departamento del Distrito Federal, Proyecto Lago de Texcoco, Ciudad de México.

Cruickshank, G. 1995. Proyecto Lago de Texcoco: rescate hidrogeológico. In: *Memoria de la evolución del proyecto que mejora en forma importante las condiciones ambientales de la ciudad de México*, Ed. G.A. Zepeda-Reyes, Comisión Nacional del Agua, Agua.

Csillag, F., Pástor, L., and Biehl, L.L. 1993. Spectral band selection for the characterization of salinity status of soils. *Remote Sensing of Environment* 43: 231–242.

Eger, H., Fleischhauer, E., Hebel, A., and Sombroek, W. 1996. Taking action for sustainable land use: Results from the 9th ISCO Conference in Bonn, Germany. *Ambio* 25: 480–483.

Everitt, J.H., Escobar, D.E., Gebermann, A.H., and Alaniz, M.A. 1988. Detecting saline soils with video imagery. *Photogrammetric Engineering and Remote Sensing* 54: 1283–1287.

FAO, ISRIC and ISSS. 1999. World reference base of soil resources. Report on World Natural Resources 84, Food and Agriculture Organization of the United Nations, International Soil Resource Information Centre, International Society of Soil Science, Rome, Italy.

Fernández Buces, N. 2006. Variabilidad espacial de la salinidad y su efecto en la vegetación en el ex-lago de Texcoco: implicaciones para su monitoreo por percepción remota. PhD dissertation. Posgea do en Ciencias Biológicas, Instituto de Geología, Universidad Nacional Autónoma de México, México D.F.

Fernández Buces, N., Siebe, C., Cram, S., and Palacio-Prieto, J.L. 2006. Mapping soil salinity using a combined spectral response index for bare soil and vegetation: A case study in the former lake Texcoco, Mexico. *Journal of Arid Environments* 65: 644–667.

ILWIS. 2005. Integrated land and watershed information system. Version 3.3. International Institute for Geo-Information and Earth Observation (ITC), Enschede, the Netherlands.

Kertéz, M. and Tóth, T. 1994. Soil survey based on sampling scheme adjusted to local heterogeneity. *Agrokémia és Talajtan* 43: 113–132.

Long, D.S. and Nielsen, G.A. 1987. Detection and inventory of saline seep using color infrared aerial photographs and video image analysis. In: *Proceedings of the 14th Workshop on Color Aerial Photography and Videography in Resource Monitoring*, Montana State University, Bozeman, MT, pp. 220–23.

Milton, E.J. 1987. Principles of field spectroscopy. *International Journal of Remote Sensing* 8: 1807–1827.

Payne, R.W. 2006. *The Guide to GenStat, Release 9—Part 2: Statistics*. VSN International, Hemel Hempstead, Great Britain.

SRH. 1971. *Estudio agrológico especial del ex-lago de Texcoco, Estado de México*. Serie Estudios, Publicación Número 2. Secretaría de Recursos Hidráulicos, Dirección General de Irrigación y Control de Ríos, Dirección de Agrología, Ciudad de México.

Szabolcs, I. 1989. *Salt-Affected Soils*. CRC Press, Boca Raton, FL.

Tanji, K.K. (Ed.). 1996. *Agricultural Salinity Assessment and Management*. American Society of Civil Engineers, New York.

Tóth, T., Csillag, F., Biehl, L.L., and Michéli, E. 1991. Characterization of semi-vegetated salt-affected soils by means of field remote sensing. *Remote Sensing of Environment* 37: 167–180.

Valeriano, M.M., Epiphanio, J.C.N., Formaggio, A.R., and Oliveira, J.B. 1995. Bi-directional reflectance factor of 14 soil classes from Brazil. *International Journal of Remote Sensing* 16: 113–128.

Verma, K.S., Saxena, R.K., Barthwal, A.K., and Deshmukh, S.N. 1994. Remote sensing technique for mapping salt-affected soils. *International Journal of Remote Sensing* 15: 1901–1914.

Webster, R. and Oliver, M.A. 2007. *Geostatistics for Environmental Scientists*, 2nd ed. John Wiley & Sons, Chichester, UK.

Webster, R., Welham, S.J., Potts, J.M., and Oliver, M.A. 2006. Estimating the spatial scales of regionalized variables by nested sampling, hierarchical analysis of variance and residual maximum likelihood. *Computers and Geosciences* 32: 1320–1333.

16 Model-Based Integrated Methods for Quantitative Estimation of Soil Salinity from Hyperspectral Remote Sensing Data

Jamshid Farifteh

CONTENTS

16.1 INTRODUCTION

The development of methods for assessing, mapping, and monitoring salt-affected soils by use of optical remote sensing data in combination with field measurements has been the objective of numerous studies during the last three decades. The approaches used range from simple visual interpretation of hardcopy satellite imagery (Verma et al., 1994), to digital analysis of broadband remote sensing data (Metternicht and Zinck, 1996; Rao et al., 1998), to quantitative analysis of hyperspectral imagery (Ben-Dor, 2002; Dehaan and Taylor, 2002), and integrated modeling (Farifteh, 2007). A review is given in Metternicht and Zinck (2003) and Farifteh et al. (2006).

    Among the monovariate techniques, various deterministic and stochastic spectral matching techniques generating similarity measures have substantial potential for quantifying salt variations in soils (Farifteh et al., 2007b). Similarity measures are usually applied to determine (dis)similarity between spectra of an unknown substance and spectra of a reference substance (Van der Meer, 2000). Deterministic techniques such as Euclidean distance and spectral angle mapper (SAM) are based on distance, angle, and correlation between two spectra. Stochastic techniques such as spectral information divergence (SID) are used to determine spectral variations by modeling spectral information as a probability distribution (Chang, 2003). Curve fitting is also a statistical technique used for modeling soil properties from soil reflectance. Absorption features observed in the visible, near-infrared (NIR), and shortwave infrared (SWIR) spectra are usually considered to be inherently Gaussian-like in shape (Clark and Roush, 1984). Successful applications of Gaussian functions in modeling absorption bands have been reported (Miller et al., 1990; Whiting et al., 2004). Inverted Gaussian (IG) function can also be used to quantify variations in salt-affected soil reflectance spectra and to establish monovariate linear regression between soil salinity levels and calculated parameters of the fitted Gaussian curve (Farifteh et al., 2007c).

    Multivariate or adaptive techniques such as partial least square regression (PLSR) and artificial neural networks (ANNs) have been widely applied to model the relationships between chemical and physical soil properties and spectral data (Udelhoven et al., 2003). Adaptive techniques refer to mathematical algorithms or computational models that have the ability to adapt or change their structure based on external or internal information that flows through the system during the learning or calibrating phase (Wikipedia, 2007). The PLSR approach is able to model several response variables simultaneously, while effectively dealing with strongly collinear and noisy independent variables (Wold et al., 2001). The PLSR algorithm, thus, has inferential capability that can be used to model a possible linear relationship between salt concentration in soils and measured reflectance spectra (Farifteh et al., 2007d). ANN is the other widely used multivariate technique that has the ability to model any linear or nonlinear relationships between a set of variables (Huang and Foo, 2002; Yang et al., 2003). An ANN consists of a large number of nodes, which simulate basic functions of biological neurons and have the ability to learn the relationship between input and output data. As the relationship between soil salinity and reflectance may not be linear, the ANN methods can be advantageous in examining this type of relationship.

The study presented in this chapter aims at developing predictive models for quantifying salt content in soils on the basis of the integrated use of soil reflectance and geophysical survey. The main question in this research is: Can a salinity warning system be developed from remote sensing imagery to detect areas that are in an early stage of salinization?

16.2 METHODS FOR SOIL SALINITY ASSESSMENT

Formation, movement, and accumulation of salts in the soil profile have substantial influence on the physical, chemical, and hydraulic properties of soil. Traditionally, scientists from different disciplines such as soil science, hydrology, geophysics, and spectrometry, have studied salt accumulation, movement, and reflectance. Usually, the effects of salt accumulation on soil properties are studied in the field and laboratory. Laboratory data on hydraulic properties were used to model solute transport in soils and predict salt distribution within the soil profile, with emphasis on the dynamics of salt movement regimes, especially in subsurface horizons (Jury, 1982). The spectral characteristics of salt minerals and the reflectance properties of salt-affected soils have been studied by remote sensing (Metternicht and Zinck, 2003; Farifteh et al., 2006). The spectral information recorded by remote sensors was used to assess and monitor soil salinity on a regional scale (De Jong, 1994; Ben-Dor et al., 1999). Near-surface geophysical sensors are now widely used to map and monitor salt-affected areas through measuring soil apparent electrical conductivity (ECa) (McNeill, 1980; George and Bennett, 2000). In practice, each of these approaches has advantages and disadvantages that limit the application in soil salinity studies. An overview of optical remote sensing, solute transport modeling, and near-surface geophysical applications in soil salinity studies is given in the following sections.

16.2.1 CONVENTIONAL METHODS

16.2.1.1 Remote Sensing and Salt-Affected Soils

Optical remote sensing has been repeatedly used as a promising tool to obtain information regarding soil properties and land degradation processes (De Jong, 1994; Dehaan and Taylor, 2002). Thousands of medium- to high-resolution images from the earth surface are available, which can be used to detect and monitor soil salinity spatially and temporally. To make use of satellite remote sensing, an appropriate scene model is needed to convert multispectral reflectance into thematic information. Potential contribution from remote sensing to salinity study can be analyzed applying the approach suggested by Verstraete et al. (1994). Dealing with remote sensing data, they distinguish three types of variables, namely, the measurable, the retrievable, and the hidden ones. The contribution of remote sensing techniques depends on whether or not the number of hidden variables is larger or smaller than the other two sets of variables. In an extreme case, if the behavior of the system under study is entirely explained by hidden variables, then there is little expectation that remote sensing data can provide useful information for this particular application. In the context of salinity, broadband remote sensing studies have often concentrated on areas affected by severe salinity, where measurable and retrievable variables were used to map salt-affected areas. The capability of broadband remote sensing in discriminating slightly affected areas is very limited, since salt concentration within the soil is not enough to show evidence at the soil surface or to influence the vegetation cover. In other words, the effect of salt in slightly saline areas is entirely explained by hidden variables.

The major factor limiting the application of optical remote sensing in salinity studies stems from the nature of the conventional satellite images, which by definition precludes extracting information about the third dimension of the 3D soil body. Furthermore, the combined reflectance of different ground covers within a single pixel and coarse spectral and spatial resolutions of satellite data are among the factors that limit the application of remote sensing in soil salinity studies.

16.2.1.2 Groundwater Modeling and Soil Salinization

The modeling of solute transport in soil both vertically and laterally provides vital information for understanding the dynamics of the salt movement regime and assessing the impact of salt accumulation under various conditions (Jury, 1982). Usually, salt-transport modeling aims at predicting salt distribution in the subsurface as a result of water percolation, groundwater level changes, and groundwater flow. Many options are available to model solute movement in the vadose zone. Mathematical models of unsaturated–saturated flow and transport of a given solute are used to simulate water flow and solute transport processes, and to determine or estimate the average concentration of salts in the soil profile (Van Dam et al., 1997). For example, Jury (1982) used a transfer function model to predict average values of solute concentration, as a function of depth and time, through highly variable field systems. In another case, a stochastic-convective model allowed to successfully predict solute concentration and estimate the leading edge of solute migration (Butters and Jury, 1989). A review of the different approaches developed for modeling preferential and nonequilibrium flow and transport in the vadose zone is given in Simunek et al. (2003).

Although solute transport modeling techniques can accurately estimate salt concentration in the soil profile, the reliability of predicted values is largely dependent on accurate information on soil hydraulic parameters (Crescimanno and Garofalo, 2005). Measurement of soil hydraulic properties is difficult. Especially, the determination of unsaturated hydraulic conductivity is costly, laborious, and time-consuming. In addition, laboratory measurements of soil hydraulic properties are only valid for the observation points where soils were sampled and, therefore, the information generated by a solute model needs to be interpolated for spatial analysis.

16.2.1.3 Geophysical Exploration and Salt-Affected Soils

Since the early twentieth century, geophysical sensors have been used in environmental studies. They are now increasingly used to extract information from the near-surface and subsurface, including the soil mantle and the regolith. The geophysical devices are designed to cover a large range of depths and are used for mapping terrain conductivity among other applications. Geophysical sensors, electromagnetic (EM) induction, and contact electrodes are techniques implemented to measure soil salinity in situ. EM surveys are conducted by inducing EM energy into the soil without physical contact (McNeill, 1980). Contact electrodes measure the apparent soil conductivity with a system of coulters in direct contact with the soil through insulated metal electrodes that penetrate the soil surface (Doerge et al., 2002). Airborne geophysical sensors are widely used in salinity studies because of advantages such as speed, cost-effectiveness, and availability of a variety of systems (Kearey and Brooks, 1994).

Geophysical sensors are capable of gathering rapidly and cheaply subsurface information with sufficient accuracy (Johnson et al., 2001). Near-surface geophysical sensing is particularly suitable for salt detection in cropland and for identifying areas of high conductivity without surface salt evidence (Howlett et al., 2001). Geophysical measurements provide average apparent conductivity values of the materials (as bulk) at a certain depth, which are influenced by changes in mineralogy, intensity of rock weathering, and water content.

16.2.2 Model-Based Integrated Method

16.2.2.1 Basic Principles

Considering the complexity of the salinization process, Farifteh (2007) proposed a model-based integrated framework to study soil salinity using a transdisciplinary concept (Klein et al., 2000). Transdisciplinarity, with the potential of linking a variety of disciplines, is inherent to the framework and provides the basis for developing a dynamic modeling approach that overcomes a number of the aforementioned limitations, while also benefitting from the advantages of conventional (monodisciplinary) methods. The framework creates the possibility to move from three different disciplines (i.e., optical/infrared remote sensing, geophysics, and solute modeling) toward a unique and new method (Figure 16.1), with prediction capabilities allowing early detection of salinization (Farifteh et al., 2004; Farifteh, 2007).

The method makes use of a knowledge-based approach for integrating surface and subsurface measurements. Knowledge-based systems offer large potential for extracting information from remotely sensed data on the basis of explicit knowledge about the objects and processes (Mulder, 1994), the use of ancillary data, and an iterative procedure to adjust the calculated results via optimization. Both multivariate PLSR and ANN modeling methods use training data sets to learn about the object and develop calibration algorithms that eventually can be used for identification and estimation of soil salinity levels. The general strategy to estimate soil salinity levels using the model-based integrated method is presented in Figure 16.2.

Integration takes place in a well-designed geographical information system (GIS) environment, with the support of geostatistics and a rule-based technique. The integrated methodology involves upscaling, because both the process and the data obtained are scale-dependent. In the case studies discussed later, a combination of hyperspectral remote sensing and in-situ geophysical measurements with statistical and mathematical modeling tools is used to demonstrate the process of upscaling from laboratory-scale data to field- and image-scale measurements. The strategy for applying the method to the determination of the relationship between the spectral reflectance of salt-affected soils measured by a spectrometer and the ECa measured through a geophysical survey is presented in Figure 16.3.

16.2.2.2 Applications of the Model-Based Integrated Method

The model-based integrated approach proposed here aims to improve the results of the techniques currently used for assessing soil salinization via integrating soil surface and near-surface

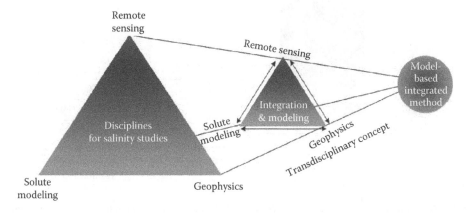

FIGURE 16.1 Conceptual model for the assessment of salt-affected soils.

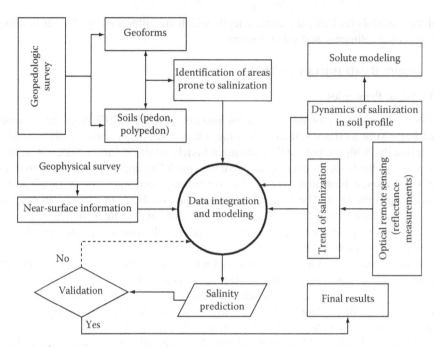

FIGURE 16.2 Methodological approach integrating remote sensing, solute modeling, and geophysical survey.

information. The application of the conceptual framework suggested has not been fully evaluated yet. However, there are several investigations in which integration approaches have been applied to improve the products derived from remotely sensed data for salinity estimation and mapping (Sah et al., 1995; Eklund et al., 1998).

GIS-based overlaying is one of the most widely used techniques to integrate information at grid-cell level. The integrated approach allows combining digital data obtained from space- or airborne

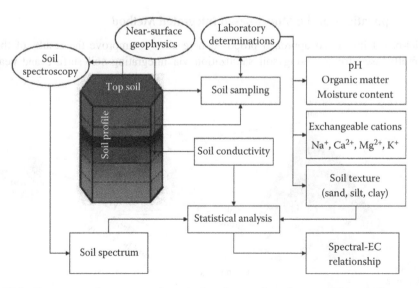

FIGURE 16.3 Strategy to integrate surface (soil reflectance) and near-surface (soil apparent ECa) measurements.

sensors with ancillary data such as digital elevation model, soil map, groundwater quality map, geological map, and land use map. This approach was found effective in classifying salt-affected areas (Sah et al., 1995; Eklund et al., 1998).

Conditional probability has been used to integrate GIS layers from databases including remotely sensed data. In this context, the Bayes' theorem can be used to assess the probability that a certain degree of salinity may occur at a given location based on a predefined piece of evidence (Duda and Hart, 1973). Such an approach was used by Schmidt et al. (2004) for mapping eucalypt forest and coastal vegetation, respectively.

Knowledge-based systems offer potential for extracting information from remotely sensed data. Together with remote sensing data, these systems use explicit knowledge about the objects and processes, ancillary data, and an iterative procedure to adjust the measurements (Mulder, 1994).

16.3 MODELING TECHNIQUES FOR ESTIMATING SOIL SALINITY

16.3.1 Spectral Similarity Measures

Two types of spectral similarity measures (spectral matching techniques) can be used to determine differences between reflectance spectra of salt-affected soils (unknown) and spectra of nonsaline soils (reference): deterministic and stochastic. Deterministic techniques are based on angle, distance, and correlation between two spectra, while stochastic techniques, in general, use sample properties and develop spectral criteria such as divergence and probability to measure dissimilarity between two spectra (Chang, 2003). In the following discussion of spectral similarity techniques, unknown spectra (for salt-affected soils) are denoted x_i and reference spectra (for nonsaline soils) are denoted y_i, both measured for i ($= 1, 2, \ldots, n$) spectral bands.

16.3.1.1 Deterministic Techniques

Several deterministic techniques were used in the case studies, including the Euclidean distance, the SAM, and the spectral similarity value (SSV).

The Euclidean distance (d_e) measures the difference in magnitude of two spectra and can be defined as

$$d_e = \sqrt{\sum_{i=1}^{n} (x_i - y_i)^2} \tag{16.1}$$

In order to remove the dependence on the number of spectral bands, d_e is modified by inserting the factor $1/n$ (Keränen et al., 2003). Low values of d_e mean high similarity and vice versa. The distance d_e can be converted to a similarity index (SI), if it is normalized, so that it lies in the range of 0 (for maximum dissimilarity) to 1 (for maximum similarity). SI is calculated as (Webster and Oliver, 1990):

$$\text{SI} = 1 - d_e \tag{16.2}$$

The distance d_e can also be scaled between 0 and 1 using Equation 16.3 (Homayouni and Roux, 2004):

$$D_e = (d_e - m)/(M - m) \tag{16.3}$$

where
   D_e is the scaled distance
   m and M are the minimum and maximum of the d_e values, respectively

The SAM calculates the angle α between an unknown spectrum and a reference spectrum by applying the following equation (Kruse et al., 1993):

$$\alpha = \arccos\left(\frac{\sum_{i=1}^{n} x_i y_i}{\sqrt{\sum_{i=1}^{n} x_i^2}\sqrt{\sum_{i=1}^{n} y_i^2}}\right) \tag{16.4}$$

The α values range between 0 and $\pi/2$. Low values of α indicate high similarity between the unknown and the reference spectra. For comparison reasons, the results can be rescaled in the range of 0–1 by dividing 2α by π (Schwarz and Staenz, 2001).

The SSV is a combination of the two above-described algorithms and is calculated as (Keränen et al., 2003):

$$\text{SSV} = \sqrt{d_e^2 + (1 - \rho^2)^2} \tag{16.5}$$

$$\rho^2 = \left(\frac{\frac{1}{n-1}\sum_{i=1}^{n}(x_i - \mu_x)(y_i - \mu_y)}{\sigma_x \sigma_y}\right)^2 \tag{16.6}$$

where
μ_x and μ_y are the means
σ_x and σ_y are the standard deviations of the unknown and reference spectra, respectively

Low values of SSV mean high similarity and vice versa.

16.3.1.2 Stochastic Techniques

Two stochastic similarity measures were used in the case studies to compare salt-affected soil and nonsaline soil reflectance spectra. The SID is defined as (Chang, 2003)

$$\text{SID}(x_i, y_i) = D(x_i \,\|\, y_i) + D(y_i \,\|\, x_i) \tag{16.7}$$

It measures the discrepancy between the spectral signatures of two vectors x_i and y_i in terms of their corresponding probability mass functions p and q, respectively (Chang, 2003). A combination of SID and SAM generates a new measure referred to as the SID–SAM. The SID–SAM between two spectral signatures x_i and y_i is given by

$$\text{SID (TAN)} = \text{SID}(x_i, y_i)\tan(\text{SAM}(x_i, y_i)) \tag{16.8}$$

$$\text{SID (SIN)} = \text{SID}(x_i, y_i)\sin(\text{SAM}(x_i, y_i)) \tag{16.9}$$

The product of SAM and SID increases spectral discriminability considerably, because it makes two similar spectra even more similar and two dissimilar spectra more distinct (Du et al., 2004). Low values of SID, SID(TAN), or SID(SIN) indicate high similarity and vice versa. The detailed description of these algorithms is available in Chang (2003) and Du et al. (2004).

16.3.2 Curve Fitting

Successful applications of IG functions in modeling absorption bands have been reported (Miller et al., 1990; Whiting et al., 2004). Using an IG function to quantify the spectral response of salt-affected soils has the advantage of making use of the general shape of the spectrum and of being independent from diagnostic absorption bands and their parameters, which are largely influenced by the amount of soil moisture and masked in an image due to atmospheric effects.

16.3.2.1 IG Function

To represent the spectral shape of an absorption band, the IG function is defined as (Miller et al., 1990):

$$R(\lambda) = R_s + (R_0 - R_s) \exp\left(\frac{-(\lambda_0 - \lambda)^2}{2\sigma^2}\right) \tag{16.10}$$

where
  R_s is the maximum reflectance
  R_0 is the minimum reflectance
  λ_0 is the central minimum or peak position wavelength
  λ is the wavelength
  σ is the Gaussian function deviation parameter describing the width of the Gaussian peak

The IG function can be fitted to portions of the EM spectrum where most spectral variations occur, by connecting the maximum reflectance wavelength to the functional center. To estimate soil salinity levels, the fundamental water absorption band at 2800 nm was used as the functional center. This peak position lies beyond the range of the FieldSpec FR spectrometer used in this study and, therefore, a fitting algorithm was needed to extrapolate the IG function to this point. The 2800 nm water absorption band was used because most of the salt minerals (e.g., hydroxyl group) exhibit a sharp absorption band in the region of 2700–2800 nm, depending on the OH type directly attached to and where it is located in the mineral. The water of hydration usually shows a sharp absorption band near 2750 nm and one or more absorption bands near 2900 nm (Nyquist and Kagel, 1971). The band at 2800 nm was also used as the function center to estimate soil moisture (Whiting et al., 2004). Furthermore, as already evidenced by Bishop et al. (1994), the IG curves are not fitted to the spectra of salt minerals, but to the spectra of the salt-affected soils, which are influenced by other soil constituents including clay minerals and water. Thus, the results can be influenced by sufficient absorption at the 2800 nm wavelength to diagnose salt minerals within soils with various clay and water contents.

16.3.2.2 Processing Steps

The parameters needed to estimate salt contents in soils, using IG functions, are obtained in three steps: preprocessing, parameter calculation, and accuracy assessment.

Preprocessing activities start with the smoothing of the reflectance spectra using convolution filters. It continues with the conversion of the relative reflectance to the natural log and the normalization of the log reflectance by the maximum log reflectance. The normalized natural log of reflectance can provide an optimum fit, and normalization of the spectra to the maximum log of reflectance is essential since it enhances the effect of a shift in position of the maximum reflectance (Whiting et al., 2004).

Parameter calculation involves the determination of the local maximum reflectance, its corresponding wavelength, and the calculation of the fitting parameters. Local maxima are used to define a general shape of the continuum, which provides the generalization of the spectrum. There are several techniques, such as first and second derivatives, wavelet transformation, and convex hull, which can be used to determine local maxima in soil spectra. Whiting et al. (2004) found that convex hull provides better results than derivatives do, since it is less susceptible to random

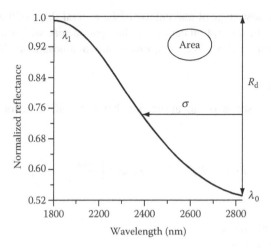

FIGURE 16.4 IG model parameters required for estimating soil salinity (λ_i is the point of maximum reflectance in the spectrum; λ_0 is the peak center; R_d is the depth; and σ is the distance to the inflection point).

variability in the spectra. A program in IDL (Interactive Data Language) was developed (Farifteh et al., 2007c) to generate convex hull boundary points from the spectral plot values and to provide a sequential list of convex hull points. Along the convex hull points, the IG function is fitted to the shortwave region of the spectrum from the point of maximum reflectance λ_i to the functional center λ_0 at 2800 nm (Figure 16.4). Other model parameters, such as functional depth R_d, distance σ to the inflection point, and area of one side of the function, were then calculated.

The third step involves the application of the derived IG model parameters, as estimation variables x_{ij} ($j = 1, \ldots, p$) in a linear regression function, to estimate soil salt contents λ_i for an unknown sample.

16.3.3 ADAPTIVE TECHNIQUES

16.3.3.1 PLSR Model

PLSR model is a method that assumes a linear relationship between a set of dependent (response) variables Y and a set of predictor variables X. The general purpose of using PLSR is to extract the orthogonal or latent predictor variables, accounting for as much of the variation of the dependent variable(s). PLSR can be used to model correlation between soil reflectance spectra (predictor variables) and soil salinity (response variable). A schematic representation of the method is given in Figure 16.5, based on the tutorial of Geladi and Kowalski (1986) and a review by Wold et al. (2001). Application of the PLSR method takes into account the matrices T, U, W, and C (Figure 16.5), which contain useful information that describes the linear relationship between the response and predictor variables. The score matrices T and U describe the similarities or dissimilarities (e.g., relationships) between response and predictor variables. The weight matrices W and C define how the variables are combined to form the quantitative relationship between X and Y.

The spectral data used for building the PLSR models need to be transformed to log ($1/R$) and are recommended to be smoothed and corrected for light scattering (Rossel, 2005). Before analysis, the transformed spectral data should be further centered or standardized (mean of zero and standard deviation of one) in order to make their distribution fairly symmetrical (transformation), focus the model on important variables (scaling), facilitate the interpretation and increase numerical stability (center), and decrease the size of the observations (normalize) if it is irrelevant (Jackson, 1991).

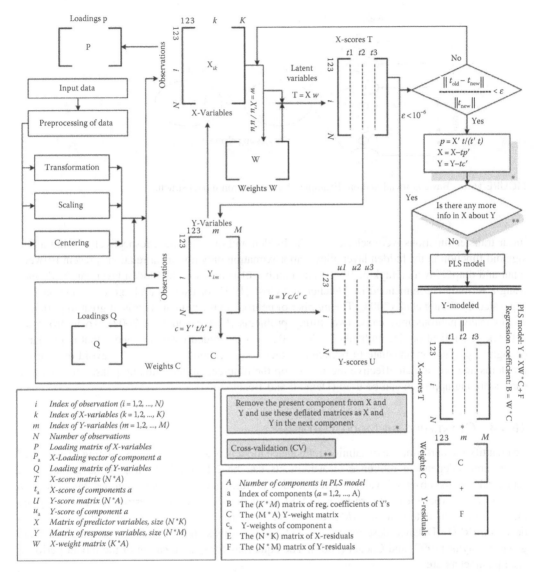

FIGURE 16.5   PLSR algorithm and data arrangement based on the procedure described in Wold et al. (2001).

16.3.3.2   ANN Model

An ANN consists of a large number of highly interconnected processing elements (nodes or units). Figure 16.6 illustrates a simple one-neuron model within an artificial network, where an input vector is passed through the neuron to provide an output value. A neural network is composed of several (sometimes thousands) such processing units, where unidirectional connections between them are provided by a corresponding weighting vector. Such networks are called multiple-layer networks and have the ability to learn the relationship between input and output data. A multiple-layer system consists of input, hidden, and output layers. The theory and mathematical basis of ANN have been extensively described in Haykin (1999).

In the case studies discussed in this chapter, a multiple-layer feed-forward backpropagation network with three (input, hidden, and output) layers was used. This type of network generally performs better than other types (Hornik et al., 1989). Nonlinear tan-sigmoid transfer functions and

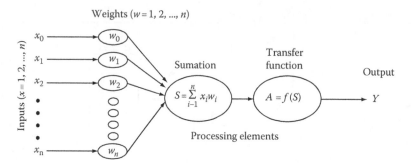

FIGURE 16.6 Basic artificial neuron: Example of a one-neuron model structure.

linear transfer functions were selected for the hidden and output layers, respectively. Using a tan-sigmoid function in the hidden layer allows approximating only nonlinear relations present between input and output layers (Haykin, 1999). The number of neurons in the hidden layer can be defined using a formula recommended by Fletcher and Goss (1993) or using a trial-and-error approach as suggested by Chang et al. (2004). The number of neurons in the hidden layer is of great importance, as too many neurons may cause over-fitting problems (Huang and Foo, 2002). To improve the network generalization, Demuth and Beale (2004) recommended using a network that is just large enough to provide an adequate fit between predictor and response variables. To avoid over-fitting problems and provide an effective means to stop the time-demanding training phase, the so-called early stopping approach can be used (Demuth and Beale, 2004).

16.3.4 QUANTIFYING THE MODEL PERFORMANCE

To quantify the performance of salinity–reflectance models, various parameters relating to estimated values and independent reference measurements can be calculated. To assess the goodness of estimated versus observed values, three model performance parameters can be calculated, in addition to the root-mean-square error (RMSE) and the mean absolute errors (MAE). They are the coefficient of determination (R^2), the model efficiency (ME), and the ratio of prediction to deviation (RPD). Detailed descriptions and definitions of these model performance parameters are given in Taylor (1997) and Cacuci (2003). The equations for quantification of the model performance parameters are

$$\text{RMSE} = \sqrt{\frac{\sum (Y' - Y)^2}{n}} \tag{16.11}$$

$$\text{MAE} = \frac{\sum_{i=1}^{n} |Y_i' - Y_i|}{n} \tag{16.12}$$

$$\text{ME} = 1 - \frac{\sum_{i=1}^{n} (Y' - Y)^2}{\sum_{i=1}^{n} (Y_i - \overline{Y})^2} \tag{16.13}$$

$$\text{RPD} = \frac{\text{SDP}}{\left(\sum (Y' - Y)^2 - \left\{ [\sum (Y' - Y)]^2 / N \right\} \big/ N - 1 \right)^{1/2}} \tag{16.14}$$

$$SDP = \left\{ \sum y'^2 - \left[\left(\sum y' \right)^2 / N \right] \Big/ (N-1) \right\}^{1/2} \qquad (16.15)$$

where

Y and Y' are measured and estimated values, respectively
SDP is the standard deviation of prediction
$\overline{Y}$ and $\overline{Y}'$ are the averages of the observed and predicted values, respectively
n is the number of observations

The R^2 and RPD indicate the strength of the statistical correlation between measured and predicted values. A predictive model is accurate if the R^2 and RPD values are higher than 0.91 and 2.5, respectively (Williams, 2001). R^2 values of 0.82–0.9 and RPD values higher than 2 indicate a good prediction, whereas R^2 values of 0.66–0.81 with a RPD value higher than 1.5 reflect a moderately good prediction. R^2 values of 0.5–0.65 indicate a poor relationship. MAE and ME values indicate the degree of agreement between measured and predicted values; hence, they provide a measure of the prediction ME (Nash and Sutcliffe, 1970). Large values of MAE and RMSE mean high prediction errors, while ME values close to 1 indicate small errors in prediction values.

16.4 CASE STUDIES FOR ESTIMATION OF SOIL SALINITY

16.4.1 Description of the Study Areas and Data Sets

16.4.1.1 Characteristics of the Test Sites

The data sets used in this study were collected from four different areas (Table 16.1). Using data sets from various field locations has the advantage of demonstrating the robustness and wide applicability of the model-based integrated method developed here. The use of data sets from different field areas and multiple scales is considered advantageous because the main objective is to

TABLE 16.1
Description of the Test Sites and Data Sets

Location	Country	Geographic Coordinates	Type of Data Sets	Soil Texture	Name of Subsets	Number of Samples
Texel Island	The Netherlands	52°99′–53°17′N 04°72′–04°48′E	Laboratory experiment	Sandy loam Sand	NaCl MgCl$_2$	48 22
Tedej	Hungary	47°39′–47°89′N 01°00′–21°43′E		Silty clay loam	MgSO$_4$	30
Muangpia	Thailand	16°00′–16°05′N 102°38′–102°43′E	Field measurements	Sand Sandy loam Sandy clay loam	Grids 1 and 6 Grids 2–5 All grids	302 316 628
Toolibin Lake	Western Australia	32°54′–32°55′S 117°36′–117°37′E	HyMap imagery and field measurements	Farifteh (2007)	Subimage 1 Subimage 2 Subimage 3	1060 3469 1237

demonstrate the capabilities of integrated modeling, rather than comparing results or addressing upscaling issues. However, this approach does not support a comprehensive model validation, as could be done with a complete data set derived from a single study area.

The experiment-scale data sets were derived from soil samples collected (a) on the island of Texel in the northwest of the Netherlands, and (b) in the Tedej area in northeast Hungary. The field data set was obtained from the Muangpia test site with an area of approximately 55 km^2, situated near Khon Kaen city in northeastern Thailand. The image data set belongs to the Toolibin lake test area in Western Australia, which is situated approximately 200 km southeast of Perth, close to the town of Narrogin. Detailed information of the test areas is given in Farifteh (2007).

16.4.1.2 Laboratory and Field Procedures for Data Collection and Analysis

The procedures for sample preparation, laboratory analysis, and spectral data acquisition are given in Farifteh (2007). The soil properties that are usually influenced by the accumulation of salts and affect the soil optical characteristics were determined for each soil sample by standard laboratory methods (Lide, 1993; Van Reeuwijk, 1993). A portable ASD FieldSpec FR spectrometer, manufactured by Analytical Spectral Devices, Inc., was employed for laboratory and field reflectance measurements. Soil apparent ECa was measured using the EM-38 geophysical exploration equipment that works on the basis of EM induction principle. The in-situ soil moisture was determined using the Thetaprobe sensor, which measures the moisture volume percentage by applying the frequency-domain technique.

16.4.1.3 Description of the Data Sets

The experimental data set includes measurements such as soil reflectance, soil particle size distribution, electrical conductivity (EC 1:2), pH, moisture content, and soluble salt content. The experimental set-up ensured that soil moisture content remained low, thus minimizing its effect on the measured spectra. The moisture content of the soil samples never exceeded 0.05%, with more than three-fourth of the samples having moisture contents less than 0.02% (Figure 16.7). The detailed description of data and laboratory procedures can be found in Farifteh et al. (2007a).

The field data set from Thailand includes soil reflectance, soil ECa, and soil moisture content, which were measured in situ in six grids. These data were distributed in two sets on the basis of textural similarity in the topsoil (0–20 cm). One set (sandy soils) included data from grids 1 and 6, and the other (sandy loam to loamy soils) included the data from grids 2–5. Details of the procedures used for field measurements and the technical specifications of the instruments are reported in Farifteh (2007).

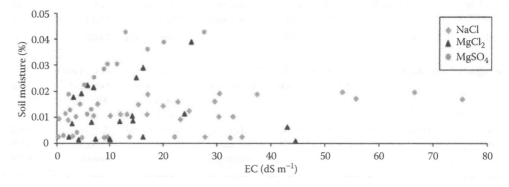

FIGURE 16.7 Scatter-plot of soil moisture content versus EC at the time of spectral measurements.

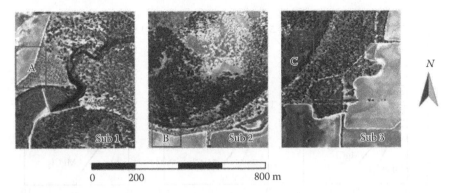

FIGURE 16.8 Hyperspectral image (HyMap) from Toolibin Lake in Western Australia. Windows A, B, and C are representative areas where soil salinity was estimated and mapped.

The image data sets from the Toolibin test area consist of image data acquired by the airborne hyperspectral sensor HyMap of the HyVista Corporation (pixel size 3.5 m) and electrical conductivity measurements. The HyMap hyperspectral scanner provides 128 bands across the 450–2500 nm wavelength region, with contiguous spectral coverage (except in the atmospheric water vapor bands) and bandwidths of 15–20 nm (HyVista, 2004). The hyperspectral image was taken on November 20, 1998, and was georeferenced and corrected for atmospheric effects by the CSIRO Office in Perth. Concurrently with the acquisition of the hyperspectral data, soil ECa was measured at intervals of approximately 3 m using the EM-38 geophysical instrument. Observation points at which ECa was measured vary from 1000 in subimage 1 to 3500 in subimage 3 (see Table 16.1).

On the basis of visual inspection, a few measured field spectra differed considerably from the rest and were thus left out. From the total number of 128 bands provided by the HyMap image, 117 bands were used for the analysis. Eleven bands (namely bands 1, 2, 3, 64, 65, 66, 95, 96, 97, 127, and 128) contained large amounts of noise and were therefore discarded. To reduce the processing load because of software limitation, the HyMap data set was split into subimages in which representative windows (Figure 16.8A through C) were selected to estimate and map soil salinity.

16.4.2 Spectral Characteristics of the Salt-Affected Soils

16.4.2.1 Salt-Induced Absorption Features in Continuum-Removed Spectra (Experiment-Scale Data Set)

Continuum-removed (CR) reflectance spectra of soils treated with different salt minerals (halite, bischofite, and epsomite) and three levels of salinity are displayed in Figure 16.9. Spectral analyses presented in Farifteh et al. (2005, 2007a) revealed that increases in soil salinity induced changes in soil reflectance for wave bands higher than 1300 nm, particularly in the water absorption bands

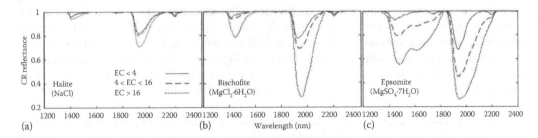

FIGURE 16.9 CR reflectance spectra of soils treated with different salt minerals (EC in dS m^{-1}).

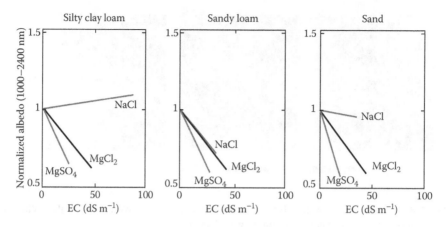

FIGURE 16.10   Normalized albedo (1000–2400 nm) as a function of soil salinity (EC) for three salt types and three soil textures.

around 1400 and 1900 nm. The absorption features beyond 1300 nm were broadened, the positions of maximum reflectance were shifted toward shorter wavelengths, and the overall reflectance changed proportionally as salt concentrations in the soil increased (Figure 16.9). The spectral analysis also revealed that soil samples with large variations in salt did not exhibit all diagnostic absorption features that can be found in the spectra of the dominant salt minerals. The number and clearness of the diagnostic bands were reduced as salt concentration in the samples decreased.

16.4.2.2   Salt-Induced Reflectance and Albedo Changes (Experiment-Scale Data Set)

The normalized albedo of saline soil samples is plotted as a function of soil salinity (EC) in Figure 16.10. The normalized albedo is the albedo of a saline sample ($\Sigma\rho\lambda$) divided by the albedo of the same soil treated with distilled water. The graph illustrates that the overall albedo may increase or decrease with the amount of salt, depending on the salt mineral and soil texture. It shows a strong decrease in soil albedo for samples treated with epsomite ($MgSO_4 \cdot 7H_2O$) and bischofite ($MgCl_2 \cdot 6H_2O$). In both cases, the decrease in soil albedo with increasing soil salinity can be observed across all soil types. The effect of halite (NaCl) on the overall albedo does not reveal a strong and consistent trend in relation to variations in soil EC and soil texture.

16.4.2.3   Soil Reflectance and Band Combinations for Salinity Indices

For a preliminary analysis, 2D correlograms of determination coefficients (R^2) were constructed by sequential regression of reflectance ratios against soil EC (Figure 16.11). The reflectance ratios include all possible pair combinations of narrow bands in the range 400–2450 nm.

   For the experiment-scale data sets, the 2D correlograms show several hot spots indicating relatively broad regions of high correlation between reflectance ratios and soil EC. The hot spots correspond to wavelength combinations with R^2 values higher than 0.7. Apparently, the best performing band combinations are those in the NIR and SWIR regions of the soil spectra. The reflectance spectra of soil samples rich in $MgSO_4$ and $MgCl_2$ provide high R^2 over large portions of the NIR and SWIR regions, indicating potential for estimation of salinity levels (Figure 16.11a). For soil samples rich in NaCl, the number of well-performing band combinations is limited and with low R^2 values.

   For the field-scale data sets, the 2D correlograms show that the reflectance spectra of salt-affected soils with sandy texture (grids 1 and 6) have better correlations with ECa than the soils with sandy loam to loam textures (grids 2–5), and that the potential spectral bands for salinity estimation

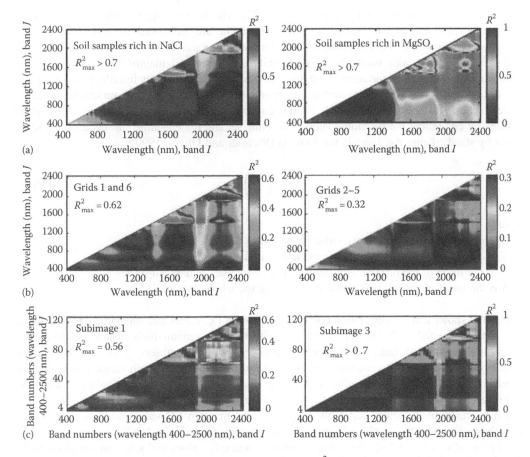

FIGURE 16.11 Examples of 2D correlograms representing the R^2 values between ratio indexes (wavelength I/wavelength J) and EC values for (a) the experiment-scale, (b) the field-scale, and (c) the image-scale data sets.

are in the NIR and SWIR regions of the soil spectra (Figure 16.11b). Because of their texture and location in lowlands where groundwater is close to the surface (1.5–2 m), the saline soils in grids 2–5 have higher moisture contents than the saline soils in grids 1 and 6, and this largely influences the relationships between ECa and reflectance.

For the image-scale data sets, the R^2 values between ECa and reflectance ratios are similar to those of the experiment-scale data sets and mostly higher than those of the field-scale data sets. The hot spot in Figure 16.11c indicates that potentially useful spectral bands for salinity estimation are those in the NIR and SWIR regions.

16.4.3 Spectral Similarity Measures for Salinity Estimation

16.4.3.1 Spectral Similarity by Deterministic and Stochastic Techniques

The reflectance spectra of the experiment-scale data sets, including soil samples with different EC values, were statistically analyzed to quantify the spectral differences and establish a relationship between soil salinity and calculated similarity values. Spectral similarity measures (Section 16.3.1) were applied to the NIR and SWIR regions (1350–2350 nm) of the EM spectrum (Farifteh et al., 2007b).

The correlation coefficients between EC values and calculated similarity values vary with soil texture, because the size and arrangement of the soil particles in relation to soil air and water influence the reflectance (Baumgardner et al., 1985). The results suggest that (a) variations in reflectance spectra of the soil samples treated with halite are influenced not only by variations in salt concentration but also by variations in texture, and (b) variations in reflectance spectra of the soil samples treated with bischofite and epsomite are influenced mainly by variations in salt concentration and less by variations in texture. In other words, variations in reflectance spectra of the soil samples treated with bischofite and epsomite are large enough to minimize the effect of texture on soil spectra. This weak relationship between texture and reflectance was further verified by combining the bischofite subsets into one set and the epsomite subsets into one set of SSVs by standardizing the data via the formula (Webster and Oliver, 1990):

$$S'_s = \left(\frac{S_{sj}}{\overline{S}_s}\right)\overline{S} \qquad (16.16)$$

where

S'_s is the standardized similarity value for sample s
S_{sj} is the original similarity value for sample s in data subset j
$\overline{S}_j$ is the mean of the original similarity values of samples in data subset j
$\overline{S}$ is the mean of the original similarity values in the whole data set

Table 16.2 shows that the correlation coefficients between EC values and standardized SSVs for all soil samples treated with bischofite and epsomite together remain high instead of decreasing as could be expected from combining the data of the three subsets. This corroborates the above conclusion that salt concentration, and not soil texture, is the main factor controlling the reflectance variations in this experiment. Similarly, the conclusion is supported by the scatter-plots in Figure 16.12 and the values in Table 16.2 in which all soil samples treated with bischofite and epsomite, respectively, show better linear relationships between EC values and similarity values than those shown by the combined subsets of soil samples treated with NaCl.

The results in Table 16.2 show that calculated similarity values and salt concentrations in the soil samples treated with either bischofite or epsomite are negatively correlated. The strong correlations between EC values and the values of d_e, SAM, and SSV indicate that the absolute reflectance, the shape of spectra, and the combination of absolute reflectance and shape of spectra are heavily influenced by the concentrations of either $MgCl_2$ or $MgSO_4$ in the soil samples. The results further suggest that the different spectral similarity measures are useful for the spectral classification of variable degrees of salinity in soils rich in $MgCl_2$ or $MgSO_4$. In the case of the soils

TABLE 16.2

Correlation Coefficients between EC Values and Calculated SSVs from Soil Samples under Laboratory Conditions (Experiment-Scale)

Dominant Salt in Soil Sample	Nr.	Soil Texture	Deterministic Methods				Stochastic Methods		
			d_e	SAM	SSV	SI	SID	SID (TAN)	SID (SIN)
NaCl	17	L	0.85	0.73	0.85	−0.85	0.64	0.57	0.57
	10	SL	0.86	0.74	0.86	−0.86	0.79	0.79	0.79
	10	S	0.29	0.75	0.29	−0.29	0.73	0.72	0.72
NaCl	37	L, SL, S	0.62	0.61	0.62	−0.60	0.56	0.50	0.50
$MgCl_2$	22	L, SL, S	0.92	0.92	0.91	−0.89	0.90	0.90	0.90
$MgSO_4$	19	L, SL, S	0.74	0.90	0.74	−0.72	0.93	0.92	0.92

L, silty clay loam; SL, sandy loam; S, sandy textures; SI, similarity index; Nr., number of samples.

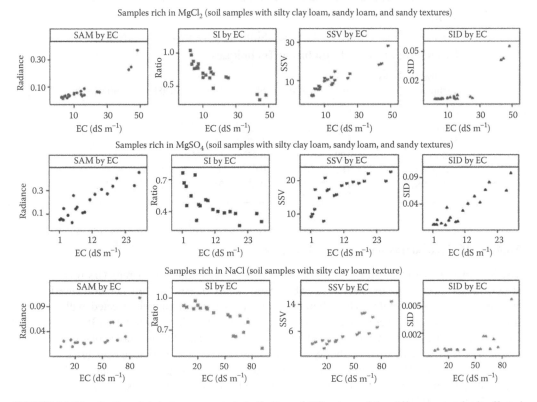

FIGURE 16.12 Scatter plots between spectral similarity and EC values of the different sets of salt-affected soil samples. The correlation coefficients for each plot are given in Table 16.2.

affected by NaCl, the spectral similarity measures help identify the properties of the reflectance spectra that could be useful for spectral discrimination of soil salinity degrees. For example, there is moderate to strong correlation between EC values and similarity values for soil samples with silty clay loam and sandy loam textures (Table 16.2). Similarly, there is significant correlation between EC and SAM values and stochastic similarity measures for soil samples with sandy texture (Table 16.2). This suggests that for NaCl-affected soils the magnitude and the shape of the reflectance spectra can be instrumental for distinguishing soil salinity levels.

16.4.3.2 Correlation between Similarity Measures

Table 16.3 shows the correlation coefficients between similarity values obtained by the deterministic and the stochastic techniques. There is high correlation among similarity values obtained using the three stochastic techniques (SID, SID(SIN), and SID(TAN)). The correlation between d_e values and SSV values is very high, whereas the correlation between SAM values and SSV values is relatively lower (Table 16.3). Considering that d_e is based on the absolute reflectance, SAM is based on the spectral angle, and SSV is a combination of d_e and SAM, then the lower correlation between SAM and SSV values, as compared to the correlation between d_e and SSV values, indicates that the SSV values are influenced more by the absolute reflectance than by the spectral angle. The results also show that the SAM values are highly correlated with the similarity values obtained by the three stochastic techniques, whereas d_e and SSV values are poorly correlated with these same similarity values (Table 16.3). This means that the similarity values are more influenced by the spectral angle than by the magnitude of the reflected energy.

TABLE 16.3
Correlation Coefficients between Similarity Values Generated
by Deterministic and Stochastic Techniques

Spectral Similarity Techniques	d_e	SAM	SSV	SID	SID(SIN)	SID(TAN)
d_e	1					
SAM	0.89	1				
SSV	1	0.89	1			
SID	0.77	0.97	0.77	1		
SID (SIN)	0.62	0.85	0.62	0.91	1	
SID (TAN)	0.62	0.85	0.62	0.91	1	1

16.4.4 IG PARAMETERS FOR SALINITY ESTIMATION

The IG function is suitable for salinity estimation, since positions and parameters (depth, width, area, and asymmetry) of the diagnostic absorption bands in soil spectra are influenced by the amount of salt concentration. The laboratory spectra of the soil samples treated with epsomite (EC: 1.6–27.7 dS m^{-1}) were used to demonstrate this relationship (Figure 16.13). It shows that the absorption features beyond 1300 nm become broader and the positions of the maximum reflectance tend to shift toward shorter wavelengths as the overall reflectance diminishes with increasing salinity. The spectral characteristics of the soil samples in Figure 16.13 suggest potential application of the IG function to model soil salinity on the basis of spectral variations (Farifteh et al., 2007c).

For the samples rich in MgSO$_4$, the amplitude R_d, the area within the Gaussian curve A, and the distance to the inflection point σ of the function appear to be good indicators of salinity (Figure 16.14 and Table 16.4). The regression fit of the amplitude R_d and the area within the Gaussian curve to the salt content have high correlation coefficients ($R^2 = 0.81$ and 0.82, respectively) (Figure 16.14 and Table 16.4). In contrast, the fit parameters calculated for samples rich in NaCl show no relationship with salt content and provide unsatisfactory estimation of soil salinity.

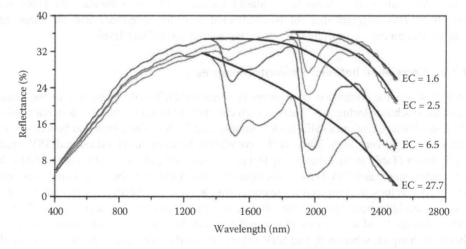

FIGURE 16.13 Laboratory spectra of salt-affected soil samples treated with epsomite (MgSO$_4 \cdot$7H$_2$O). The bold lines illustrate the general pattern in the SWIR region (EC values in dS m^{-1}).

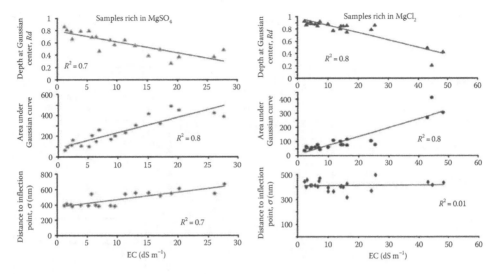

FIGURE 16.14 Linear relationships between soil salinity levels and IG model parameters calculated from soil spectra obtained from laboratory samples treated with epsomite (left) and bischofite (right).

Overall, the laboratory results suggest that the application of the IG function to salinity assessment depends on the salt types.

Validation of the predictive models of salt contents in soils based on IG parameters is summarized in Table 16.4. The linear spectral models for estimating salinity levels in terms of EC gave RMSE between 0.8 and 2.9 dS m^{-1} and R^2 values between 0.68 and 0.82. Among the three salt types, only for $MgCl_2$ and $MgSO_4$ could the salinity levels be estimated with cross-validated RMSE (RMSE$_{CV}$) lower than 1.2 dS m^{-1}. In most cases, predictive models using area under fitted curve gave better results than models using distance to the inflection point σ or depth at Gaussian center R_d.

16.4.5 ADAPTIVE TECHNIQUES FOR SALINITY ESTIMATION AND MAPPING

16.4.5.1 PLSR Method

Application of the PLSR algorithm to the various data sets resulted in calibration models. Examples of PLSR calibration model parameters are illustrated in Figure 16.15. The optimal number of latent

---

TABLE 16.4

Coefficients of Determination (R^2) and RMSE between Measured and Estimated Soil Salinity Values (EC 1:2)

Dominant Salt in Soil Sample	Soil Salinity Gaussian Model Accuracy											
	R^2			Slope (β)			Intercept (ε)			RMSE		
IGMP	R_d	σ	A	R_d	σ	A	R_d	σ	A	R_d	σ	A
NaCl	0.1	0.11	0.1	-0.001	-0.7	0.8	0.9	464	44	3.4	3.4	3.4
$MgCl_2$	0.8	0.01	0.8	-0.01	0.14	6.3	1.0	411	8.3	1.2	2.9	1.2
$MgSO_4$	0.7	0.7	0.8	-0.02	9.6	15	0.8	371	82	1.0	1.0	0.8

Notes: Numbers of samples used are 37, 22, and 19 for samples rich in NaCl, $MgCl_2$, and $MgSO_4$, respectively.
IGMP, IG model parameters; σ, distance to inflection point; R_d, depth; A, area.

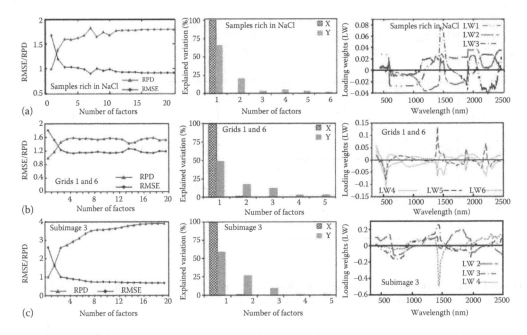

FIGURE 16.15 Parameters of the PLSR calibration model. Letters a, b, and c represent three examples selected from the experiment-, field-, and image-scale data sets, respectively.

variables (or orthogonal factors) that can be used for salinity estimation can be defined through the analysis of the $RMSE_{CV}$ and RPD values of the calibration models. The predictive ability of individual X and Y variables as a function of the number of PLSR factors can be assessed by the amount of variation in both X and Y data explained by the latent variables.

For all data sets, more than 95% of the variations are explained by the first five latent variables (Table 16.5). The contribution of each wavelength band can be visualized by the analysis of the

TABLE 16.5
Performance Statistics of the PLSR Calibration Models

Data Scales	Data Sets	Nr.	RMSE	RPD	Variations Explained by Model Variables (%)				
					X_1	X_2	Y_1	Y_2	Total Y
Experiment-scale	NaCl	6	15	2.0	99.8	0.14	65.2	19.8	97.6
	$MgCl_2$	12	18	1.0	99.4	0.64	58.0	31.7	100
	$MgSO_4$	6	4	2.0	97.7	2.25	72.1	18.5	99.8
Field-scale	Grids 1 and 6	11	114	1.6	99.2	0.7	48.1	17.4	86.4
	Grids 2–5	19	59	1.5	99.8	0.1	85.1	3.8	97.8
	All grids	12	87	1.6	99.4	0.4	64.7	7.9	90.4
Image-scale	Subimage 1	20	97	2.4	99.9	0.02	61	11	96
	Subimage 2	30	101	2.7	99.9	0.05	64	14	95.6
	Subimage 3	19	67	3.9	99.9	0.06	58	26	98

Nr., number of latent variables used by calibration model; X_1, X_2, explained X variations (%) by latent variables 1 and 2, respectively; Y_1, Y_2, explained Y variations (%) by latent variables 1 and 2, respectively; Total Y, total explained Y variations by latent variables used for estimation. RPD is a ratio showing the accuracy of the PLSR calibration model (not reliable $1.5 < RPD > 2.5$ reliable).

computed loading weights (LWs). The computed LWs support the results of the 2D correlogram analysis stating that spectral bands useful for salinity estimation are those in the NIR and SWIR regions. Strong covariation between soil ECa and soil reflectance in the NIR and SWIR regions can also be visualized in the form of positive and negative peaks of the regression coefficients B (Figure 16.5). The statistics of the estimated Y variable given in Table 16.5 show how well soil salinity can be estimated from reflectance data with the PLSR method.

For the experiment-scale data, the RMSE and RPD values together with the total variation in Y explained by the model variables indicate that the calibration models for all soil samples (treated with NaCl, $MgCl_2$, and $MgSO_4$) performed very well (Table 16.5). For the image-scale data sets, the RPD and RMSE values and the explained variation in Y derived from all the calibration models suggest reliable estimation ability (Table 16.5). In contrast, the performance of the calibration models concerning the field-scale data sets resulted to be relatively poor.

The PLSR calibration models developed for each data set at each scale were then used to estimate soil salinity. The accuracy of the estimated salinity values and the level of performance of the predictive models were assessed on the basis of the performance statistics derived from measured and estimated salinity values and on the basis of the calculated model performance indices (Table 16.6).

For the experiment-scale data, accurate EC estimations can be made based on PLSR models developed from samples rich in $MgSO_4$ and $MgCl_2$ (Figure 16.16a and Table 16.6). For the samples rich in NaCl, the EC estimations can be considered to be good, because the R^2 values are higher than 0.81 and the RPD of the estimations varies between 2.1 and 6.3. This is also indicated by the low RMSE and MAE values and the high ME values (Table 16.6). In the case of the soil samples treated with $MgSO_4$ and $MgCl_2$, the high accuracy of salinity estimation can be attributed to (a) salt-induced absorption bands at around 1488, 1630, and 1946 nm for samples treated with $MgSO_4$, and 1451, 1556, and 1952 nm for samples treated with $MgCl_2$; (b) enlargement of the width and area of absorption bands at around 1400 and 1900 nm due to the increase in salt concentration in the soil; (c) strong decrease in soil albedo with increasing soil salinity; and (d) shift of the maximum reflectance shoulder toward shorter wavelengths as the salt concentration increases (Farifteh et al., 2007a–c). For soil samples salinized with NaCl, the estimated ECa values are on average higher than the measured ones. The high RMSE and MAE values, together with low ME values, confirm the large differences between measured and estimated ECa values (Table 16.6).

In general, the field-scale data sets show poor relationships between ECa values and soil reflectance (Table 16.6 and Figure 16.16). High RMSE and MAE values and low ME values indicate large variations between measured and estimated ECa values. The field data from grids

TABLE 16.6
Performance Statistics of the PLSR Predictive Models

Data Scales	Data Sets	Nr.	R^2	Slope	Intercept	RMSE	MAE	ME	RPD
Experiment-scale	NaCl	6	0.81	0.712	16.5	12.8	10.4	0.75	1.7
	$MgCl_2$	12	0.92	1.01	− 1.21	4.4	3.3	0.90	3.5
	$MgSO_4$	6	0.98	1.01	0.142	1.3	1.0	0.97	6.2
Field-scale	Grids 1 and 6	10	0.8	0.883	25.1	70	50	0.78	2.1
	Grids 2–5	23	0.6	0.784	32.7	60	47	0.50	1.4
	All grids	17	0.6	0.83	36.1	88	68	0.56	1.5
Image-scale	Subimage 1	22	0.86	0.897	22.8	87.4	66.6	0.86	2.6
	Subimage 2	22	0.87	0.875	31.2	101	76.6	0.86	2.5
	Subimage 3	16	0.94	0.927	36.9	63.2	40.0	0.94	4.2

Nr., optimal number of latent variables used by predictive model.

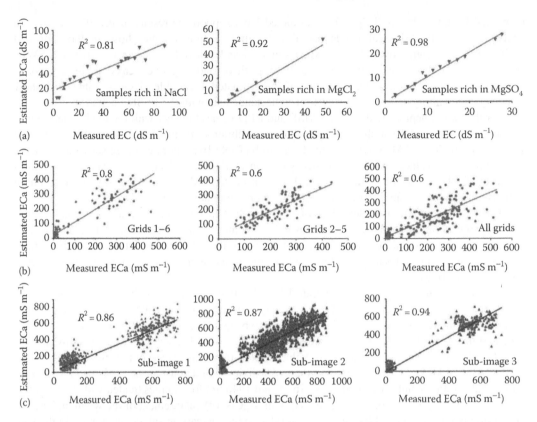

FIGURE 16.16 Scatter-plots of measured versus estimated ECa values derived from the PLSR analysis of soil reflectance.

1 and 6 provide slightly better estimation results ($R^2 = 0.8$ and RPD $= 2.2$) than the data from grids 2–5 ($R^2 = 0.6$ and RPD $= 1.3$). Poor soil salinity estimation from field reflectance spectra is most likely due to high soil moisture, as the field-scale data sets were collected in situ from an area with subtropical climate, while the experimental-scale data sets were obtained from laboratory-dried soil samples and the image-scale data sets pertain to an area with semiarid climate.

For the image-scale data sets, the R^2, RPD, and ME values of the predictive models of the subimages 1, 2, and 3 are satisfactory (Table 16.6). Even though the PLSR analysis reveals high correlation between soil spectra and ECa values, spectral variations at the image level can also be due to variations in vegetation density, land cover type, soil moisture, soil type, and reflectance of other scene components within a pixel area. It must be noted that the measured ECa values represent the salinity of the soil profile to a depth of 1.5 m, whereas the image reflectance data refer only to the soil surface. Despite this important difference, the covariation between the measurements is remarkably high (Table 16.6 and Figure 16.16).

Using the HyMap hyperspectral imagery, the PLSR calibration models developed for the image-scale data sets were employed to map variations in soil salinity in three small representative windows (70×100 pixels) of subimages 1, 2, and 3 (Figure 16.8). For each window, a salinity map was created from estimated ECa values recoded into seven salinity classes (Figure 16.17: maps a–c). As part of a postclassification processing, prior to map presentation, each map was smoothed with a median filter (window size 7×7) to suppress high frequency variations (i.e., noise) and portray only the general soil salinity pattern. Descriptive statistics derived from the salinity maps show that the estimated ECa values in the selected areas have a range similar to that of the measured ECa values (Farifteh et al., 2007d).

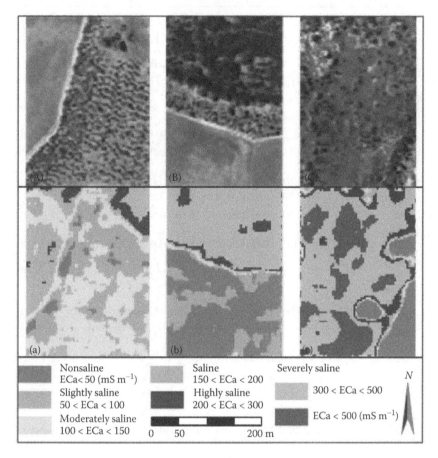

FIGURE 16.17   Salinity maps (a–c) resulting from the PLSR analysis applied to windows A, B, and C (70 × 100 pixels) of the hyperspectral subimages. Location of maps a–c is indicated in Figure 16.8.

16.4.5.2   ANN Method

For the experiment-scale data sets, three ANN models were developed, one for each salt mineral. In general, R^2 was high and RMSE low for the training models (Table 16.7). This is probably because the weights and bias were optimized during training. RMSE values at various phases of learning or training, validation data, and test data sets are illustrated in Figure 16.18. Overall, ANN models for soil samples salinized with $MgCl_2$ and $MgSO_4$ have higher R^2 values and lower RMSE and MAE values than those of the ANN models for soil samples salinized with NaCl. The R^2 and RMSE values of the training models and the poorer R^2 and RMSE values of the validation and test models indicate that the ANN models learned quite well but have poor generalization ability, probably because of the small size of the data sets.

For the field-scale data sets, the R^2 values of the validation and test data sets (i.e., grids 1 and 6) are mostly lower than those of the training data set (Table 16.7). This suggests that the ANN models of the field-scale data did not learn quite well and thus have poor generalization ability (Figure 16.18). In general, soil salinity estimation accuracy using the field-scale data sets is relatively poor, presumably related to the field and climate conditions and accuracy of the field equipments.

For the image-scale data sets, the ANN models work satisfactorily. The high R^2 values obtained from most of the subimage data sets indicate that the ANN models learned quite well and mostly have good generalization ability (Figure 16.18). The ECa estimates in subimages 1, 3, and 6 are very good ($0.86 < R^2 < 0.96$ and $2.5 < RPD < 4.9$). This is also supported by the RMSE, MAE, and ME

TABLE 16.7

ANN Model Parameters for Soil Salinity (ECa) Estimation

Data Scales	Data Sets	RMSE Tra.	Val.	Test	R^2 Tra.	Val.	Test	R^2	ME	RMSE	MAE	RPD	Nr. ite.	Nr. neu.
Experiment-scale	NaCl	3.45	8.5	9.4	0.98	0.86	0.86	0.87	0.84	9.4	7.7	2.5	14	9
	MgCl$_2$	1.8	3.0	5.4	0.98	0.98	0.88	0.89	0.79	5.3	4.3	2.3	13	5
	MgSO$_4$	0.55	2.3	2.8	0.99	0.87	0.91	0.97	0.96	1.4	0.9	5.0	20	5
Field-scale	Grids 1 and 6	105	82	89	0.69	0.72	0.60	0.65	0.56	91	65	1.5	11	9
	Grids 2–5	66	61	62	0.42	0.41	0.42	0.46	0.17	59	46	1.1	16	9
	All grids	89	78	78	0.67	0.54	0.50	0.59	0.39	79	61	1.3	12	9
Image-scale	Sub image1	86	87	90	0.87	0.86	0.85	0.86	0.84	90.3	59.7	2.5	25	7
	Sub image2	100	101	102	0.86	0.86	0.86	0.86	0.89	89.3	63.4	2.5	89	13
	Sub image3	45	52	51	0.97	0.96	0.96	0.96	0.96	61.4	35.7	4.9	58	11

Tra., training; Val., validation; Nr. ite., number of iterations; Nr. neu., number of neurons.

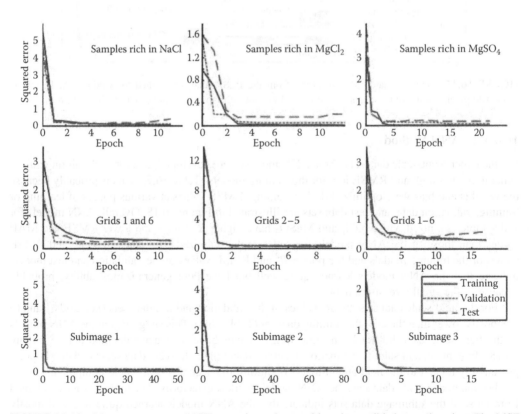

FIGURE 16.18 Mean square error (MSE) at various phases of learning, validation, and testing. The MSE values are calculated from experimental, field, and image data sets.

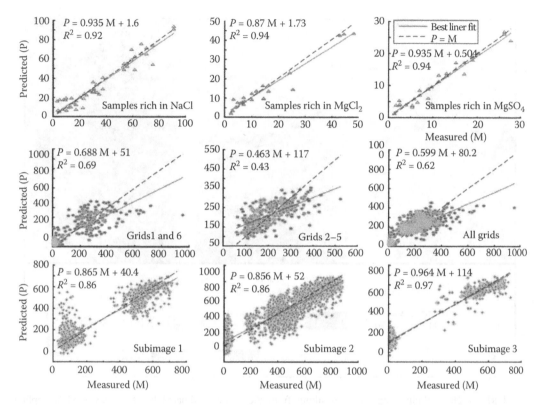

FIGURE 16.19 Scatter-plots of measured versus estimated ECa values derived from the ANN analysis of soil reflectance.

values. The correlations between measured and estimated ECa values at the various phases of the ANN modeling (training, validation, and testing) are illustrated in Figure 16.19. The results in Figure 16.19 and Table 16.7 indicate that hyperspectral imagery provides suitable data for soil salinity mapping over large areas, if an accurate salinity model from ground data in a representative training site is available.

The image reflectance data of the windows (70×100 pixels) selected from the hyperspectral image (see Figure 16.8) were used in the ANN models, previously trained with the image-scale data sets, to map the spatial distribution of soil salinity. For each window, a salinity map was created with a maximum of seven salinity classes (Figure 16.20: maps a–c). A median filter (window size 7×7) was applied to the maps of estimated ECa values as a postclassification technique to reduce high frequency variations and smooth the boundaries between soil salinity classes.

The descriptive statistics derived from the salinity maps show that the estimated ECa values in the selected areas are quite lower than the observed ECa values. For example, the maximum observed ECa value in subimage 2 is 880 mS m^{-1}, while the maximum estimated ECa value is 615 mS m^{-1} (Farifteh et al., 2007d). The tendency of the ANN models to underestimate ECa is not clear but can be probably related to data distribution, spectral characteristics of the imagery (e.g., mixed pixels, especially because of variations in the vegetation cover), or ANN generalization.

16.4.5.3 Comparison of the PLSR and ANN Modeling Approaches

To compare the robustness of PLSR-estimated ECa values vis-à-vis that of the ANN-estimated ECa values, with respect to the corresponding measured ECa values, the statistical parameters used earlier (Sections 16.4.5.1 and 16.4.5.2) to assess the performance of individual models cannot be

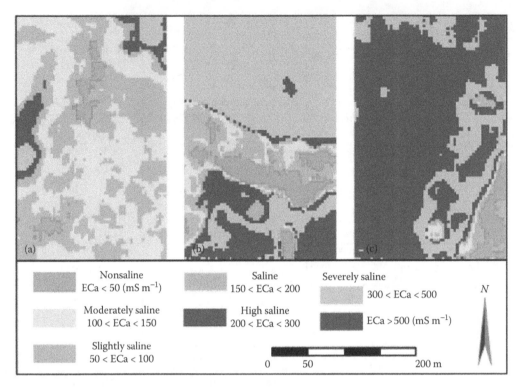

FIGURE 16.20 Salinity maps (a–c) resulting from the values estimated by the ANN models applied to windows A, B, and C (70 × 100 pixels) of the hyperspectral subimages.

used appropriately because of the difference in data treatment in each of the two modeling techniques. In order to have an uniform basis on which to judge whether soil salinity estimations based on reflectance data are robust when using PLSR or when using ANN, the aforementioned three sets of ECa values were assumed to be replicate measurements obtained by different methods, and an one-way analysis of variance (ANOVA) was performed on them. The one-way ANOVA was applied to two data sets: one set including the measured and PLSR-estimated ECa values and another set including the measured and ANN-estimated ECa values. The null hypothesis in either set of the one-way ANOVA was that a set of measured ECa values and a corresponding set of estimated ECa values had, with respect to variance, identical means. The null hypothesis was tested by calculating an F-value and comparing this with a critical F-value at a certain significance level. A calculated F-value not only allows further assessment of individual PLSR or ANN salinity–reflectance models but also provides a basis, in terms of magnitude of calculated F-values and their corresponding p-values, to judge if a PLSR model is better than its corresponding ANN model or vice versa.

For the experimental-scale data sets, the F-values calculated between measured and PLSR-estimated ECa values and between measured and ANN-estimated ECa values are all statistically smaller than the corresponding critical F-values (Table 16.8). These results suggest that the sets of estimated ECa values, either by PLSR or ANN, are statistically similar to their corresponding sets of measured ECa values. This supports an earlier observation that soil salinity estimation based on reflectance data would be dependent on salt types. However, as the size of the experimental-scale data sets is small, the above suggestion as to the performance of the individual models might not be really conclusive. In addition, the p-values of the calculated F-values do not show a clear pattern that could suggest that PLSR-estimated ECa values are better than ANN-estimated ECa values or vice versa.

TABLE 16.8

Results of One-Way ANOVA between Measured and PLSR-Estimated ECa Values and Between Measured and ANN-Estimated ECa Values

Data Scales	Data Sets	PLSR		ANN		F Critical	df
		F-value	p-value	F-value	p-value		
Experiment-scale	NaCl	0.5755	0.4525	0.0019	0.9658	4.08	1, 41
	MgCl$_2$	0.0173	0.8972	0.0612	0.8082	4.60	1, 14
	MgSO$_4$	0.0099	0.9216	0.0015	0.9699	4.20	1, 29
Field-scale	Grids 1 and 6	0.0909	0.7633	0.0083	0.9274	3.89	1, 198
	Grids 2–5	1.3460	0.2473	0.0093	0.9234	3.89	1, 214
	All grids	2.6568	0.1039	0.2830	0.5950	3.86	1, 414
Image-scale	Subimage 1	0.6543	0.4188	0.0166	0.8976	3.85	1, 1058
	Subimage 2	0.1498	0.6988	0.0043	0.9478	3.84	1, 3492
	Subimage 3	0.7960	0.3725	0.0027	0.9582	3.85	1, 1234

Note: p-value significant at $p < 0.05$.

For the field-scale data sets, the calculated F-values between measured and PLSR-estimated ECa values and between measured and ANN-estimated ECa values (Table 16.8) suggest that the sets of estimated ECa values, either by PLSR or ANN, are statistically similar to their corresponding sets of measured ECa values. The F-values are lower and the corresponding p-values are higher, for either PLSR or ANN models, in grids 1 and 6 (lower soil moisture due to sandy texture) than the values of the same parameters in grids 2–5 (higher soil moisture due to sandy loam to loam textures). This suggests that high soil moisture content renders soil salinity estimation from reflectance data ineffective. The p-values of the calculated F-values between measured and PLSR-estimated ECa values are consistently much smaller than the p-values of the calculated F-values between measured and ANN-estimated ECa values. These results indicate that, based on the field-scale data sets, modeling the relationship between soil salinity and soil reflectance via ANN would result in slightly better soil salinity estimation than via PLSR. In the specific case of areas with tropical climate, soil salinity estimation based on reflectance data would be better via ANN modeling than via PLSR modeling.

For the image-scale data sets, the results of the one-way ANOVA (Table 16.8) indicate that the sets of estimated ECa values, either by PLSR or ANN, are statistically similar to their corresponding sets of measured ECa values. Thus, modeling the salinity–reflectance relationship in soils and soil salinity mapping based on reflectance would be presumably more robust through ANN modeling than through PLSR modeling. The salinity maps estimated from the image-scale data sets (Figures 16.17 and 16.20) show good agreement between PLSR and ANN estimated values in the salinity maps a–c (Table 16.8). In the histograms of these subimages, the estimated values of the ANN models compare to those of the PLSR models, presenting smaller range and distributions and being closer to the average estimated ECa values (Figure 16.21). The summary statistics (Farifteh et al., 2007d) also show that the minimum and maximum values estimated by PLSR models are often higher than the ones estimated by the ANN models. The differences between the salinity maps produced by PLSR and ANN are illustrated in Figure 16.22. The maps have been prepared by subtracting the PLSR estimation maps (Figure 16.17) from the ANN estimation maps (Figure 16.20). The resulting difference maps were then normalized to obtain an index between −1 and 1. Positive values show areas where PLSR estimations are better, while negative values show areas where ANN estimation values are better. Values close to 0 represent areas of similar estimations, and pixels with values close to 1 or −1 represent maximum differences between the two methods.

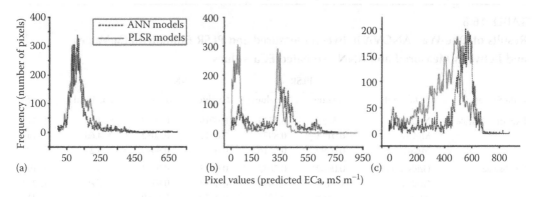

FIGURE 16.21 Histograms calculated from the salinity maps a–c (see Figures 16.17 and 16.20) generated using the PLSR and ANN estimated values.

In summary, for the calibration of salinity–reflectance relationships and for the estimation of soil salinity using reflectance data, the performance statistics calculated between measured and PLSR-estimated EC values and between measured and ANN-estimated EC values show that, for the experiment- and field-scale data sets, both methods perform similarly, whereas for the image-scale data sets the ANN modeling slightly outperforms the PLSR modeling. However, the performance differences between these two methods are not significant and, therefore, soil salinity based on soil reflectance can be estimated and mapped by either PLSR or ANN.

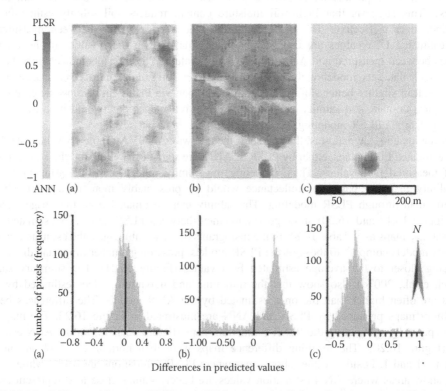

FIGURE 16.22 Differences between PLSR and ANN estimated salinity values (maps a–c). The calculated histograms show the data distribution in each map.

16.5 DISCUSSION AND CONCLUSIONS

Soil salinity has a large impact in arid and semiarid regions where arable land is needed to supply food and fibers to rapidly increasing populations, while human activities and natural conditions contribute to the expansion of saline areas. This conflictive situation is a major concern at local and global scales where enormous efforts are geared toward preventing or slowing down the land salinization threat. As a part of these efforts, scientific studies are carried out to help combat salinization by providing comprehensive understanding of the process, developing methods and algorithms, and implementing a variety of application tools. Optical remote sensing, geophysics, and solute modeling are the three most used approaches for detection, estimation, and mapping of salt-affected soils. Each approach has not only advantages but also constraints that make their use limited for the identification of salinization problems, particularly early-stage detection of saline areas. More accurate estimation and mapping of soil salinity are feasible when using predictive modeling tools with integration capability that combines surface and near-surface measurements.

In this chapter, the following issues and features were addressed: (a) a new conceptual model for salinity assessment; (b) spectral behavior of salt-affected soils under laboratory conditions; (c) quantifying similarities between spectra of salt-affected soils; (d) relationship between soil properties and reflectance spectra; (e) field application of experimental results; (f) upscaling by using airborne imagery combined with field data; and (g) estimating salinity and mapping saline areas using modeling tools and evaluation of the results.

16.5.1 Spectroscopy of Salt-Affected Soils

From the spectral analysis of the experiment-scale data sets the following conclusions were reached:

- Study of salt-affected soils, using hyperspectral data, should focus on the spectral variations in the NIR and SWIR regions of the soil reflectance spectra rather than on individual diagnostic absorption features.
- Quantification of salt abundance in soils, in the upscaling phase, should be based on the general shape of the reflectance spectra rather than on absorption band parameters (depth, area, etc.), since most of the spectral features that diagnose the presence of salt minerals or salt-affected soils are masked in an image due to atmospheric effect.

16.5.2 Modeling Algorithms: Monovariate Techniques

Comprehensive analysis of the visible, NIR, and SWIR regions of the reflectance spectra of salt-affected soil samples using spectral matching techniques indicates considerable differences in reflectance magnitude, absorption strength, and spectral angle. The results show significant correlations between soil EC and spectral similarity measures, indicating that the similarity between sample spectra decreases as salt concentration in the soil increases. Spectral similarity measures can be used as diagnostic indicators to differentiate among salt-affected soil spectra obtained from laboratory experiments.

The fitting of an IG function to the SWIR region of the soil spectra showed that the parameters of the fitted curve such as functional depth, distance to the inflection point and area, can be used as predictors in regression analysis to estimate soil salinity depending on the dominant salt types in soil. This means that the application of the IG modeling approach is limited in salinity studies. The approach was further used to test field data sets where the in-situ measured soil reflectance spectra were largely influenced by both salt and moisture (Farifteh, 2007). The poor correlation between soil ECa and IG model parameters for the field spectral data confirmed the influence of moisture on soil reflectance spectra and its contribution to causing incongruity in model calculations. The results suggest that the estimation of soil moisture using the same approach can be undermined by the accumulation of salts in the soil.

16.5.3 Modeling Algorithms and Integrated Approach: Multivariate Techniques

Adaptive methods such as PLSR and ANN are very effective for establishing relationships between soil reflectance and soil EC. The performance indexes calculated from the values estimated by both methods suggest large similarity between the two approaches, even though the comparison of the individual values estimated by each model showed much less agreement between estimated values, especially for the image data set.

The significant correlation between soil EC and soil reflectance obtained at experiment scale for both methods provided the basis for upscaling toward the airborne hyperspectral data. In the upscaling process, the laboratory experiment data were replaced by pixel reflectance data derived from hyperspectral imagery and in-situ ECa values measured by geophysical instruments. The accurate estimation of soil salinity by both methods (PLSR and ANN), at upscaled level, supports the possibility of integrating subsurface geophysical EM measurements and surface reflectance derived from airborne hyperspectral images.

16.5.4 Future Research on Remote Sensing and Soil Salinity

This chapter illustrates ways for detecting, estimating, and monitoring soil salinity, especially at the early stage. However, the application of the model-based integrated method can be extended if further research is carried out on the following topics:

- The possibility to integrate remote sensing and solute modeling, or remote sensing, geophysical survey, and solute modeling, needs to be further explored. More accurate estimation of salinity and wider applications can be expected from such an integrated approach (Farifteh, 2007).
- An area that needs additional attention concerns soil moisture and its influence on reflectance spectra of salt-affected soils. Further studies are required to determine the calibrating coefficients that can be used to eliminate the background spectra caused by soil moisture. In this regard, the potential of ground-penetrating radar and high-resolution thermal infrared data needs to be explored.
- Lastly, the extension of soil salinity research needs an optimal integration approach. Even though there are several techniques such as overlaying, conditional probability, and evidential reasoning that have already been employed for data integration, further research is recommended to explore the possibilities that offer an optimal integration of surface and near-surface measurements (Farifteh, 2007).

REFERENCES

Baumgardner, M.F., Silva, L.F., Biehl, L.L., and Stoner, E.R. 1985. Reflectance properties of soils. *Advances in Agronomy* 38: 1–44.

Ben-Dor, E. 2002. Quantitative remote sensing of soil properties. *Advances in Agronomy* 75: 173–243.

Ben-Dor, E., Irons, J.R., and Epema, G.F. 1999. Soil reflectance. In: *Remote Sensing for the Earth Sciences: Manual of Remote Sensing*, ed. A.N. Rencz. Wiley and Sons, New York, pp. 111–188.

Bishop, J.L., Pieters, C.M., and Edwards, J.O. 1994. Infrared spectroscopic analyses on the nature of water in montmorillonite. *Clays and Clay Minerals* 42(6): 702–716.

Butters, G.L. and Jury, W.A. 1989. Field scale transport of bromide in an unsaturated soil, 2. Dispersion modelling. *Water Resources Research* 25: 1583–1589.

Cacuci, D.G. 2003. *Sensitivity and Uncertainty Analysis, Volume 1: Theory.* Chapman & Hall/CRC, New York.

Chang, C.I. 2003. *Hyperspectral Imaging: Techniques for Spectral Detection and Classification.* Springer, New York.

Chang, Y.M., Chang, L.C., and Chang, F.J. 2004. Comparison of static feed-forward and dynamic-feedback neural networks for rainfall runoff modeling. *Journal of Hydrology* 290: 297–311.

Clark, R.N. and Roush, T.L. 1984. Reflectance spectroscopy: Quantitative analysis techniques for remote sensing applications. *Journal of Geophysical Research* 89(B7): 6329–6340.

Crescimanno, G. and Garofalo, P. 2005. Application and evaluation of the SWAP model for simulating water and solute transport in a cracking clay soil. *Soil Science Society of America Journal* 69: 1943–1954.

Dehaan, R.L. and Taylor, G.R. 2002. Field-derived spectra of salinized soils and vegetation as indicators of irrigation-induced soil salinization. *Remote Sensing of Environment* 80: 406–417.

De Jong, S.M. 1994. Applications of reflective remote sensing for land degradation studies in a mediterranean environment. Nederlandse Geografische Studies 177. Utrecht University, Utrecht, the Netherlands.

Demuth, H. and Beale, M. 2004. *Neural Network Toolbox for Use with MatLab. User's Guide Version 4.* The Math Works, Natick, MA.

Doerge, T., Kitchen, N.R., and Lund, E.D. 2002. Soil electrical conductivity mapping. Site-specific management guidelines, SSMG-30. Potash and Phosphate Institute (PPI), South Dakota State University (SDSU), Brookings, SD.

Du, Y., Chang, C.-I., Ren, H., Chein-Chi, C., Jensen, J.O., and D'Amico, F.M. 2004. New hyperspectral discrimination measure for spectral characterization. *Optical Engineering* 43(8): 1777–1786.

Duda, R.O. and Hart, P.E. 1973. *Pattern Recognition and Scene Analysis*. John Willey & Sons, New York.

Eklund, P.W., Kirkby, S.D., and Salim, A. 1998. Data mining and soil salinity analysis. *International Journal of Geographical Information Science* 12: 247–268.

Farifteh, J. 2007. Imaging spectroscopy of salt-affected soils: Model-based integrated method. PhD dissertation 143, International Institute for Geo-Information Science and Earth Observation (ITC), the Netherlands.

Farifteh, J., Bouma, A., and Van der Meijde, M. 2004. A new approach in the detection of salt affected soils: Integrating spectroscopy and geophysical measurements (abstract). In *IGARSS: Proceedings of the IEEE International Geoscience and Remote Sensing Symposium: Science for Society, Exploring and Managing a Changing Planet*. Anchorage, AK.

Farifteh, J., van der Meer, F., and Van der Meijde, M. 2005. Spectral characteristics of salt-affected soils; impact on imaging spectroscopy (abstract). In: *Proceedings of the 4th Workshop on Imaging Spectroscopy: New Quality in Environmental Studies*, Warsaw, Poland, p. 48.

Farifteh, J., Farshad, A., and George, R.J. 2006. Assessing salt-affected soils using remote sensing, solute modeling and geophysics. *Geoderma* 130: 191–206.

Farifteh, J., van der Meer, F., van der Meijde, M., and Atzberger, C. 2007a. Spectral characteristics of salt-affected soils: A laboratory experiment. *Geoderma*, 145, 3–4: 196–206.

Farifteh, J., van der Meer, F., and Carranza, E.J.M. 2007b. Similarity measures for spectral discrimination of salt-affected soils. *International Journal of Remote Sensing* 28(23): 5273–5293.

Farifteh, J., Tolpekin, V., van der Meer, F., and Sukchan, S. 2007c. Salinity modeling of soil reflectance spectra. *International Journal of Remote Sensing* (under review).

Farifteh, J., van der Meer, F., Atzberger, M.C., and Carranza, E.J.M. 2007d. Quantitative analysis of salt-affected soil reflectance spectra: a comparison of two adaptive methods (PLSR and ANN). *Remote Sensing of Environment* 110: 59–78.

Fletcher, D. and Goss, E. 1993. Forecasting with neural networks: An application using bankruptcy data. *Information and Management* 24: 159–167.

Geladi, P. and Kowalski, B.R. 1986. Partial least-squares regression: A tutorial. *Analytica Chimica Acta* 185: 1–17.

George, R.J. and Bennett, D.L. 2000. Airborne geophysics provides improved spatial information for the management of dryland salinity. In: *Land Degradation*, Ed. A.J. Conacher. Kluwer Academic Publisher, Perth, Australia, pp. 305–319.

Haykin, S. 1999. *Neural Networks: A Comprehensive Foundation*. 2nd ed. Prentice-Hall, Upper Saddle River, NJ.

Homayouni, S. and Roux, M. 2004. Hyperspectral image analysis for material mapping using spectral matching. Available at http://www.isprs.org/istanbul2004/comm7/papers/10.pdf.

Hornik, K., Stinchcombe, M., and White, H. 1989. Multilayer feedforward networks are universal approximators. *Neural Networks* 2: 359–366.

Howlett, A., Roach, M.J., and Reid, J.E. 2001. Geophysical characteristics of salinization at cape Portland, NE Tasmania. *Exploration Geophysics* 32: 214–218.

Huang, W. and Foo, S. 2002. Neural network modeling of salinity variation in Apalachicola River. *Water Research* 36: 356–362.

HyVista. 2004. Airborne hyperspectral surveys and information products. HyVista Corporations. Available at http://www.hyvista.com/main.html.

Jackson, J.E. 1991. *A User's Guide to Principal Components*. John Wiley & Sons, New York.

Johnson, C.K., Doran, J.W., Duke, H.R., Wienhold, B.J., Eskridge, K.M., and J.F. Shanahan. 2001. Field-scale electrical conductivity mapping for delineating soil condition. *Soil Science Society of America Journal* 65: 1829–1837.

Jury, W.A. 1982. Simulation of solute transport using a transfer function model. *Water Resources Research* 18: 363–368.

Kearey, P. and Brooks, M. 1994. *An Introduction to Geophysical Exploration*, 2nd ed. Blackwell Science, Malden, MA.

Keränen, P., Kaarna, A., and Toivanen, P. 2003. Spectral similarity measures for classification in lossy compression of hyperspectral images. In: *Proceedings of the 9th International Symposium on Remote Sensing: Image and Signal Processing for Remote Sensing VI*, Ed. A.B. Serpico. SPIE Vol. 4885, pp. 285–296.

Klein, J.T., Grossenbacher-Mansuy, W., Häberli, R., Bill, A., Scholz, R.W., and Welti, M. (Eds.). 2000. *Transdisciplinarity: Joint Problem Solving among Science, Technology, and Society; An Effective Way for Managing Complexity*. Birkhäuser, Basel, Berlin, Boston.

Kruse, F.A., Lefkoff, A.B., Boardman, J.B., Heidebrecht, K.B., Shapiro, A.T., Barloon, P.J., and Goetz, A.F.H. 1993. The spectral image processing system (SIPS)—Interactive visualization and analysis of imaging spectrometer data. *Remote Sensing of Environment* 44: 145–163.

Lide, D.R. 1993. *Handbook of Chemistry and Physics*, 74th ed. CRC Press, Boca Raton, FL.

McNeill, J.D. 1980. *Electrical Conductivity of Soils and Rocks*. Geonics Ltd., Technical note TN-5. Mississauga, Ontario, Canada.

Metternicht, G.I. and Zinck, J.A. 1996. Modeling salinity-alkalinity classes for mapping salt-affected topsoils in the semi-arid valleys of Cochabamba (Bolivia). *ITC Journal* 1996–2: 125–135.

Metternicht, G.I. and Zinck, J.A. 2003. Remote sensing of soil salinity: Potentials and constraints. *Remote Sensing of Environment* 85: 1–20.

Miller, J.R., Hare, E.W., and Wu, J. 1990. Quantitative characterization of the vegetation red edge reflectance: 1. An inverted-Gaussian reflectance model. *International Journal of Remote Sensing* 11: 1755–1773.

Mulder, N.J. 1994. A theory of knowledge-based image analysis with application to SAR data of agriculture. In: *Proceedings of the European Optical Society and International Society of Optical Engineering Symposium, VIII*, Ed. A.B. Serpico. SPIE Vol. 4885, pp. 200–212.

Nash, J.E. and Sutcliffe, J.V. 1970. River flow forecasting through conceptual models: 1. A discussion of principles. *Journal of Hydrology* 10: 282–290.

Nyquist, R.P. and Kagel, R.O. 1971. Infrared spectra of inorganic compounds (3800–45 cm^{-1}). Academic Press, New York.

Rao, B.R.M., Dwivedi, R.S., Sreenivas, K., Khan, Q.I., Ramana, K.V., Thammappa, S.S., and Fyzee, M.A. 1998. An inventory of salt-affected soils and waterlogged areas in the Nagarjunsagar Canal Command Area of Southern India, using space-borne multispectral data. *Land Degradation and Development* 9: 357–367.

Rossel, R.A.V. 2005. ParLeS version 2.1a. Pre-processing of data, principal component analysis and partial least squares regression with leave-one-out cross validation. Available at http://www.usyd.edu.au/su/agric/acpa/people/rvrossel/soft01.htm.

Sah, A.K., Eiumnoh, A., Murai, S., and P. Parkpian. 1995. Mapping of salt-affected soils using remote sensing and geographic information systems: A case study of Nakhon Ratchaslma, Thailand. In: *Proceedings of the 16th Asian Conference of Remote sensing*, G-3-1 to G-3-6.

Schmidt, K.S., Skidmore, A.K., Kloosterman, E.H., van Oosten, H., Kumar, L., and Janssen, J.A.M. 2004. Mapping coastal vegetation using an expert system and hyperspectral imagery. *Photogrammetric Engineering and Remote Sensing* 70: 703–715.

Schwarz, J. and Staenz, K. 2001. Adaptive threshold for spectral matching of hyperspectral data. *Canadian Journal of Remote Sensing* 27(3): 216–224.

Simunek, J., Jarvis, N.J., Van Genuchten, M.T., and Gardenas, A. 2003. Review and comparison of models for describing non-equilibrium and preferential flow and transport in vadose zone. *Journal of Hydrology* 272: 14–35.

Taylor, J.R. 1997. *An Introduction to Error Analysis: The Study of Uncertainties in Physical Measurements*, 2nd ed. University Science Books, Sausalito, CA.

Udelhoven, T., Emmerling, C., and Jarmer, T. 2003. Quantitative analysis of soil chemical properties with diffuse reflectance spectrometry and partial least-square regression: A feasibility study. *Plant and Soil* 251: 319–329.

Van Dam, J.C., Huygen, J., Wesseling, J.G., Feddes, R.A., Kabat, P., van Walsum, P.E.V., Groenendijk, P., and Diepen, C.A.1997. Theory of SWAP version 2.0. Simulation of water flow, solute transport and plant growth in the soil-water-atmosphere-plant environment. DLO Winand Staring Centre, Technical Document 45. Wageningen, the Netherlands.

Van der Meer, F. 2000. Spectral curve shape matching with a continuum removed CCSM Algorithm. *International Journal of Remote Sensing* 21: 3179–3185.

Van Reeuwijk, L.P. 1993. Procedures for soil analysis. Technical Paper No. 9. International Soil Reference and Information Centre (ISRIC). Wageningen, the Netherlands.

Verma, K.S., Saxena, R.K., Barthwal, A.K., and Deshmukh, S.N. 1994. Remote sensing technique for mapping salt-affected soils. *International Journal of Remote Sensing* 15: 1901–1914.

Verstraete, M.M., Pinty, B., and Myneni, R. 1994. Understanding the biosphere from space: Strategies to exploit remote sensing data. In: *Proceeding of the 6th ISPRS International Symposium on Physical Measurements and Signatures in Remote Sensing*, CNES, Val d'Isere, France, pp. 993–1004.

Webster, R. and Oliver, M.A. 1990. *Statistical Methods in Soil and Land Resource Survey*. Oxford University Press, New York.

Whiting, M.L., Li, L., and Ustin, S.L. 2004. Predicting water content using Gaussian model on soil spectra. *Remote Sensing of Environment* 89: 535–552.

Wikipedia. 2007. *On-Line Encyclopedia*. Available at http://En.Wikipedia.Org/Wiki/.

Williams, P.C. 2001. Implementation of near-infrared technology. In: *Near-Infrared Technology in the Agricultural and Food Industries*, Ed. P.C. Williams and K.H. Norris. American Association of Cereal Chemists. St. Paul, MN, pp. 145–171.

Wold, S., Sjöström, M., and Eriksson, L. 2001. PLS-regression: A basic tool for chemometrics. *Chemometrics and Intelligent Laboratory Systems* 58: 109–130.

Yang, H., Griffiths, P.R., and Tate, J.B. 2003. Comparison of partial least squares regression and multi-layer neural networks for quantification of nonlinear systems and application to gas phase Fourier transform infrared spectra. *Analytica Chimica Acta* 489: 125–136.

17 Data Mining for Soil Salinity Modeling

Peter Eklund and Stephen D. Kirkby

CONTENTS

17.1 INTRODUCTION

The application of artificial intelligence to spatial data processing, including both remotely sensed and digitized spatial data, was not well explored before this paper was published in the *International Journal of Geographical Information Systems* in 1998 (Eklund et al., 1998). It was common at that time to apply expert or knowledge-based system techniques to spatial data, as shown by Wang and Newkirk (1988) and Johnsson and Kanonier (1991). Knowledge-based system approaches were used to interrogate various attributes in data layers to color a map or perform feature extraction, but ours is the first paper in the geospatial sciences to use knowledge and data discovery techniques for decision support in secondary salinization. An excellent survey of work done using knowledge discovery in the geosciences in general can be found in Yuan et al. (2004). The contribution of this chapter lies in the application of techniques from artificial intelligence to the problem domain of secondary salinization. The original paper published in the *International Journal of Geographical Information Systems* was revised, updated, and improved for inclusion in this book. Permission for this purpose was sought and granted for by Taylor & Francis.

The presentation of the chapter is organized by three experiments using environmental spatial data and machine-learning algorithms. It subsequently draws a comparison of the suitability of machine-learning techniques for decision support in the analysis of secondary soil salinity, following the style of presentation in the machine-learning literature (Mooney et al., 1989). An introduction to the Salt Manager knowledge-based decision support system (Kirkby, 1994) and to machine learning as a data analysis and acquisition technique precedes our experimental descriptions and the comparison.

The first experiment describes how the popular inductive machine-learning program C4.5 (Quinlan, 1993) was used to restructure a rule-base from a knowledge-based decision support system and how the learning program could be used to improve the classification and decision support system performance by reorganizing the rules. In this chapter, we call expert-derived rules *classification rules* and rules derived by machine *association rules*.

The second experiment describes how the same symbolic machine-learning algorithm (C4.5) can aid in data mining or knowledge discovery, revealing how important classification desiderata present in patterns of spatial data can be reused to better model secondary salinization.

The third experiment measures the effect of additional electromagnetic (EM) spatial data on predicting secondary salinization. The additional EM layer allows us to pose the question whether, and in what way, a new spatial EM layer can assist in predicting secondary salinization and, in doing so, improve the model. What new domain knowledge results and how does an additional map layer impact the decision support system? The answers to these questions provide not only important clues about the domain knowledge for secondary salinization but also conditions data acquisition, mapping, and remote sensing activities for managing and preventing secondary salinity.

In our knowledge discovery experiments, spatial data sets are separated into training and test sets and use the knowledge-based system classification as synthetic ground truth. The Salt Manager knowledge-based system has itself been ground-truthed by field survey with good correspondence. We measure the relative performance of the machine-induced classifiers on unseen data generated by several machine-learning algorithms, including C4.5 (Quinlan, 1993), based on the creation of decision trees using the principle of minimum entropy: IB1 and IB2 (Aha et al., 1991), which are two algorithms that generate classification predictions using only training instances; MPIL1 and MPIL2 (Romaniuk, 1993), which are multipass instance-based learning (IBL) algorithms that create nearest neighbor-like classifiers (Hart, 1968); and feed-forward backpropagation neural networks, using various unit and layer topologies. Classification performance for each machine-learning technique is varied against the training set size to produce learning curves for each method and, in doing so, measure the classification performance of each technique for predicting secondary salinity.

17.2 BACKGROUND FOR MODELING AND PREDICTING SECONDARY SALINITY

The aim of our research is to define the conditions that prevent or minimize secondary salinization and lead to sustainable land practices for a range of land uses, including irrigation, agriculture, pastoralism, forestry, and conservation. In order to establish such practices, decision support systems need to provide evidence-reinforcing predictions to land management practices and also explain the logic that underlies decision-making about land use. For this reason, a knowledge-based system is an appropriate platform for this style of decision support. The knowledge-based approach provides the level of explanation necessary to reinforce recommendations and accommodates heuristic and informal knowledge about the conditions that lead to secondary salinization. Classification rules (IF condition(s), THEN effect) can be formulated by human domain experts, and these encode heuristics (formal and informal) about the causal relationships between environmental attributes such as depth to groundwater, slope, elevation, soil type, remote sensing spectral data, and (in combination) the likelihood of secondary salinization at any given point in Cartesian space. These classification rules can iterate over millions of data points, interrogating attributes in multiple spatial data layers, and be used to color a map showing the degrees of anticipated secondary salinization. The result is a geographical information-based decision support system (called Salt Manager) that can aid in land management and salinity prediction, prevention, and treatment.

Our case study is to model, measure, and predict the effect and occurrence of secondary salinization on pastoral land in an area located 200 km NNW of Adelaide in South Australia. The hypothesis tested and reinforced by Salt Manager is that systematic measurement of the depth to

TABLE 17.1
Spatial Data Layers: Description, Sources, and Format Transportation

Data Types	Acquisition	Conversion
Soil	Digital GIS coverage	Vector/raster
Geology	Digital GIS coverage	Vector/raster
Slope	Topographic survey	DEM/raster
Elevation	Topographic survey	DEM/raster
EM31	Field survey	Kriging/raster
Groundwater	Field survey	Kriging/raster
Cadastral	Digital GIS coverage	Not applicable
Roads	Digital GIS coverage	Not applicable
Drainage lines	Digital GIS coverage	Not applicable
Landsat Band7 image	October 1991 (late flush imagery)	Raster

groundwater can be used to underwrite a knowledge-based decision support system for deciding secondary salinization. Such a system can in turn be used to predict the presence (or absence) of secondary salinization at any given point. Specifically, the objective of the Salt Manager is to identify the area's hydrology in terms of groundwater recharge and discharge, information that can be used to initiate land management practices to prevent, minimize, and treat the effects of secondary salinization.

The case study area includes Jamestown in the mid-north of South Australia. Spatial data are represented in raster format, because this allows easy manipulation of map layers as Boolean operations, and remotely sensed imagery can be integrated with geo-based data taken at various times and from various sources. In addition, some vector layers are also used, namely, for roads and cadastral data. A description of the spatial data layers and original sources is given in Table 17.1, and a more detailed description of the attribute values for each data layer is presented in Table 17.2.

TABLE 17.2
Spatial Data Layer, Data Type, and Range of Values

Attribute Name	Data Type	Value Range
Soil	Discrete/Enumerated	$\{1, 2, \ldots, 13\}$
Geology	Discrete/enumerated	$\{1, 2, \ldots, 8\}$
Slope	Continuous	$1°–7°$
Elevation	Continuous	1–6 m
EM31	Continuous	0–300 mS m^{-1}
Groundwater	Continuous	1–14 m
TMBand1	Continuous	64–159 μm
TMBand2	Continuous	24–90 μm
TMBand3	Continuous	22–145 μm
TMBand4	Continuous	28–190 μm
TMBand5	Continuous	16–233 μm
TMBand6	Continuous	125–154 μm
TMBand7	Continuous	4–145 μm

Spatial resolution is fixed at 30 m^2 to normalize to remotely sensed spectral data. Other data sets are generalized to this resolution and some other minor data transformations are performed. For instance, geology and soil data nomenclatures are interchanged with integers to improve processing speed, e.g., soil value 3 = Yanga Silt Loam, geology value 3 = Mintaro Shale, etc. Accuracy and precision between the data sets varied. For instance, a Geonics EM31 ground conductivity-meter was used to measure the bulk electrical conductivity (EC) of the soil profiles. The measurement unit is milli-Siemens per meter (mS m^{-1}). Data were collected in both horizontal and vertical dipole modes. EM31 data were collected at 894 points, compared to only 17 piezometer installations for groundwater monitoring within a 1 × 3 km catchment. Spatial error estimates were calculated for the groundwater and EM31 data sets, but not for the entire data set. Nevertheless, the features of interest in the study, namely, current and potential salt-affected groundwater discharge and recharge areas, are significant enough to be identified within 30 m^2 cells. Further, the spatial data coverages are not instantaneous snapshots at a particular point in time, as the occurrence of secondary salinity is a problem that manifests over long periods and bringing together disparate data sources from different time-points is standard practice for decision support in land management and geographical information system (GIS).

Having collected data from multiple sources, we concentrate on encoding both formal and informal decision knowledge regarding secondary salinity. It is well known that secondary salinity occurs only in groundwater discharge areas that are locations where groundwater rises through the soil profile. Therefore, we need a decision procedure that determines the degree of groundwater discharge. We can build a useful hydrological model of the region by classifying recharge areas where water enters the soil profile and recharges groundwater. An example of such informal knowledge is that secondary salinity never occurs along a ridgeline with shallow skeletal soils, because (1) water rapidly runs off and (2) the depth to the groundwater through the soil profile is large. Such a heuristic, coarse as it is, and operating over all the available data, can be formulated as the following classification rule:

if slope > 4%, then

    if groundwater > 4 meters, then

    conclude low potential discharge

In Salt Manager, there are three groundwater discharge and three recharge classes. These were adopted on the advice of a panel of five domain experts. The recharge and discharge classes are illustrated in Table 17.3.

As with the information about recharge and discharge classes, the same five salinity experts were interviewed to obtain heuristic knowledge about secondary salinization. The Salt Manager study used the focused interview method of knowledge acquisition (de Greef and Breuker, 1985), a standard methodology in the field. Interviews were recorded over a 3-month period in 1992. A rule base was constructed and three of the five experts reconsulted. Finally, two other local experts were asked to comment on the classification rules provided by the panel of domain experts.

TABLE 17.3
Groundwater Recharge and Discharge
Classes Defined by Domain Experts

Groundwater Discharge	Groundwater Recharge
Existing (ED)	High potential (HPR)
High potential (HPD)	Medium potential (MPR)
Low potential (LPD)	Low potential (LPR)

RuleX: if ground water ≤ 3 then
if soil type in {3, 5, 7, 12, 13} then
if geology = {1, 2, 3, 4, 5, 6, 7} then
if elevation ≤ 2 then
if slope ≤ 3 then
→ class 2 (HPD)

FIGURE 17.1   Typical groundwater discharge classification rule.

This resulted in several additional rules. A typical production rule is shown in Figure 17.1. Sixteen production rules of this type were defined for the six classes of recharge and discharge described in Table 17.3. The rules forward chain over the spatial map database to derive the salinity map shown in Figure 17.2.

The discharge–recharge map (called a salinity map for brevity) defines a data set of attribute values along with discharge classes. It suggests areas where (1) effective high water usage management strategies require implementation (groundwater recharge or potential groundwater discharge zones) and (2) where salt-tolerant cropping practices should be initiated (existing or potential groundwater discharge zones). The classification map has a high correspondence to surveyed ground-truth as shown in the confusion matrices of Tables 17.4 and 17.5. This partially vindicates the time-independence of the spatial data. The Salt Manager-derived classes conform to ground survey despite the layers coming from different points in time and sources. Table 17.4 indicates groundwater recharge classification accuracy of 75% for the Salt Manager system when compared to data collected from eight nested piezometer point locations (calculated by dividing the sum of the diagonals by the total number of cases). The eight points in Table 17.4 were multiple nested piezometers, monitored by the Department of Agriculture of South Australia, and were the field validation points for this test. Overall we had 17 piezometers, but on the advice of the South

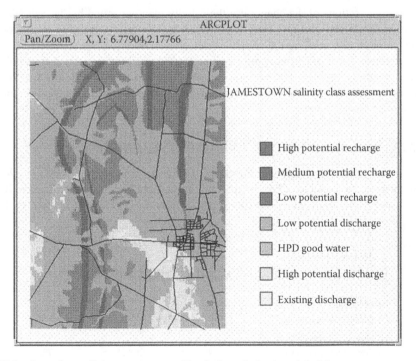

FIGURE 17.2   Secondary salinity map generated by the knowledge-based decision support system called Salt Manager (approximate scale 1:30,000).

TABLE 17.4

Salt Manager Confusion Matrix for Recharge Zones

Salt Manager	Classification (Recharge)			
Field Data	HPR	MPR	LPR	Total
High potential (HPR)	1			1
Medium potential (MPR)		3		4
Low potential (LPR)			2	3
Total	1	4	3	8

The accuracy of the Salt Manager system is read by examining the diagonal where it can be seen that three of the four matrix piezometer sites for ground-truthed field data for MPR are correctly predicted by Salt Manager and two of three piezometer sites for LPR are correctly predicted by salt manager. Therefore, the correspondence between Salt Manager and the eight piezometer sites for recharge is 6/8 or 75%.

Australian Department of Agriculture we chose these nested piezometer sites as they characterized the groundwater systems beneath the various soil-landform associations in the catchment areas and had been monitored on a daily basis. The eight piezometer nests enabled vertical gradients at each site to be calculated and the direction of the groundwater flow determined, thus indicating if the site is acting as a recharge or discharge area. The nested piezometers were distributed evenly over the entire catchment area and were monitored on a daily basis since February 1991. In contrast, Table 17.5 compares the Salt Manager discharge classification measured in hectares to the ground-truth field data. The correspondence between the classifier and the surveyed ground-truth is 94.3%. Detailed discussions on these results, the benefits, and limitations of the model, can be found in Kirkby (1994).

The remaining sections of the chaper use the salinity map generated by the Salt Manager as a synthetic ground truth and examine how data mining techniques can be used to improve system performance through restructuring the classification rules. In addition, we comment on knowledge discovery outcomes that result from using machine-learning techniques. We also examine several other machine-learning alternatives for the secondary salinization domain. Before we do, we give a brief tutorial on symbolic machine learning, since its application to land management and decision support in soil salinity is novel.

17.3   BACKGROUND TO INDUCTIVE MACHINE LEARNING

In symbolic logic deduction is a sound process of inference. If we know $\forall x\, P(x) \rightarrow Q(x)$ and we know $P(a)$, then we can conclude $Q(a)$. Induction is a plausible but unsound inference. If we know $P(a) \rightarrow Q(a)$ and $P(b) \rightarrow Q(b)$, then we can plausibly infer $\forall x\, P(x) \rightarrow Q(x)$. However, there is no

TABLE 17.5

Salt Manager Classification Matrix for Discharge (ha)

Salt Manager	Classification (Discharge)		
Field Data	ED	HV	Total
Existing discharge (ED)	20.5	11.7	32.2
Healthy vegetation (HV)	5.4	268.2	273.6
Total	25.9	279.9	305.8

guarantee that $\forall x\ P(x) \rightarrow Q(x)$ holds. Inferring $P(x) \rightarrow Q(x)$ from $P(a) \rightarrow Q(a)$, $P(b) \rightarrow Q(b)$, $P(c) \rightarrow Q(c)$, etc., is an example of unsound inference, though a plausible explanation for $P(a) \rightarrow Q(a)$ and $P(b) \rightarrow Q(b)$ and $P(c) \rightarrow Q(c)$. Inductive machine learning is a process of acquiring knowledge by drawing inductive inferences from proven facts: In the simplest case, from positive and negative examples of some observed events but more generally from multiple classes of observed events. Such a process involves operations of generalizing, specializing, transforming, correcting, and refining knowledge representations. The study and modeling of this form of machine learning is one of the central topics in artificial intelligence (Dieterich and Michalski, 1983; Dieterich, 1990) and has also become a standard technique in data analysis (Maimon, 2005).

The earliest application of inductive learning is in automated construction of association rules for expert or knowledge-based systems (Quinlan, 1979). With the popular focused-interview method of Hayward et al. (1987), constructing knowledge bases was found to be a time-consuming process of formalizing expert knowledge through encoding it in some knowledge representation system, such as classification rules and semantic networks. Inductive machine-learning programs provided both an improvement on the then current interview-based acquisition techniques and a basis for data mining methods (Quinlan, 1983). In selected domains, inductive learning systems are able to determine decision rules by inductive inference from expert decisions, and there are many successful generic systems that employ these techniques (Quinlan, 1979; Michalski and Chilausky, 1980; Quinlan, 1986; Cestnik et al., 1987). In doing so, these systems demonstrated a considerable increase in the speed of constructing sets of rule bases, thus accelerating the capacity to transfer knowledge from domain expert to machine. Further, the machine-produced association rules often classify data more efficiently than the classification rules derived from human experts.

An important application of inductive machine-learning programs is in experimental sciences. Induction can be used to detect interesting conceptual patterns that repeat or reveal the inherent structure of data in a collection of observations. The world cannot necessarily be explained in all cases using well-understood theories or mathematical formulas, and so one method of generating explanations for data is by observing thousands or millions of examples. For this reason, much of the training data and representational schemes of inductive machine-learning programs are framed as logical descriptions: usually combinations of conditions based on propositional logic, unary, and binary relations. These descriptions can be used to characterize data in terms of high-level, human-oriented concepts and relationships. Examples of such inductive learning algorithms are the ID3 system (Quinlan, 1986) and its descendants including the Assistant Professional (Cestnik et al., 1988) and the C4.5 system (Quinlan, 1993), but there are other systems and techniques that perform similar tasks (Breiman et al., 1984).

C4.5 is a system of automated knowledge acquisition for knowledge base and other artificial intelligence applications. The main vehicle for automated construction of knowledge bases is the induction of decision trees from examples. An example of a decision tree generated by C4.5 is shown is Figure 17.3. C4.5 is noise-tolerant and therefore able to deal with unreliable measurement information. It can also work with incomplete descriptions of the world. The machine-learning system points out important (or irrelevant) data for a decision problem. If data do not appear in a decision tree, then it is often the case that the decision tree carries no information relevant for the decision problem, or it is redundant (circumscribed by some other conditional(s)) with respect to the decision outcome. C4.5 can be used as an association rule generator, since each traversal from root to leaf node represents an association rule of the "IF condition(s), THEN effect" form. The association rules can subsequently be used to solve new decision problems on previously unseen data.

A decision tree, such as the one shown in Figure 17.3, is a tree where nodes stand for attributes, branches stand for attribute values, and leaves stand for classes. A decision tree is read from top to bottom. Every branch, from the root of the tree to the leaf, represents one decision rule. The skill in constructing a decision tree is to select an attribute for the current node so that the final decision tree

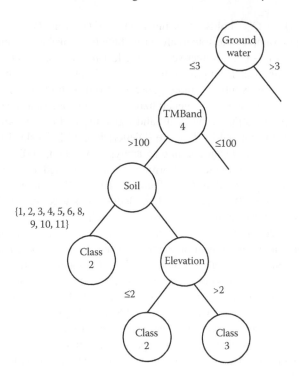

FIGURE 17.3   Partial view of a decision tree produced by C4.5 to predict secondary salinization.

is in some way minimal. A decision tree is constructed from the root node of the tree down. The tree is built by choosing an attribute that minimizes the amount of disorder in the subtree rooted at a given node, as we wish to minimize disorder or maximize gain. The disorder associated with an attribute partition is the probability-weighted sum over all training instances, which can be partitioned by the attribute test, and is given by the following equation:

$$\text{Disorder} = \sum_c -\frac{n_{bc}}{n_b} \log_2 \frac{n_{bc}}{n_b} \qquad (17.1)$$

where
   n_{bc} is the number of instances in a branch of class c
   n_b is the number of instances in the branch in total
   c is the number of classes in the branch

Let us suppose for simplicity we have only two classes of observations, A and B. If the number of members in class A is identical to the number of members in class B, then there is maximum disorder that is from the above formula: $-1/2 \log_2 1/2 - 1/2 \log_2 1/2 = 1/2 + 1/2 = 1$. On the other hand, if there are only A's and no B's (or vice versa), there is perfect order that is: $-1 \log_2 1 - 0 \log_2 0 = 0$. As the partition over the set of instances moves from homogeneity to equal balance of each, disorder varies from 0 to 1. This gives a way of computing the disorder in each partition (or subbranch) of the decision tree. To obtain the average disorder of the entire decision tree, each subbranch is weighted with the percentage of instances that belong to it, as follows:

$$\text{Average disorder} = \sum_b \frac{n_b}{n_t} \sum_c -\frac{n_{bc}}{n_b} \log_2 \frac{n_{bc}}{n_b} \qquad (17.2)$$

where

n_t is the total number of observations in the entire training set
b is the number of branches in a given partition

Informally, at every decision level, C4.5 selects the most informative attribute, an attribute that contributes the most information for the classification of the learning observations in the current node. When the best attribute is selected, the set of training observations is split into subsets of observations, one for each attribute value combination. Each subset of the learning observations is then used to construct further subtrees. When a termination condition is satisfied in the current subset of observations, a leaf node is generated. The termination conditions are given as follows: (1) all observations from a subset belong to a single class; (2) no more attributes are available to select; or (3) a pruning factor limiting the depth and number of divisions of subtrees is satisfied.

As mentioned, a decision rule can be constructed by following the path from the root of the decision tree to a leaf node. For instance, in Figure 17.3, consider the leftmost branch of the decision tree, *if Ground Water ≤ 3 and TMB and 4 > 100 and Soil in {1, 2, 3, 4, 5, 6, 8, 9, 11} then Class = 2 (HPD)*. Tracing the choices at each node in the decision tree recovers one association rule, which is sometimes identical to a classification rule from the knowledge base that can be derived from human domain experts.

Often, in real-world classification tasks, the class of observations in the training set cannot be expressed as a function of the attribute values. This occurs because either the attribute values contain errors, or the attributes collectively provide insufficient information to classify an object. In these circumstances, continuing to divide the training set until all subsets containing members of a single class may be impossible. Even when such perfect division is possible, it may be inadvisable because the lower levels of the tree may model idiosyncrasies like noise in the training data rather than provide useful structure (Quinlan, 1990). One way to deal with this problem is to let the tree grow and then remove unimportant or unsubstantiated portions of the decision tree by pruning it. Decision tree pruning involves replacing a subtree by one of its branches or, more commonly, by a leaf node. It usually results in a smaller tree and has the additional advantage of increased accuracy when classifying unseen objects (Quinlan, 1990).

17.4 THREE EXPERIMENTS IN KNOWLEDGE AND DATA DISCOVERY OF SECONDARY SALINITY

There are three experiments that organize this section. The first is to use a machine-learning algorithm on spatial data classified by expert-derived classification rules to improve classification efficiency and system performance. The task is to use the machine-learning algorithm to generate association rules that optimally reconstruct the human-derived classification rules. As mentioned earlier, the automatically derived rules are often called association rules to distinguish them from classification rules, derived by human experts. The second experiment explores new patterns of attribute value combinations in the association rules. Exploring association rules can reveal new patterns that can lead to new discoveries about secondary salinization. This process is referred to as knowledge and data discovery. The third experiment measures the effect of an additional EM spatial data layer on the knowledge-based decision support system.

17.4.1 RESTRUCTURING THE RULE BASE: FROM CLASSIFICATION RULES TO ASSOCIATION RULES

There are several problems with rule bases generated through the focused-interview process and these can be generalized to any similar environmental and geospatial knowledge-based approach. Firstly, the knowledge-based system analyses attribute information for 12 spatial data layers stored in a GIS and will potentially need to process millions of data points to build a salinity map. In such a model, an inefficient rule structure can dramatically affect performance. For example, consider the simple rule *"if p and q then r."* Each of the propositions *p* and *q* represents antecedents or conditionals and *r* is the consequent. Imagine now that *"if p and q then r"* is a rule that applies over a database of millions of cases where *p* is satisfied but *q* fails in a majority of cases. Because *q* fails more often than *p*, it is a better discriminator. The classifier can be reorganized and optimized so that the rule reads *"if q and p then r."* C4.5 has the effect of producing such an optimized rule structure by generating an information theoretic ordering on conditionals of the associative rules. As an illustration of the effect of C4.5 on the structure of expert rules, consider the following classification rule:

> *Rule A: if soil in {12, 13, 3, 5, 7} then*
>
>> *if slope ≤3 then*
>>
>> *if elevation ≤2 then*
>>
>> *if groundwater ≤3 then*
>>
>> *if TMBand4 > 100 then*
>>
>> → *class 2 (HPD)*

The C4.5 gain criterion selects the *elevation* attribute as the dominant discriminator. The training set is partitioned into two sets: one for training instances with *elevation* ≤2 and the other for instances with *elevation* >2. Each of the partitioned training sets is considered for each subtree of the decision tree. In the subtree where *elevation* ≤2, groundwater depth is selected as the next node in the subtree. The effect of this observation is to restructure *Rule A* so that the dominant attribute values are tested first. The resulting restructured association *Rule A'* is given below:

> *Rule A': if elevation ≤2 then*
>
>> *if groundwater ≤3 then*
>>
>> *if slope ≤3 then*
>>
>> *if soil in {12, 13, 3, 5, 7} then*
>>
>> *if TMBand4 > 100 then*
>>
>> → *class 2 (HPD)*

Rule A' is called an association rule, derived from a decision tree, the principal output of C4.5. The decision tree for this domain is large. A portion of the tree is given in Figure 17.3.

There are several metrics that are suitable for assessing the relative quality (or minimality) of a decision tree (Fayyad and Irani, 1990). The general idea of Occam's razor (Churchland, 1984) is an important one in science and it is appropriate when considering decision trees. Is the decision tree the simplest explanation of the data? There are a number of ways to quantify this idea of minimality:

- *M1*: percentage errors on classifying unseen examples (⇓). Once a decision tree is built, how many examples (as a percentage), not used to construct the tree, cannot be accounted for by the tree?
- *M2*: number of rules in the rule base (leaves in the decision tree) (⇓). This idea is to reduce the number of leaf nodes; this in turn minimizes the number of rules that cover or account for the data.
- *M3*: number of nodes in the tree (⇓). In some situations there will be as many leaf nodes in one potential decision tree as in another. Both may explain the data, but the number of

intermediate nodes that cover the data should be minimized in order to reduce the complexity of the association rules.

- *M4*: total number of preconditions in the rule base ($\Downarrow$). This measures the generality of the entire set of rules. It is a more accurate measure of generality than the average number of preconditions per association rule, since the latter fails to indicate if the number of rules is relatively large when compared to the number of training examples.
- *M5*: average example support per association rule (per leaf) ($\Uparrow$). This is the average number of examples in the test set to which an association rule is applicable and correctly predicts the class. It is a measure of the applicability (generality or utility) of the association rules.

The ($\Downarrow$) and ($\Uparrow$) symbols indicate that it is desirable (intuitively) for the corresponding measure to be minimized or maximized, respectively. Applied to Salt Manager, the salinity map is partitioned such that 80% is used as a training data set to construct a decision tree and the remaining 20% is withheld as a test data set. The classification accuracy for secondary salinity indicates a high degree of confidence in the decision tree produced by C4.5 on the 20% of the data that is withheld, called unseen data. These results are presented in Table 17.11. This table also shows that the accuracy of the generated association rules for secondary salinity is almost as good as the decision tree, with the advantage of having fewer and more readable association rules.

There are two issues concerning the decision rules generated by C4.5 in the secondary salinity domain. The first is the logical structure or readability of the association rules. Are they understandable by the domain expert? Does the structure of the association rules present the most efficient means of determining a class at the expense of the domain logic? To test this, the domain experts were shown the association rules. The experts could read and also recognize their knowledge. The order of the conditionals was different (as earlier discussed), but attribute value tests were mostly identical.

The second issue is the efficiency of the association rule base in terms of the metrics *M1–M5* described above (Table 17.6). Table 17.7 summarizes the rule numbers associated with each class for the induction-derived association rules and expert-derived classification rules. There are more association rules than classification rules. However, more association rules have fewer attribute value tests (conditionals) and therefore classify the data more efficiently. Further efficiency can also be introduced. For instance, some of the rules are general and represent a natural default class that can be assumed when no other rule applies. In the case of the default class LPD, general rules for concluding LPD can be removed, namely, if no rule matches the data, then it can be assumed to be a member of the default or background class. Introducing a default class can prove a substantial saving; in our case study, 60% of the area of the classification map belongs to the default class (see Figure 17.1 and the *Low Potential Discharge* class).

TABLE 17.6
Decision Tree Metrics

Performance Metrics for Decision Trees

	M1	*M2*	*M3*	*M4*	*M5*
Decision tree	4	48	36	84	506
Pruned decision tree	5	35	27	62	694
Association rules	16	19	—	67	1277

M1, Percentage errors on classifying unseen examples; *M2*, number of rules in the rule base (leaves in the decision tree); *M3*, number of nodes in the tree; *M4*, total number of preconditions in the rule base; *M5*, average example support per rule (per leaf).

TABLE 17.7

Association and Classification Rules Associated with Each Salinity Class

Class Description	Number of Classification Rules (Expert-Derived)	Number of Association Rules (Machine-Derived)
Existing discharge (1)	1	4
High potential discharge (2)	1	5
Low potential discharge (3)	3	3
Low potential recharge (4)	4	4
Medium potential recharge (5)	3	5
High potential recharge (6)	3	1
Total	15	22

Association rules reconstituted from the decision tree were recoded in the Salt Manager knowledge-based decision support system and used to regenerate the classification map. The reconstruction of the salinity map (see Figure 17.2) was achieved in 50% of the time it took with the expert-derived classification rules. When a default class was introduced, this process was reduced to 30% of the original classification time. The performance of Salt Manager is therefore improved by 85% by the introduction of association rules and default map class.

17.4.2 KNOWLEDGE DISCOVERY USING C4.5 FOR SECONDARY SALINITY

An important aspect of restructuring the rule base using C4.5 is to verify that the association rules correspond to the logic used in the initial expert-derived classification rules. The decision rules were shown to the panel of domain experts who were asked to identify any implausible rules. The experts were initially puzzled by the combination of C4.5 rules shown in Figure 17.4, believing they represented a counterfactual.

*Rule*1: *if ground water* ≤ 3 *then*
   if slope ≤ 3 *then*
   *TMBand*3 ≤ 39 *then*
   if soil type in {3, 5, 7, 12, 13} *then*
   *TMBand*4 ≤ 100 *then*
      → *class* 2 (*HPD*)

*Rule*14: *if elevation* ≤ 2 *then*
   if ground water ≤ 3 *then*
   if slope ≤ 3 *then*
   if soil type in {3, 5, 7, 12, 13} *then*
   *TMBand*3 > 58 *then*
      → *class* 2 (*HPD*)

*Rule*16: *if elevation* ≤ 2 *then*
   if ground water ≤ 3 *then*
   if slope ≤ 3 *then*
   if soil type in {3, 5, 7, 12, 13} *then*
   *TMBand*4 > 100 *then*
      → *class* 2 (*HPD*)

FIGURE 17.4 Conflict and incoherence between rule logic suggest a revised *HPD* class.

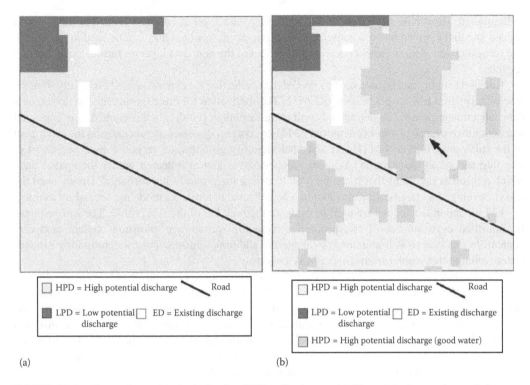

(a) (b)

FIGURE 17.5 Comparison of maps indicating HPD split classes. (a) The original secondary salinity map shows the majority of pixels are thought to be HPD. (b) The HPD is split into two, showing that the darker areas are flushed HPD pixels (HPD$_f$) and these points are indicative of an area that had been contoured to implement a land management strategy. The pixel size is 30 m^2.

Stressed vegetation expected for *HPD* (class 2) is consistent with and satisfied by *rule 14* (high spectral response *TMBand3*). However, *rules 1* and *16* cover vegetation that is experiencing flushed growth and is identified on the basis of the following conditionals: *TMBand3* low spectral response and *TMBand4* high spectral response. *Rules 1* and *16* are inconsistent with *HPD* (class 2), as the healthy vegetation implies a high groundwater table with low secondary salt content.

The rules in Figure 17.4 suggest two subclasses of *HPD*. The original *HPD* class was split into *HPD$_f$* indicating healthy vegetation (a good water class) and *HPD$_s$* indicating stressed vegetation (a salt water class). Given the explanation of the two rules, it is now necessary to reclassify the data using the new *HPD* classes. The above rules were included in the Salt Manager and a new salinity map created. The differences between the two maps are illustrated in Figure 17.5. The new class *HPD$_f$*, as indicated by the arrow, is clearly defined by Figure 17.5. The resulting map shown in Figure 17.5 is interesting because the new class *HPD$_f$* occurs in an area where contouring as a method of surface water control was implemented, an indication that this land management strategy was preventing or managing salinization. This discovery, mined from the data, is independent evidence that water management using contouring controls saline discharge. Importantly, the *HPD$_f$* class only occurred within the vicinity of the engineering and contouring works.

17.4.3 How an EM Data Layer Can Be Used to Determine Groundwater Recharge

The third experiment describes a methodology for the addition of GIS data layers to the rule-based model. In particular, it reports the effect of adding an EM (EM31) data layer to the existing GIS spatial database. Machine-learning approaches should not be used as a surrogate for statistical techniques to determine the effect of an additional data layer (Cressie, 1993). The EM31 data arrived late to the research and our approach tests the extent to which they can be used in the model. We reuse and derive knowledge concerning the effect of the new data layer on future data collection regimes.

Both field survey and remote sensing spatial data collection are expensive, and especially so in the case of geo-based EM coverage such as EM31. The motivation for this experiment is to decide on a data collection regime for modeling and combating secondary salinity. In this context, it is important to assess the relevance of a newly acquired EM31 coverage to groundwater recharge in the catchment of the study area. Can an EM31 data layer help identify groundwater recharge areas and assist in mapping secondary salinization? Is there independent evidence reflected within the spatial data, which reinforces intuitions that it can be used to determine groundwater recharge? Do we need the EM31 coverage for classification and should it be collected in future for modeling secondary salinity?

EM induction techniques are used to measure the bulk EC of the soil profile. The EM response is controlled by a number of soil properties including salt storage, moisture content, and clay content. In this case, it is important to evaluate the additional information for determining groundwater recharge that results from using EM31 coverage.

Geophysical surveys carried out in the catchment area used a Geonics EM31 ground conductivity-meter. The surface EM31 data layer was collected in both the vertical and horizontal modes in order to map the spatial variability of the bulk EC of the landscape and, in so doing, determine the potential groundwater recharge. Field groundwater recharge coverage for secondary salinization was added to a GIS database with three classes of groundwater recharge, namely, low, medium, and high potential. This was supplemented by a partial EM31 coverage. The resulting database was input to the C4.5 and a decision tree classifier was produced.

The purpose was to discover (1) if the resultant decision tree encapsulates any significant attribute value discriminators in the new EM31 coverage layer; (2) if data contained within existing coverage subsume the new EM31 coverage, making it irrelevant or redundant to the classification task (if so, in what way); and (3) if any conclusions are forthcoming about the secondary salinity classification task without prior knowledge of the effect of the EM31 coverage in the domain classification task.

Henschke et al. (1994) produced an extensive report on groundwater discharge and recharge in the Jamestown area. The effects of the EM31 layer are discussed at length in this report. Most importantly, the EM31 frequency distributions for each soil type are presented (Figure 17.6) and a summary of the expected response for discharge in each soil type is given in tabular form (Table 17.8). Measured in mS m^{-1} in Table 17.8, the discharge categories describe the apparent EC (ECa) of the surface.

The frequency distribution of ECa suggests that it can be used to distinguish soil textural groups to a limited degree. For instance, an ECa of 200 mS m^{-1} indicates either Belalie Clay Loam or Belalie Loam, and this value excludes the soil types Yanga Silt Loam and Canowie Loam. Although there is a definite overlap, the distributions are unique. This is summarized in Table 17.8. Although responses of 140–300 mS m^{-1} are indicative of Belalie Loam, the range of 60–140 mS m^{-1} includes both Belalie Loam and Belalie Clay Loam. Additionally, a recharge map of the catchment area accompanied the EM31 coverage (Henschke et al., 1994). This map is reproduced in Figure 17.6 and can be used to check our salinity map results.

The main test is whether the knowledge of recharge classes that follows from Table 17.8 can be recovered from the GIS database using a survey recharge map of the catchment area and inductive machine learning. This has the effect of independently verifying expert knowledge on the usefulness of the EM31 coverage to classify zones of recharge. It also includes the possibility of discovering additional knowledge on EM31 effects in the recharge classes. The obvious assumption is that the EM31 has a positive impact on the recharge classification task and, therefore, can be used as an attribute test within one or more association rules. However, this was unknown to the experimenters who conducted this machine-learning experiment without any prior disclosure.

Expert assessment of groundwater recharge zones   EM31 Verticle dipole Jan 1992

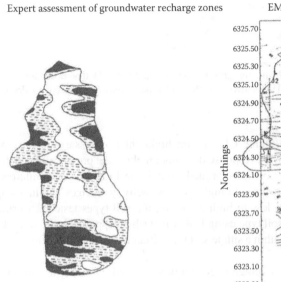

■ High potential groundwater recharge
▣ Medium potential groundwater recharge
☐ Low potential groundwater recharge

FIGURE 17.6 Frequency distribution of EM31 data (image on right) and categories of ECa values derived by human expert (image on left) for each soil unit. (From Henschke, C.J., McCarthy, D.G., Richardson, S.B., and Evans, T.D., Dryland salinity catchment investigation: Jamestown catchment mid-north, South Australia. Technical Report 193, Primary Industries Department, South Australia, 1994.)

In knowledge and data discovery terms, the C4.5 program produced a decision tree that contained a number of interesting rules. Consider the Figures 17.7 and 17.8. The occurrence of Yanga Silt Loam (soil value = 3 in Figure 17.7) verifies expert knowledge from Table 17.8, but the absence of any EM31 conditional test in *Rule B* (Figure 17.8) indicates the subsuming role of this soil type for high potential recharge (HPR). In contrast, *Rule B* and *Rule C* (Figure 17.9) illustrate the importance of EM31 data (the *em31ver* and *em31hor* attribute tests in Figures 17.8 and 17.9, respectively) to determine a recharge class. In light of *Rule A, Rule B* substantiates the expert opinion for Yanga Silt Loam in Table 17.8, namely, that this soil type is indicative of the recharge class.

TABLE 17.8
Catchment Recharge Zones for EM31 Values and Soil Types

Recharge Potential	Area of Catchment (ha)	Dominant Soils	ECa zone (mS m^{-1})
High	71.2 (23.7%)	Yanga Silt Loam	0–60
		Canowie Loam	
		Yarcowie Loam	
Moderate	116.9 (38.8%)	Belalie Loam	60–140
		Belalie Clay Loam	
Low	112.9 (37.5%)	Belalie Loam	140–300

Source: From Henschke, C.J., McCarthy, D.G., Richardson, S.B., and Evans, T.D., Dryland salinity catchment investigation: Jamestown catchment mid-north, South Australia. Technical Report 193, Primary Industries Department, South Australia, 1994.

Rule A: if ground water ≤ 12 then
   if slope > 1 then
      if soil type in {3} then
         → *class 6 (HPR)*

FIGURE 17.7   Rule A: An association rule for groundwater recharge (HPR) that does not involve any conditionals using the EM31 coverage. This shows that the soil type is mostly sufficient to determine the HPR class.

Interestingly, Belalie Loam and Belalie Clay Loam highlight a necessary expansion of the expert's recharge classifier, as a low EM31 response does not preclude the presence of Belalie Loam and Belalie Clay Loam soil types, as it could be argued from Table 17.8. A second inference from *Rule B* is that an EM31 response can be used to highlight conductivity changes within a single soil type, because EM31 values vary from low to high across all the soil types tested. Discriminating recharge between cells with the same soil type using EM31 is further supported by *Rule C*. This rule separates high and medium recharge within a single soil type (Belalie Loam = soil value = 13) using the EM31 response (*em31hor*).

The conclusion is that the EM31 coverage is generally subsumed by soil type. Although little additional knowledge results from the EM31 coverage for classifying groundwater recharge, we argue that EM31 data can be used to distinguish recharge within Belalie Loam (soil value = 13). In general, these findings increase confidence in the quality of the soil data layer and show that the EM31 data can be used as a surrogate for soil type. However, the cost involved in the collection of the EM31 coverage is relatively high and, given the quality of the soil type data layer, mostly unnecessary.

Furthermore, it is the exception to this general rule that is interesting and provides new insight. Additional HPR cells are located using the EM31 coverage in the Belalie Loam and Belalie Clay Loam soil types. Variations in EM31 response indicate that cells, previously considered to be low or medium recharge, are actually high recharge cells, and the EM31 value is the only discriminating attribute in those cells. Therefore, it is only in Belalie Loam and Belalie Clay Loam that EM31 is of any use in predicting groundwater recharge. This is a significant result because HPR cells can be identified using EM31 and because identifying high recharge is far more important than discriminating between low and medium recharge cells. High groundwater recharge cells are important to identify because they provide 36% of all groundwater recharge within the catchment area. A proportion of 28% of all groundwater recharge derives from outside the boundary of the catchment area and the balance (36%) comes from the other recharge classes (Henschke et al., 1994).

The data mining results indicate that an EM31 data collection regime should be initiated on the basis of the soil types and in particular be restricted to the Belalie Loam and Belalie Clay Loam subset of the soil map. Because these soil types cover 58% of the catchment area, the EM31 coverage will prove useful in identifying groundwater recharge only for those soil types. EM31 data for the remaining 42% of the area, including all other soil types present, are of no use and should not be collected.

Rule B: if elevation > 2 then
   if ground water ≤ 6 then
   if slope > 3 then
   if em31ver < 4 then
   if soil type in {3, 12, 13} then
         → *class 6 (HPR)*

FIGURE 17.8   Rule B: An association rule for groundwater recharge (HPR) involving conditional tests for EM31 attributes. An EM31 response can be used to highlight conductivity changes within a single soil type.

Rule C: if ground water ≤ 8 then
if soil type in {13} then
if em 31hor ≤ 3 then
if geology in {5, 8} then
→ class 6 (HPR)

FIGURE 17.9 Rule C: An association rule for groundwater recharge (HPR) involving conditional tests for EM31 attributes that separates high and medium recharge within a single soil type (Belalie Loam = soil value = 13).

The finding that EM31 is a surrogate for certain soil types and that it can be used to distinguish between medium and high potential groundwater recharge cells for Belalie Loam and Belalie Clay Loam soil types is an important result. For other soil types in the catchment area, EM31 is subsumed by soil data and cannot be used to discriminate groundwater recharge. Therefore, EM31 collection should be administered with close reference to the soil map and only for these soil types. This knowledge can be used to condition the data collection regimes in future similar studies.

17.5 ANALYSIS OF VARIOUS MACHINE-LEARNING CLASSIFIERS FOR SECONDARY SALINITY

Geoscientists try to construct accurate GIS-based decision support systems and models from the smallest possible field survey data. This is because of the computational overhead involved in processing large amounts of data, but more importantly because of the prohibitive cost of ground-based surveys and data acquisition.

Hereafter, we test the classification accuracy of several machine-learning techniques to verify that the selection of the classification method C4.5, used in the three experiments described above, was the appropriate choice. The three techniques we examined are inductive learning, backpropagation, and IBL. We have tested the classification accuracy of these techniques against a ground-truth classification map derived from the Salt Manager system (Kirkby, 1994), discussed in the previous section and whose output is illustrated in Figure 17.2.

Only a limited amount of ground-truth data exists for the study area, as shown in Tables 17.4 and 17.5. A set of 17 piezometers was installed in the study catchment area and monitored for 27 months to determine groundwater recharge rates, and so the performance of the various machine-learning classifiers was measured against the output from the Salt Manager system.

Each learning technique is a mechanism to measure the accuracy of the classifiers tested in relation to the synthetic ground truth from Salt Manager. There is a natural bias toward C4.5 given that the synthetic ground-truth map is symbolic in nature and was generated from the Salt Manager rule-base that is similar to the output from C4.5. This is supported by the high correspondence between the rules generated by C4.5 and the expert-derived classification rules described in the experiment presented in Section 17.4.1 and shown in Table 17.11. However, the question remains whether any one machine-learning technique can produce at least as efficient a classifier as C4.5 using the knowledge-based system classification as a synthetic ground truth. In other words, are the other machine-learning methods capable of producing a classifier that works as well as C4.5 on unseen data? Our approach is a standard method in the machine-learning literature and provides some confidence in the knowledge discovery results for modeling secondary salinity (Mooney et al., 1989).

17.5.1 IBL AND ITS PERFORMANCE IN SECONDARY SALINITY CLASSIFICATION

All IBL algorithms derive from the well-known nearest-neighbor classifier (Cover and Hart, 1967). These algorithms determine the similarity of two instances as their Euclidean distance in an n-dimensional feature space. The nearest-neighbor method classifies an instance as a member of the same concept as its most similar instance (Aha et al., 1991). Several IBL algorithms are given in Kibler et al. (1989) and Aha et al. (1991). These include the IB1 and IB2 algorithms and also a MPIL algorithm (Romaniuk, 1993).

The IB1 algorithm is the simplest IBL algorithm. It is equivalent to the classic nearest-neighbor algorithm, except that it normalizes the attribute ranges and has the capability to deal with missing values. The IB2 algorithm is similar to IB1, except that it saves only misclassified instances. It displays a high degree of similarity with the condensed nearest-neighbor algorithm (Hart, 1968), since it saves and uses only selected instances for prediction. By being selective in the instances it saves, the storage requirements of the IB2 algorithm are significantly smaller than those of IB1, especially when instances are adequately separated in the feature space. On the other hand, IB2 appears to be more sensitive to noise than IB1.

The MPIL algorithm allows multiple passes over the complete set of instances, as opposed to a single pass in the IB1 and IB2 algorithms. A set of features represents an instance, while a single feature represents a continuous or discrete attribute value. Associated with every instance is a neighborhood, spanned by the instance, which can be either a sphere (MPIL1) or a k-dimensional box (MPIL2). The details of these two algorithms are given in Romaniuk (1993).

All supervised learning classifiers involve training and testing. IB1 stores all training data and does not involve any training phase, as all the training instances are used to define the machine-learning classifier. For IB2, MPIL1, and MPIL2, training involves storing prototypical cases that summarize training data. All four IBL algorithms have the same testing phase. Each test case is scanned through the stored data and selects the most similar case. If the classes of two cases are identical, classification is considered correct, otherwise incorrect.

The first experiment attempts to compare the performance accuracy of the four IBL algorithms and results are reported in Table 17.9. The performance accuracy is split into three parts: (1) the performance of the training set only; (2) the performance of the unseen test set to determine the generalization capabilities of the algorithms; and (3) the storage requirements of the different approaches. IB1 is used as a base case, since it requires storage of all training instances.

Table 17.9 gives insight into the performance of the IBL algorithms in the salinity and EM31 domains. Both domains are partitioned into training set and test sets. The training set size is 800 and the test set size is 500 for both domains. The most striking result is the good performance of MPIL1 and MPIL2 over IB1. Both MPIL1 and MPIL2 show the most significant improvement in performance for the two data sets, compared to IB1 and IB2. They achieved perfect classification results on the training set and a very high degree of classification accuracy on the unseen test set. Both require

TABLE 17.9

Performance of IBL Algorithms (% Accuracy)

Algorithm	Training	Test	Storage
Performance on secondary salinity data			
IB1	100	89	100
IB2	89.2	85.6	13.9
MPIL1	100	89	12.5
MPIL2	100	95.8	9.9
Performance on EM31 data			
IB1	100	60.1	100
IB2	56.3	56.9	19.7
MPIL1	100	85.2	18.5
MPIL2	100	84.9	19.2

only a small fraction of IB1's storage. MPIL1 achieves highest accuracy for the EM31 data and MPIL2 outperforms other IBL algorithms for salinity data.

17.5.2 PERFORMANCE OF BACKPROPAGATION FOR SECONDARY SALINIZATION

The backpropagation algorithm is central to much current work on machine learning using neural network approaches (Hertz et al., 1991). Its discovery in relation to neural networks is generally accredited to Rumelhart, Hinton, and Williams (Rumelhart et al., 1986a, b). However, in different contexts it was invented independently several times (Bryson and Ho, 1969). The algorithm gives a prescription for changing the weights in any feed-forward network to learn a training set of input–output pairs.

Input nodes record observed features from the training data set and pass activation forward through an intermediate layer(s) of hidden nodes to the output layer. A neural network is structured hierarchically. It has an input layer consisting of a number of nodes, representing the input values of training and test instances, likewise, an output layer whose node values represent the output class, and any number of hidden layers between the input and output layers, each layer containing a fixed number of nodes. Each node in the neural network is linked to every node at the next layer above and below via a weighted connection. The activation levels of the nodes in the output layer determine the output of the network and the signal across the weighted connection. The existence of hidden nodes allows the network to develop complex feature detectors.

Our research implements a backpropagation neural network in the C programming language. Discrete attributes are encoded using a set of binary input nodes, one for each value. For example, geology has eight input nodes or attribute values. If an instance has geology "5," then the fifth input node for geology is set to 1.0 and the other seven are set to 0. Each continuous attribute is modeled by one input unit and scaled between 0 and 1. Output nodes correspond uniquely to classes, 1 node for each class. The network was tested with several hidden nodes and both one and two hidden layers. Optimal results were obtained with 10 hidden nodes and 1 hidden layer. Extending the model to two hidden layers had no effect on the classification performance. Therefore, for the salinity data, 31 input, 10 hidden, and 6 output nodes were employed. The network was tested with various learning rates (0.05, 0.1, 0.5, 1) and momenta (0.0, 0.1, 0.5, 0.9). Of these, a 0.05 learning rate and a momentum value of 0.9 gave the best classification results against the test data. The initial weights w_{ij} were constrained to lie in the range:

$$\frac{-1}{\sqrt{k_i}} \leq w_{ij} \leq \frac{1}{\sqrt{k_i}} \tag{17.3}$$

where k_i is the number of inputs to nodes i, incrementally updated with one pattern at a time chosen in a random order from the training set. This approach of setting and updating weights avoids local minima.

Table 17.10 records the results achieved by backpropagation for two different activation functions: sigmoid and hyperbolic tangent. The number of hidden nodes varied from none (in one layer only) to 20. The results of the two data sets are given for 0, 5, 10, and 15 hidden nodes. For both domains, the network gives good results and high classification accuracy with ten hidden nodes using the sigmoid activation function. The results do not show improvement neither when increasing the number of hidden nodes from 10 to 20 nor when introducing a second hidden layer.

17.5.3 INDUCTIVE MACHINE LEARNING AND ITS PERFORMANCE RELATIVE TO OTHER APPROACHES

After pruning the decision tree, inductive learning achieved 99.8% and 96% accuracy on secondary salinity and EM31 data, respectively. These results are discussed in the first experiment above and summarized in Table 17.11.

TABLE 17.10

Percentage Classification Accuracy Using Different Activation Functions

No. of Hidden Nodes	Sigmoid Function		Hyperbolic Tangent	
	Salinity Data	EM31 Data	Salinity Data	EM31 Data
0	92.6	82.4	91.0	80.1
5	96.9	78.5	94.7	78.8
10	97.6	91.5	97.1	88.2
15	95.3	90.6	94.8	85.0

The plots in Figure 17.10 show the learning curves of the three machine-learning techniques for the secondary salinity data. The graphs show the speed at which the classification accuracy increases during training. The learning curve is a useful indicator of the resolution of the conglomerate data set and the performance of the classifiers on the data used in the domain. From Figure 17.10, we conclude that C4.5 produces classification accuracy better than 95%, even with a small training set (more than 235 instances). To achieve equivalent accuracy, more than 500 training instances are required when using a backpropagation neural network (BP in Figure 17.10), and more than 700 training instances when using the IBL program MPIL2. Figure 17.10 also suggests that a backpropagation neural network and the IBL MPIL2 perform well for large training sets, but performance deteriorates markedly for smaller training sets in this domain.

C4.5 outperforms both backpropagation and instance-based MPIL2. The reasons in general are that the domain favors the symbolic attribute-value treatment. Certain attributes are more significant than others for salinity-class discrimination. For instance, groundwater depth is the prime discriminator for saline discharge, because salt-affected water close to the earth's surface characterizes secondary saline discharge. Salt-affected or not, water close to the soil surface belongs to the HPD class. Likewise, vegetation stress can be inferred from the spectral signatures of the Landsat5 TMBand3 and TMBand4.

In domains where attributes have a more uniform influence on classification, the performance of backpropagation neural network and instance-based MPIL2 may be significantly improved. It is worth noting that the ground-truth data have been synthesized using a symbolic rule-based classifier (Salt Manager). Rule-based classifiers also precisely resemble the output vocabulary of C4.5. Consequently, the bias deriving from both these observations results in favor of C4.5 against the other approaches.

In general, no conclusions can be drawn from this study about the performance of the instance-based, backpropagation neural network and inductive learning algorithms. However, on the basis of the results obtained in this domain, C4.5 emerges as the most effective classifier on the smallest training set against the synthesized ground truth produced by the Salt Manager, and its use for knowledge and data discovery purposes is therefore vindicated in our experiments through the presentation of such a comparison.

TABLE 17.11

Performance of C4.5 (% Accuracy)

Domain	Decision Tree	Pruned Tree	Rules Generated
Secondary salinity	99.4	99.8	99.8
EM31	94	96	95.5

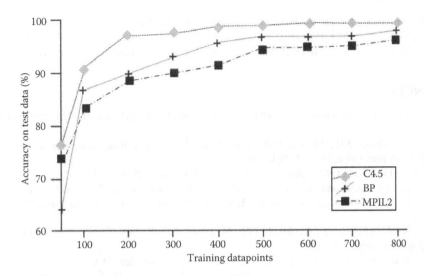

FIGURE 17.10 Classifier accuracy/performance as a function of training points.

Different domain data and circumstances require different machine-learning classifiers. The method of building the learning curves shown in Figure 17.10 gives a way of selecting the best classifier for the domain data in question. Our result is limited to modeling secondary salinization. However, the approach demonstrates a method to build effective machine-learning classifiers in similar spatial data sets. The recovery of the minimum learning set is strongly related to the original data granularity. The data layers have disparate resolution and the size of the minimum learning set reflects the aggregated view of the data.

17.6 CONCLUSION

The results presented in this chapter are measured in terms of performance enhancements of a knowledge-based decision support system and conclusions are drawn about secondary salinity using knowledge discovery techniques. In particular, our experience recommends coupling inductive machine learning with knowledge-based GIS decision support for predicting secondary salinization.

The chapter introduces a knowledge-based decision support system called Salt Manager, which uses mixed remotely-sensed and GIS data for environmental planning and monitoring. Salt Manager can be used to derive a map of secondary salinity from mapping groundwater recharge and discharge, which, because it is rule-based, can be used to explain it's reasoning and decision outcomes.

This chapter analyzes three experiments dealing with salinity recharge and discharge data that involve symbolic machine-learning techniques. In the first experiment, classified GIS data are passed to an inductive learning program that has the task to reconstruct the classification rules from the Salt Manager knowledge-base. The resulting rules were used to improve Salt Manager's classification performance by 85%. Further, the knowledge and data discovery process was able to independently verify the effects of contouring engineering on managing secondary salinization in the study area.

The second experiment predicts the effect of a new EM31 data layer on salinity classification and reveals that, contrary to conventional wisdom, EM31 has limited utility in identifying recharge areas. In our case study, EM31 can be used to predict recharge only in very restricted soil types. This outcome would not have been known without mining the data and is a useful result that can be applied in data acquisition and mapping exercises for future studies.

Finally, we compare the classification viability of different machine-learning algorithms for modeling secondary salinization and, by constructing learning curves for each candidate machine-learning technique, we are able to validate the soundness of our method and approach.

REFERENCES

Aha, D.W., Kibler, D., and Albert, M.K. 1991. Instance-based learning algorithms. *Machine Learning* 6(1): 37–66.

Breiman, L., Freidman, J.H., Olshen, R.A., and Stone, C.J. 1984. *Classification and Regression Trees.* Wadsworth International Group, Belmont, CA.

Bryson, A.E. and Ho, Y.C. 1969. *Applied Optimal Control.* Blaisdell, New York.

Cestnik, B., Kononenko, I., and Bratko, I. 1987. Assistant 86: A knowledge-elicitation tool for sophisticated users. In *Progress in Machine Learning*, Eds. I. Bratko and N. Lavrac. Sigma Press, Wilmslow, UK, pp. 31–45.

Cestnik, B., Kononenko, I., and Bratko, I. 1988. *Assistant Professional Systems, User's Manual.* Edward Kardelj University, Slovenia.

Churchland, P.M. 1984. *Matter and Consciousness.* MIT Press, Cambridge, MA.

Cover, T.M. and Hart, P.E. 1967. Nearest neighbor pattern classification. *IEEE Transactions on Information Theory* 13: 21–27.

Cressie, N.A.C. 1993. *Statistics for Spatial Data.* Wiley-Interscience, UK.

de Greef, P.A. and Breuker, J. 1985. A case study in structured knowledge acquisition. In *Proceedings of the Third International Conference on Artificial Intelligence (IJCAI'85).* Morgan Kaufmann, Los Angles, CA, pp. 390–392.

Dietterich, T.G. 1990. Machine learning. *Annual Review of Computer* 4: 255–306.

Dietterich, T.G. and Michalski, R.S. 1983. A comparative review of selected methods for learning from examples. In *Machine Learning: An Artificial Intelligence Approach*, Eds. R.S. Michalski, J.G. Carbonell, and T.M. Mitchell. Tioga Publishing Company, Wellsboro, PA, pp. 41–81.

Eklund, P.W., Kirkby, S.D., and Salim, A. 1998. Data mining and soil salinity analysis. *International Journal of Geographical Information Science* 12(3): 247–268.

Fayyad, U. and Irani, K. 1990. What should be minimized in a decision tree? In *Proceedings of the National Conference on Artificial Intelligence (AAAI).* Morgan Kaufmann, San Francisco, CA, pp. 749–754.

Hart, P.E. 1968. The condensed nearest neighbor rule. *IEEE Transactions on Information Theory* 14: 515–516.

Hayward, A.A., Wielinga, B., and Breuker, J. 1987. Structured analysis of knowledge. *Journal Man-Machine Studies* 26: 487–497.

Henschke, C.J., McCarthy, D.G., Richardson, S.B., and Evans, T.D. 1994. Dryland salinity catchment investigation: Jamestown catchment mid-north, South Australia. Technical Report 193, Primary Industries Department, South Australia.

Hertz, J.A., Palmer, R.G., and Krogh, A.S. 1991. *Introduction to the Theory of Neural Computation.* Addison Wesley, Boston, MA.

Johnsson, K. and Kanonier, J. 1991. Knowledge based land-use classification. *IEEE Geoscience and Remote Sensing (IGARS)* 3(3–6): 1847–1850.

Kibler, D., Aha, D.W., and Albert, M.K. 1989. Instance-based prediction of real valued attributes. *Computational Intelligence* 5: 51–57.

Kirkby, S.D. 1994. Managing dryland salinisation with an integrated expert system/geographic information system. PhD dissertation, Department of Geography, The University of Adelaide, Australia.

Maimon, O. 2005. *Data Mining and Knowledge Discovery Handbook.* Springer-Verlag, New York.

Michalski, R.S. and Chilausky, R.L. 1980. Learning by being told and learning from examples: An experimental comparison of the two methods of knowledge acquisition in the context of developing an expert system for soybean diagnosis. *International Journal of Policy Analysis and Information Systems* 4(2): 125–161.

Mooney, R., Shavlik, J., Towell, G., and Gove, A. 1989. An experimental comparison of symbolic and connectionist learning algorithms. In *Proceedings of the Eleventh International Conference on Artificial Intelligence (IJCAI'89).* Morgan Kaufmann, Detroit, MI, pp. 775–780.

Quinlan, J.R. 1979. Discovering rules by induction from large collection of examples. In *Expert Systems in the Microelectronic Age*, Ed. D. Michie. University Press, Edinburgh, UK.

Quinlan, J.R. 1983. Learning efficient classification procedures and their application to chess games. In *Machine Learning: An Artificial Intelligence Approach*, Eds. R.S. Michalski, J.G. Carbonell, and T.M. Mitchell. Tioga Publishing Company, Palo Alto, CA, pp. 463–482.

Quinlan, J.R. 1986. Induction of decision trees. *Machine Learning* 1(1): 81–106.

Quinlan, J.R. 1990. Decision trees and decision-making. *IEEE Transactions on Systems, Man and Cybernetics* 20(2): 339–346.

Quinlan, J.R. 1993. *C4.5: Programs for Machine Learning*. Morgan Kaufmann, San Mateo, CA.

Romaniuk, S.G. 1993. Multi-pass instance-based learning. Technical Report TRH3/93, Department of Information System and Computer Science, National University of Singapore, Singapore.

Rumelhart, D.E., Hinton, D.E., and Williams, R.J. 1986a. Learning representations by backpropagating errors. *Nature* 323: 533–536.

Rumelhart, D.E., Hinton, D.E., and Williams, R.J. 1986b. Learning representations by backpropagating errors. In *Parallel Distributed Processing*, Vol. 1, Eds. D.E. Rumelhart and J.L. McClelland. MIT Press, Cambridge, MA.

Wang, F. and Newkirk, R. 1988. A knowledge based system for highway network extraction. *IEEE Geoscience and Remote Sensing (IGARS)* 26(5): 525–531.

Yuan, M., Buttenfield, B., Gahegan, M.N., and Miller, H. 2004. Geospatial data mining and knowledge discovery. In *Research Challenges in Geographic Information Science*, Eds. R. McMaster and L. Usery, pp. 365–388. John Wiley & Sons, New York.

Zhang, K. and Giardino, J. 1992. Expert system based knowledge extraction from GIS. *IEEE Geoscience and Remote Sensing (IGARS)* 1: 319–321.

Conclusions

The traditional approach to soil salinity mapping from sole field observations and laboratory determinations is challenged, arguing that it is expensive and time-consuming. It must be recognized that it is now quite some time that such an approach is no longer implemented at local and regional scales, and that remote sensing has been integrated to varying degrees in salinity studies over the last three decades. There are, however, some key aspects where advances in remote sensing technology contributed to largely overhaul the more traditional ways of salinity mapping, with for instance aerial photographs. Improvements in temporal, spatial, and spectral resolutions have considerably increased the impact of remote sensing in salinity assessment, mapping, and monitoring.

Significant conclusions drawn from the collective expertise assembled in this book are given below:

- Provision of (relatively cheap) multispectral data at short, repeated intervals has allowed detecting changes in soil salinity and assessing the speed of salinization.
- Use of hyperspectral data has helped improve the identification of salt types, on the basis of their characteristic absorption features in narrow spectral bands, and the identification of pure endmembers in selected imagery, allowing the unmixing of feature associations at subpixel level.
- Implementation of electromagnetic induction techniques has permitted to solve the depth issue, from the identification of salinity features occurring at the terrain surface to the measurement of apparent electrical conductivity in the subsurface and substratum layers.
- Imagery with higher spatial resolution has contributed to more accurate salinity mapping, using for instance fuzzy logic to account for gradual changes across the landscape by replacing strict information classes with crisp boundaries by fuzzy information classes with transitional boundaries.

Other relevant conclusions that can be drawn from the cross-examination of the various contributions are as follows:

- Data synergy, combining various types of remotely sensed data, provided by ground-based (surface and subsurface), airborne, and satellite-borne sensors, together with expert knowledge, field observations, and a minimum set of laboratory determinations, is advocated.
- Satellite-based sensors covering the optical-infrared and microwave parts of the spectrum (e.g., Landsat and SPOT series, Terra-ASTER, LISS, IKONOS, Radarsat, ERS, ALOS, Envisat, CHRIS-PROBA, Hyperion) are cost-effective for covering large areas and for utilization on a multitemporal basis (e.g., monitoring tasks). Airborne hyperspectral sensors (e.g., AVIRIS, HyMap, AISA-ES) are generally used over relatively small areas. This is mainly due to higher cost of data acquisition, as compared to that of Landsat-like satellite imagery, and more complex image calibration and data processing.
- Modeling spatial distribution and temporal changes in soil salinity increases the accuracy of salinity mapping in time and space and helps assess salinity hazard.
- Application-oriented remote sensing provides base-line data for land management and conservation in salt-affected and salt-prone areas, including information for the improvement of farming practices in dryland areas, water application and drainage in irrigation schemes, and preservation of coastal marshes and inland lagoons and playas that are vital to migratory birds in semiarid and arid regions.

- Further research is needed in developing approaches to detect and map early stages of soil salinity. Most contributions show success in mapping and monitoring saline versus nonsaline areas, or identifying moderate to high salinity. However, farmers and land managers alike appreciate information on incipient salinity problems for timely and cost-effective decisions on adequate ways to manage and remediate, wherever possible, this degradation process.

J. Alfred Zinck
Graciela Metternicht

Index